Effect of Disorder and Defects in Ion-Implanted Semiconductors: Electrical and Physicochemical Characterization

SEMICONDUCTORS
AND SEMIMETALS
Volume 45

Semiconductors and Semimetals

A Treatise

Edited by R. K. Willardson
CONSULTING PHYSICIST
SPOKANE, WASHINGTON

Eicke R. Weber
DEPARTMENT OF MATERIALS SCIENCE
AND MINERAL ENGINEERING
UNIVERSITY OF CALIFORNIA AT
BERKELEY

*In memory of Dr. Albert C. Beer, Founding Co-Editor in 1966
and Editor Emeritus of Semiconductors and Semimetals.
Died January 19, 1997, Columbus, OH.*

Effect of Disorder and Defects in Ion-Implanted Semiconductors: Electrical and Physicochemical Characterization

SEMICONDUCTORS
AND SEMIMETALS

Volume 45

Volume Editors

GÉRARD GHIBAUDO

LABORATOIRE DE PHYSIQUE DES COMPOSANTS À SEMICONDUCTEURS/ENSERG
GRENOBLE, FRANCE

CONSTANTINOS CHRISTOFIDES

DEPARTMENT OF NATURAL SCIENCES
UNIVERSITY OF CYPRUS
NICOSIA, CYPRUS

ACADEMIC PRESS
San Diego   London   Boston
New York   Sydney   Tokyo   Toronto

This book is printed on acid-free paper.

COPYRIGHT © 1997 BY ACADEMIC PRESS

ALL RIGHTS RESERVED.
NO PART OF THIS PUBLICATION MAY BE REPRODUCED OR TRANSMITTED IN ANY FORM OR BY ANY MEANS, ELECTRONIC OR MECHANICAL, INCLUDING PHOTOCOPY, RECORDING, OR ANY INFORMATION STORAGE AND RETRIEVAL SYSTEM, WITHOUT PERMISSION IN WRITING FROM THE PUBLISHER.
The appearance of the code at the bottom of the first page of a chapter in this book indicates the Publisher's consent that copies of the chapter may be made for personal or internal use, or for the personal or internal use of specific clients. This consent is given on the condition, however, that the copier pay the stated per copy fee through the Copyright Clearance Center, Inc. (222 Rosewood Drive, Danvers, Massachusetts 01923), for copying beyond that permitted by Sections 107 or 108 of the U.S. Copyright Law. This consent does not extend to other kinds of copying, such as copying for general distribution, for advertising or promotional purposes, for creating new collective works, or for resale. Copy fees for pre-1997 chapters are as shown on the chapter title pages; if no fee code appears on the chapter title page, the copy fee is the same as for current chapters. 0080-8784/97 $25.00

ACADEMIC PRESS
525 B Street, Suite 1900, San Diego, CA 92101-4495, USA
1300 Boylston Street, Chestnut Hill, Massachusetts 02167, USA
http://www.apnet.com

ACADEMIC PRESS LIMITED
24-28 Oval Road, London NW1 7DX, UK
http://www.hbuk.co.uk/ap/

International Standard Book Number: 0-12-752145-3

Printed in the United States of America
97 98 99 00 01 BB 9 8 7 6 5 4 3 2 1

Contents

LIST OF CONTRIBUTORS xi
FOREWORD xiii
PREFACE xvii

Chapter 1 Ion Implantation into Semiconductors: Historical Perspectives
Heiner Ryssel

I. Introduction 1
II. Early History 2
III. Ion Implanters 5
   1. Development of Ion Implanters 5
   2. Acceleration and Mass Separation 8
   3. Beam Scanning and Dosimetry 11
   4. Contamination 13
   5. Implanter Trends 14
IV. Metal-Oxide Semiconductor Devices 15
   1. Self-aligned Gate Process 15
   2. Threshold Adjust 16
   3. Well Doping 18
   4. Drain Engineering 20
   5. Metal-Oxide Semiconductor Trends 21
V. Bipolar Devices 22
   1. Arsenic Emitter 23
   2. Boron Base 24
   3. Bipolar Trends 25
VI. Conclusions 26
   References 26

Chapter 2 Electronic Stopping Power for Energetic Ions in Solids
You-Nian Wang and Teng-Cai Ma

I. Introduction	32
II. General Theory	33
1. Electron-Gas Model	33
2. Dielectric Function	34
3. Local Density Approximation	35
III. Electronic Stopping Power for Protons	36
1. Method of Numerical Calculation	36
2. Low- and High-Velocity Approximations	37
3. Fitted Formula and Comparison with Experimental Data	38
IV. Electronic Stopping Power for Heavy Ions	42
1. The Brandt-Kitagawa Model for Charge Distribution of Projectile	42
2. Effective Stopping Charges	44
3. Comparison with Experimental Data	45
V. Electronic Stopping Power for Molecular Ions	47
1. Coulomb Explosion	47
2. The Vicinage Effect in the Stopping Power	49
VI. Summary	52
References	53

Chapter 3 Solid Effect on the Electronic Stopping of Crystalline Target and Application to Range Estimation
Sachiko T. Nakagawa

I. Introduction	55
II. Local Density Approximation for Binary Collision	58
1. Electron Density of Solid-State Target Atoms	58
2. Local Density Approximation for the Nuclear Stopping Power	59
3. Local Density Approximation for the Electronic Stopping Power	61
III. Impact Parameter-Dependence of the Electronic Stopping Power in Crystalline Solids	64
1. Original Oen-Robinson Model and Innovation	64
2. Determination of the Impact Parameter-Dependence of the Electronic Stopping Power in a Cluster	65
3. Solid Effects on the Electronic Stopping Power from Cluster Calculation	68
IV. Electronic Stopping Power of Chemical Compounds	69
1. Axial Channeling in Zinc-blende	69
2. Combination Rule as an Alternative to the Bragg Rule for Electronic Stopping Power of Compounds	71
V. Electronic Stopping Power and Range Profiles *via* Computer Simulations	72
1. Influence of the Electronic Stopping Power on Range Profiles	73
2. Solid-Effects on Electronic Stopping Power from Computer Simulations	76
VI. Concluding Remarks	78
References	81

Chapter 4 Ion Beams in Amorphous Semiconductor Research
G. Müller, S. Kalbitzer and G. N. Greaves

I. Introduction	85
II. Ion Beam Production of Amorphous Silicon	87
1. Impact of Disorder on Doping, Electronic Transport, and Optical Properties	87
2. Irreversible Ordering Phenomena in Undoped Amorphous Material	95
3. Structural Relaxation and Thermal Crystallization of Doped Material	99
4. Effects of Hydrogenation and Fluorination	105
III. Ion Beam Doping of Plasma-Deposited Amorphous Silicon	106
1. Gas Phase and Ion Implantation Doping	106
2. Doping Mechanism in Hydrogenated Material	110
3. Generation and Annealing of Implantation Damage	115
IV. Structural and Configurational Changes in Amorphous Silicon	121
1. Changes in Pure Material	121
2. Changes in Hydrogenated Material	122
References	123

Chapter 5 Sheet and Spreading Resistance Analysis of Ion Implanted and Annealed Semiconductors
Jumana Boussey-Said

I. Introduction	129
II. Sheet Resistance Measurement	130
1. Collinear Four-Point Probe Method	131
2. Sheet Resistance of Arbitrarily Shaped Samples	133
III. Spreading Resistance Probes Profiling	135
1. Principle of the Spreading Resistance Profiling Measurements	136
2. Extraction of Resistivity Profiles from Spreading Resistance Raw Data	138
IV. Applications	143
1. High-Dose and High-Energy Arsenic-Implanted Silicon Layers	143
2. Low-Energy Boron and Boron Fluoride Implantation in a Preamorphized Silicon Substrate	152
V. Summary	161
References	162

Chapter 6 Studies of the Stripping Hall Effect in Ion-Implanted Silicon
M. L. Polignano and G. Queirolo

I. Introduction	165
II. Survey of Basic Theory	166
1. Resistivity Measurements	168
2. Mobility Measurements	170
3. Carrier and Mobility Profiling	172
III. Applications of the Technique	175
1. Experimental Details	175
2. High Fluence Boron Fluoride-Implanted Layers	175
3. Measurements on Shallows, Heavily Arsenic-Doped Layers: Solubility and Mobility Data	186

	4. Active Dopant Concentration in Phosphorus-Doped Polycrystalline Layers	189
	Conclusions	192
	References	193

Chapter 7 Transmission Electron Microscopy Analyses
J. Stoemenos

I.	Introduction	195
II.	Transmission Electron Microscopy	196
	1. Imaging Ray Path in Transmission Electron Microscopy	196
	2. Crystallographic Structure and Chemical Composition	199
	3. Specimen Preparation	199
III.	Defects Produced by Ion Implantation, Transmission Electron Microscopy	201
	1. Microelectronics and Microscopy	201
	2. Evolution of End-of-Range Defects	206
	3. End-of-Range Defects and Electrical Properties	208
	4. Kinetics of End-of-Range Defects	208
IV.	Implantation Using Molecular Ions	210
V.	Implantation Conditions Inhibiting the Formation of End-of-Range Defects	212
VI.	Material Modification by Ion Beam Synthesis	216
	1. Silicon Separation by Implanted Oxygen (SIMOX)	216
	2. β-Silicon Carbide Formed by Carbon Implantation into Silicon	227
VII.	Ion-Beam-Induced Epitaxial Crystallization	230
VIII.	Closing Remarks	235
	References	235

Chapter 8 Rutherford Backscattering Studies of Ion Implanted Semiconductors
Roberta Nipoti and Marco Servidori

I.	Introduction	239
II.	Measurement of Disorder Depth Profiles by RBS-Channeling	240
III.	Typical Results for Silicon and Silicon Carbide	248
	1. Silicon	248
	2. Silicon Carbide	255
	References	258

Chapter 9 X-ray Diffraction Techniques
P. Zaumseil

I.	Introduction	261
II.	Basic Aspects of Crystal Lattice Modification Due to Ion-Implantation	262
	1. Implantation-Induced Variation of the Lattice Parameters	263
	2. Dopant Incorporation	265
	3. Defect Generation During Annealing	266
III.	X-ray Methods to Analyze Implanted Samples	268
	1. Simulation of the Reflection Curve of a Disturbed Crystal	268
	2. Experimental Techniques	271
	3. Limitations of X-ray Diffraction Techniques	273

IV. Examples of Special Applications 274
   1. Determination of Dislocation Loop Size and Density in Implanted and
      Annealed Silicon 274
   2. Arsenic Implantation and Annealing 276
   3. Characterization of Boron-Implanted and Annealed Silicon 278
V. Summary 280
   References 281

   INDEX 283

   CONTENTS OF VOLUMES IN THIS SERIES 287

List of Contributors

Numbers in parenthesis indicate the pages on which the authors' contributions begin.

JUMANA BOUSSEY-SAID (129), *Laboratoire de Physique des Composants à Semiconducteurs, ENSERG-INPG-UMR CNRS 5531, BP 257, 38016 Grenoble Cedex 1, France*

G. N. GREAVES (85), *Daresbury Laboratory, Warrington WA4 4AD, England*

S. KALBITZER (85), *Max Planck Institut Für Kernphysik, 69029 Heidelberg, Germany*

TENG-CAI MA (31), *Department of Physics, Dalian University of Technology, Dalian, 116023, P.R. China*

G. MÜLLER (85), *Daimler Benz AG, Forschung und Technik, 81663 München, Germany*

SACHIKO T. NAKAGAWA (55), *Department of Applied Physics, Okayama University of Science, Okayama 700, Japan*

ROBERTA NIPOTI (239), *Consiglio Nazionale Delle Ricerche Instituto LAMEL, 40129 Bologna, Italy*

M. L. POLIGNANO (165), *SGS-Thomson Microelectronics, 20041 Agrate Brianza, Italy*

G. QUEIROLO (165), *SGS-Thomson Microelectronics, 20041 Agrate Brianza, Italy*

HEINER RYSSEL (1), *Lehrstuhl Für Elektronische Bauelemente, Universitat Erlangen-Numberg, D-91058 Erlangen, Germany and Fraunhofer Institut Für Integrierte Schaltungen, D-91058 Erlangen, Germany*

MARCO SERVIDORI (239), *Consiglio Nazionale Delle Ricerche Instituto LAMEL, 40129 Bologna, Italy*

J. STOEMENOS (195), *Department of Physics, Aristotle University of Thessaloniki, 54006 Thessaloniki, Greece*

YOU-NIAN WANG (31), *Department of Physics, Dalian University of Technology, Dalian, 116023, P.R. China and Department of Physics, University of Waterloo, Waterloo, N2L 3G1, Ontario, Canada*

P. ZAUMSEIL (261), *Institute for Semiconductor Physics, D-15230 Frankfurt (Oder), Germany*

Foreword

Implantation of impurity atoms for doping semiconductor wafers offers many advantages such as rapidity, mass separation for purity requirements, accuracy and a wide range of doses, flexibility of profile depth, and a control over the amount of ions in a specific region. The bombardment of solids by ion implantation also has been widely used for the study of amorphization and physical analysis of structural disorder introduced by irradiation.

One main disadvantage of the ion implantation process is damage introduced into the semiconductor resulting from the energetic character of the process. The consequence of ion bombardment is the amorphization of the semiconductor surface and, at high ion concentrations, the presence of electrical defects such as interstitial impurities, dislocations, grain boundaries, and inhomogeneities. This implantation-induced disorder leads to strong degradation of the electrical features of the materials. It is this degradation that must be dealt with during the device fabrication process.

Ion implantation therefore must be followed by one or more annealing processes for the semiconductor to recover its crystallinity and for the doping impurity to become active. In general, thermal annealing in a conventional furnace is currently used to activate the doping such that thermal diffusion takes place, leading to a large redistribution of the impurity atoms. This redistribution then causes enlargement of the junctions, which may be prohibitive for the optimal operation of short-channel complementary metal-oxide semiconductor (CMOS) devices. Moreover, these implantation-induced defects strongly modify the recombination properties of the semiconductor that may affect the operation of bipolar devices and p-n junctions. Rapid thermal annealing (RTA) methods have been introduced to minimize redistribution of the impurities. To this end, an annealing procedure using light or electron beams has been proposed to reduce the annealing duration while maintaining a sufficiently high temperature to activate the doping species. The most commonly used technique is the so-called lamp RTA process in which the thermal heating of the sample is ensured by the heat dissipation of halogen lamps surrounding the wafers. These lamp RTA machines provide very fast heating pulses (approximately 1 to 10 sec) that allow the temperature to range from 1000 to 1100°C without significant thermal diffusion.

All chapters of Volumes 45 and 46 attempt to analyze the experimental results obtained by the four main families of techniques of characterization on implanted wafers: electrical, physicochemical, optical, and thermal wave. It is important to note that this is the first book in the field of ion implantation to review thermal wave studies on implanted materials. Each chapter of these focuses on the following important effects of ion implantation and the annealing processes:

- Damage induced (short-range disorder and long-range disorder) by ion implantation;
- Spatial distribution profiles of the damage;
- Kinetics of annihilation mechanisms (the defect layer and the amorphous layer);
- Electrical activation of the implanted impurities.

Chapter 1 of Volume 45, is an introduction to ion implantation into semiconductors with the aim of tracing the historical perspectives and demonstrating the impact of ion implantation on the fabrication of integrated circuits (ICs) in modern microelectronics. The importance of the ion implantation step in the technologic processes for ICs is emphasized and illustrated by various examples dealing with CMOS technologies and bipolar device fabrication.

Chapters 2 and 3 provide the reader with the basic principles of ion implantation into semiconductors. The mechanisms of interaction between the implanted ions and the target material are presented. Chapter 2 concentrates on the theoretical developments for calculating the electronic stopping power for energetic ions in solids. More specifically, the electronic stopping power investigates the linear response dielectric theory and the local density approximation method. Empirical formulas for evaluating the stopping power of protons are exemplified. Chapter 3 deals with the influence of solid effects on the electronic stopping power for ions implanted into a solid target. These solid effects, which cause the material dependence of the radiation effects, are: (a) the phase effects related to the nature of the irradiated medium and (b) the directional effect due to the crystallography. This issue was mainly devoted to the study of the impact parameter dependence of the stopping power, which concerns the different contributions arising from valence and core electrons. Furthermore, Monte Carlo simulations are used to calculate the range profiles of ions implanted into binary III-V semiconductors. Chapter 4 concerns ion implantation into amorphous semiconductors. The main issues addressed concern the ion beam technique experiments used to produce amorphous silicon and to dope plasma-deposited amorphous silicon. Moreover, Chapter 4 focuses on the physics underlying the structural changes occurring in a target material in which irradiation-induced disorder is produced at the same time. The influence of disorder on the dopant configuration is also discussed, with the help of electrical and optical experiments.

Chapters 5 and 6 of Volume 45 are devoted to the electrical characterization of ion-implanted semiconductors. More specifically, Chapter 5 presents a review of the techniques and results obtained by sheet and spreading resistance measurements. In particular, the principles and practical precautions of the spreading resistance method are critically given with special attention to the resistivity extraction procedure and calibration routines. Afterward, the experimental results obtained using these techniques on implanted silicon are presented. It is therefore important to note that proper combination of such electrical analysis with structural characterization methods, such as transmission electron microscopy (TEM) or X-ray diffraction, can provide reliable information about the electrical nature of ion-implantation–induced damage. Chapter 6 deals with the technique of carrier profiling by differential sheet resistance and sheet Hall coefficient measurements obtained by anodic stripping. This technique is distinctive because it can measure the carrier concentration and mobility profiles. This method therefore appears to be a complementary characterization tool, compared with the spreading resistance technique, because it provides crucial information about the changes in the scattering mechanisms in ion-implanted semiconductors and their evolutions with the annealing processes.

Chapters 7, 8, and 9 of this volume are dedicated to the physicochemical studies of ion-implanted semiconductors. Chapter 7 summarizes the capabilities of TEM to analyze the defects and the crystallographic changes after ion implantation. First the basic principles of TEM are traced back, along with the details for sample preparation. Then, experimental results demonstrating the ability of the technique to analyze the implantation-induced damage and its annihilation after annealing are presented in the context of silicon material. In particular, the kinetics of end-of-range (EOR) defects are revealed because of the sensitivity of TEM analysis. The formation and the evolution of amorphous layers after high dose implantation are emphasized.

The TEM studies are also exemplified as being powerful physical characterization tools for evaluating the quality of new materials under development, for example, separation by implementation of oxygen (SIMOX) or silicon carbide. Chapter 8 provides Rutherford backscattering (RBS) studies of ion-implanted semiconductors. The principles of the RBS-channeling technique are presented first. The RBS-channeling technique is a useful tool for studying the evolution of the disorder, which is a function of the irradiation conditions and postirradiation thermal treatments. This channeling technique gives physical information about the isolated scattering center and extended defect distribution within the material. Examples of results obtained on implanted silicon and silicon carbide are then given. These examples focus on the annihilation of damage after thermal treatment. Chapter 9 provides X-ray diffraction techniques for the study of ion-implanted semiconductors. The different measuring techniques available, such as the triple-crystal and double-crystal diffractometries, are described. The theoretic models necessary to analyze and interpret the reflection curves are presented. Typical examples for the application of X-ray techniques are then given in the context of implantation-induced defects.

GÉRARD GHIBAUDO
CONSTANTINOS CHRISTOFIDES

Preface

It is most likely that the historian of the future will characterize our century as the "Silicon Era." Nearly 50 years after the great discovery of the transistor, which constitutes the basis of modern technology, the end of the 1980s marked the passage from the microelectronics to the nanoelectronics of the twenty-first century.

Without question, the semiconductor has served as the foundation for the progress in microelectronics. In the 1970s and 1980s, various techniques for the fabrication and treatment of semiconducting materials were developed. Ion implantation is a key technological process in modern microelectronics that was introduced as an alternative to diffusion for the semiconductor doping process. Extensive research in the field of ion implantation since the early 1960s has increased our knowledge of the physics of semiconductors and the technology of materials and devices.

Despite these great strides, there remain problems and questions regarding ion-implantation that are technically and academically challenging. As a result, ion implantation is a fundamental topic of thousands of articles, hundreds of PhD theses, and many books. The question always arises for each new book in such a popular field: Why another book on ion implantation? A general review of the literature shows that, to date, most publications have focused on the effects that ion implantation has on the material and in the modification of the implanted layer after high-temperature annealing. Unfortunately, no attention has been given to the annealing of defects in the implanted layer. All the more, the annealing kinetics of the damaged layer and the dynamics of the changing layer present an important field of interest, and probably, therein lies the most exciting physics. We hope that the present approach by several experts satisfies the needs of at least one generation of new researchers.

This book presents the history of the development, the theoretical basis, and the main experimental works of ion-implantation physics and technology. An overview is given of most of the techniques that have been used for the characterization of implanted semiconductor materials, and for the first time, these techniques are presented and compared critically. In each chapter, the experimental and theoretical aspects of the techniques are presented pedagogically; however, the main emphasis is on the physics of both the disorder and the defects generated by ion implantation. Furthermore, this book is our ambitious attempt to concentrate on the kinetics of annealing and to highlight such aspects as the construction mechanisms of defects and the various annihilation processes that occur during both the phase change and the recrystallization of the implanted layer. We also try to provide information concerning the modification of the implanted layer, not only on the surface but also under it and up to the interface between the surface and the crystalline substrate. To achieve these objectives, we have obtained contributions from specialists worldwide, from both academic and research institutions. Furthermore, to give this work a multidimensional character, we have included contributions from researchers in industry.

All of this information is compiled in two separate volumes, Volumes 45 and 46, in the well-known Semiconductors and Semimetals Series. Volume 45 focuses on the physics of implantation and the creation of defects and their characterization by electrical and physicochemical techniques. Volume 46 is devoted to optical and thermal wave analyses of ion-implanted semiconductors. In addition, this second volume deals with specific questions arising in ion implantation into quantum wells and compound semiconductors.

The whole contribution can be regarded as a synthesis of chapters written by well-established professors and researchers, as well as by younger scientists who have provided a remarkable contribution to the field in recent years. In most chapters, a brief historical review presents the evolution of relevant concepts. Most of the information in this compilation concerns the basis of some theoretical and experimental aspects and is of permanent value. Several experimental results—from the late 1970s to the present— are given. Moreover, both volumes are important guides to the modern literature in this field.

We thank all the contributors to these volumes for their enthusiasm and dedication to the project. It has been a great pleasure to work with them. We also thank all the reviewers of the invited chapters. Special thanks are due to M. Marcou for her skillful and patient secretarial assistance. Thanks are also expressed to Dr. Z. Ruder, Executive Editor at Academic Press for his guidance during this effort. Last, but not least, we are grateful to our families for their patience during the many stages of writing and editing of the manuscript.

Dedicated to our parents, Anthoula, Christos, Jeanne, and Jean

CHAPTER 1

Ion Implantation into Semiconductors: Historical Perspectives

Heiner Ryssel

LEHRSTUHL FÜR ELEKTRONISCHE BAUELEMENTE,
UNIVERSITAT ERLANGEN-NUMBERG
AND FRAUHOFER INSTITUT FÜR INTEGRIERTE SCHALTUNGEN,
ERLANGEN, GERMANY

I. INTRODUCTION	1
II. EARLY HISTORY	2
III. ION IMPLANTERS	5
1. Development of Ion Implanters	5
2. Acceleration and Mass Separation	8
3. Beam Scanning and Dosimetry	11
4. Contamination	13
5. Implanter Trends	14
IV. METAL-OXIDE SEMICONDUCTOR DEVICES	15
1. Self-aligned Gate Process	15
2. Threshold Adjust	16
3. Well Doping	18
4. Drain Engineering	20
5. Metal-Oxide Semiconductor Trends	21
V. BIPOLAR DEVICES	22
1. Arsenic Emitter	23
2. Boron Base	24
3. Bipolar Trends	25
VI. CONCLUSIONS	26
References	26

I. Introduction

Ion implantation was first used in manufacturing integrated circuits in the early 1970s. There is, however, a much longer tradition of using heavy ion beams. It is not easy to write a review of ion implantation into semiconductors. Should one start with Goldstein's "Kanalstrahlen" (channel rays) in glow discharges in 1886 or with the fundamental work of Bohr who calculated the range of heavy ions in matter as early as 1913 (Bohr, 1913).

Based on this, many years later Lindhard, Scharff, and Schiøtt (1963) developed the theory for range and range straggling of low-energy ions. The well-known LSS theory is still used for most range calculations necessary for practical applications of ion implantation.

I decided to focus on the development of ion implanters, which was the prerequisite to implantation into semiconductors, and on the use of ion implantation to manufacture metal-oxide semiconductor (MOS) and bipolar devices. This is a strong limitation but otherwise this chapter would have become much too lengthy. Implantation experiments were carried out with virtually all known semiconductors. Real production, however, exists only for silicon devices.

Several books have been written on the subject of ion implantation into semiconductors. Amazingly early, the famous book by Mayer, Eriksson, and Davies (1970) was published. It focuses more on fundamental aspects. Later, a more practical book was published by Dearnaley et al. (1973). The book by Wilson and Brewer (1973) deals mainly with implanter-oriented topics. Carter and Grant (1976) treat mainly semiconductors but also many basic aspects. A more general book was written by Ryssel and Ruge (1986). Detailed information on all aspects of ion implantation can be found in the series of IBMM (Ion Beam Modification of Material) conferences[1] and IIT (Ion Implantation Technology) conferences[2] as well as in the proceedings of numerous other conferences.

II. Early History

Presumably, Ohl (1952) was the first person who tried to change the electrical properties of semiconductor devices with ion irradiation. In 1952, Ohl implanted silicon point-contact diodes with various ions and found improvement in the reverse current-voltage curve. Hydrogen, helium, nitrogen, and argon ions were used. The improvement was probably due to damage and not to chemical doping. Later, Cussins (1955) implanted germanium with a wide variety of ions, including boron. He annealed the samples for the first time but the achieved p-type conductivity disappeared.

In the meantime, Shockley (1957) had filed a patent in 1954, which was granted in 1957. The title of the patent is "Forming Semiconductive Devices by Ionic Bombardment." After the bipolar transistor, this was his second essential invention in the field of semiconductors. Although its goal was to

---

[1]Reference IBBM 1971a, 1971b, 1973, 1975, 1977, 1979, 1981, 1983, 1985, 1987, 1990, 1991, 1993, 1995.
[2]References IIT 1981, 1982, 1985, 1987, 1989, 1991, 1993, 1995.

implant the base of bipolar transistors, it covered nearly all important aspects of ion implantation. In the following, parts of the patent are cited (Shockley, 1957):

... a semiconductive body of one conductivity type is bombarded with a monoenergetic beam of ions of a significant impurity element characteristic of a conductivity type opposite to that of the body in a manner to convert the conductivity type of a thin layer in the interior of the body ....
In the process of the invention, the energy level of the bombarding beam is adjusted so that the projected beam will penetrate into the interior of the semiconductive body and be localized there for converting that region to opposite conductivity type. By utilization of a monoenergetic beam, the ions are all made to penetrate to a fairly uniform depth whereby the thickness of the region of converted conductivity type is kept small. The thickness of this region can be controlled by variations in the energies of the bombarding ions. Thereafter, the semiconductor body is treated to repair the radiation damage done to the surface region penetrated. This is done advantageously by annealing at a temperature sufficiently low that there results little migration from the region of deposition of the significant impurity ions introduced by bombardment ....
The junction may be formed in accordance with a preselected pattern either by the use of a deflection system which sweeps an ion beam focused to an appropriate cross section over the semiconductive body or else by interposing a suitable apertured mask between the ion source and the semiconductive body. In this latter technique, it is advantageous to employ a mask aperture on a scale considerably enlarged over the size of the junction pattern desired and thereafter to condense, in the manner familiar to workers in ion optics, the ions passing through the mask to reduce the scale of the pattern to that desired for the junction.

Of course, he did not yet know the approximately Gaussian shape of the implantation profile, however, he clearly stated the necessity of annealing and masking (now, however, implantation is not done by using an aperture, but at the time of the patent planar technology had not yet been invented!). Working with a focused ion beam never made it out of the laboratory, and using demagnifying lenses is still in the experimental phase but exclusively for ion beam lithography (Stengl, Löschner, Muray, 1986).
Not much happened until the early 1960s when Alväger and Hansen (1962) made the first nuclear detectors by implanting phosphorus into high-resistivity silicon at 10 keV and annealing at 600° C. Obviously, they were not aware of Cussins' work and Shockley's patent. McCaldin and Widmer (1963) prepared p-n junctions by cesium implantation. Still more exotic at that time, Gibbons (1968) and Gibbons, Moll, and Meyer (1965) implanted rare earth elements into semiconductors to produce electroluminescent layers. Afterward, King et al. (1965) and Martin, Harrison, and King (1955) also made nuclear detectors and, for the first time, solar cells.

Very important was that in the mid-1960s several nuclear research centers (Chalk River Nuclear Laboratories, Oak Ridge National Laboratory, and the Atomic Energy Research Establishment at Harwell) began to become interested in ion implantation. This was probably because they had ion accelerators at hand, which were used originally for the large nuclear programs of the United States and England in the 1940s and 1950s. In addition, two groups in California—at the California Institute of Technology and Stanford University—led by Mayer and Gibbons, were very active at the same time. In these centers, many of the basic investigations concerning range distributions in crystalline and amorphous semiconductors, channeling, low-temperature and high-temperature implantation, annealing, and epitaxial recrystallization were performed.

Since about 1965, an increasing number of investigations were performed and it is impractical to list all these experiments here. In Figure 1, the number of papers dealing with ion implantation into semiconductors for the first 30 years of ion implantation is shown. The data are taken from the bibliographies of Mazzio (1971), Seager (1973), and Agajanian (1974) and through a computer search. Since that time, hundreds of papers have been published yearly, not only with ion implantation as a main subject but just with implantation as a standard technique used to manufacture devices. I personally entered ion implantation research in 1970 when it began to boom, and I saw the rise and the fall of many implanter companies which tried to make a fortune from this new technique. Since that time, I have made implanted devices in germanium (bipolar transistors and avalanche photodiodes), silicon (bipolar transistors, MOS transistors, and nuclear

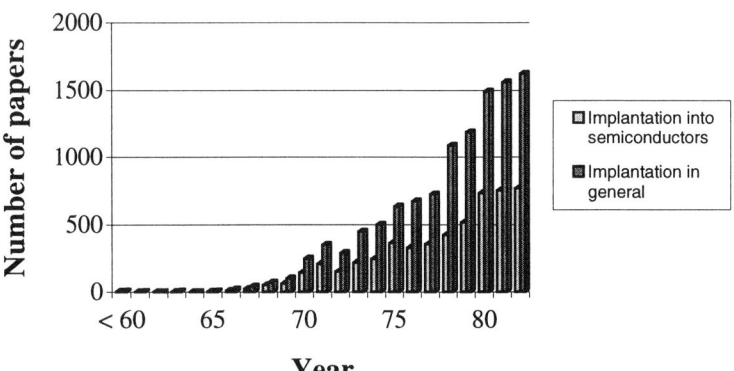

FIG. 1. Number of papers dealing with ion implantation in general from the first article published by Ohl (1952) until 1982, and with ion implantation into semiconductors.

1 ION IMPLANTATION INTO SEMICONDUCTORS 5

detectors), gallium arsenide (GaAs), gallium arsenidephosphide (GaAsP) and gallium phosphide (GaP) (light emitting diodes), indium antimonide (InSb), lead tin telluride (PbSnTe), cadmium mercury telluride (HgCdTe) (infrared detectors). I have also studied, quite generally, the surface modification of metals and polymers by ion implantation, but which are not the subject of this chapter.

In the beginning of ion implantation it was not clear, despite Shockley's patent, that semiconductors would become the main application of ion implantation. This can be seen clearly from the number of papers published. It is remarkable, however, that among the first experiments with semiconductors, III–V compounds (Mayer *et al.*, 1967) and silicon carbide (SiC) (Makarov and Petrov, 1966) found interest.

III. Ion Implanters

Ion implantation as an industrial process depends heavily on the proper equipment. Early ion accelerators were far from suitable for mass production of devices. Instead, they were typical "physicist" types of machines, that is, great, bulky, and complicated to operate with high downtime.

1. DEVELOPMENT OF ION IMPLANTERS

Ion implanters, as they are used today, have their origins in two roots. Nuclear physicists were using "low-energy ion accelerators" to study the interaction of ions with atoms since the early 1920s. For scientists working with ion implanters, these machines, with their MeV energies, would be classified as high-energy implanters. The ion beams had a circular cross section. Pre-analysis acceleration was used, as were large magnets for mass analysis and beam switching into different beamlines for different experiments. Low- and medium-current implanters were developed from these machines. High Voltage Engineering Corp. (HVEC), founded in 1946, developed into a big business by selling Van de Graaff and tandem accelerators to numerous research institutes and universities worldwide. These were typically low-current machines, operating in the microampere range.

The second origin stems from the big effort in the early 1940s to manufacture substantial quantities of highly enriched Uranium 235. More than 1000 low-energy ion separators were built in the United States until 1945, each with an ion beam capacity of tens or even hundreds of milliamperes (Aitken, 1989). Some of these so-called calutrons were used later to supply isotopically enriched nuclides. In Europe, several of these

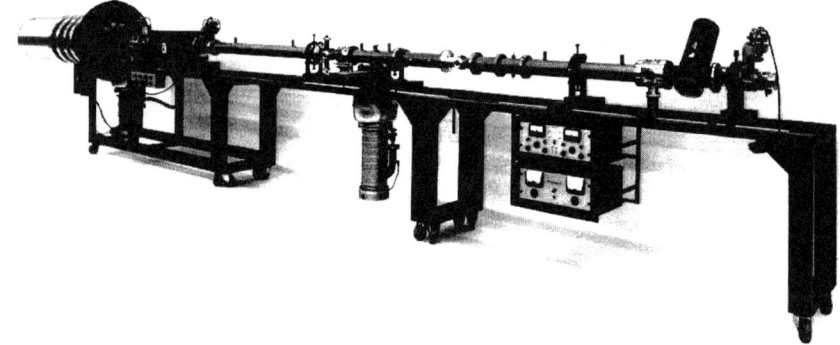

FIG. 2. Ion implanter built in 1970 by Accelerators, Inc.

heavy-ion accelerators also were developed. The ion beams had a rectangular shape and the extraction voltage was the same as the acceleration voltage. Together with mechanical scanning of the samples, the high-current implanters were developed from these isotope separators. A scaled-down calutron ion source was the so-called Freeman source, which is still the workhorse in most industrial implanters (Freeman 1963).

HVEC also had the opportunity to lead the implanter business, but failed. In 1971, a group around Peter Rose left HVEC to form a new company named Extrion. At the same time, Accelerators, Inc. of Austin, Texas, had already sold implanters for two to three million dollars a year. In our laboratory, we bought the first Accelerators machine in 1970. In fact, this machine was very similar to a Texas Nuclear neutron generator equipped with an $E \times B$ mass separator and xy scanner—but with no target chamber. At that time, the customer himself had to build it. Figure 2 shows this basic machine, which had no lead shielding against X-rays, no protection against touching high voltage, and no vented gas box, not to mention many other shortcomings. My first target chamber was a cassette-type design with 14 slides loaded with 1-in. to 2-in. wafers, similar to a target chamber of the Harwell group (Goode, 1976), and later by Extrion (Rose, 1985).

Extrion developed very successful implanters and was acquired in 1975 by Varian. It is one of the major implanter companies of today. In 1973, key people from Accelerators, Inc. left and founded Eltek Inc. and the same took place in 1978 with Varian-Extrion when Nova was founded. Eltek and Nova were acquired in 1980 by Eaton to form the Ion Beam System Division of Eaton. The company is now a leading manufacturer of implantation equipment.

1 ION IMPLANTATION INTO SEMICONDUCTORS

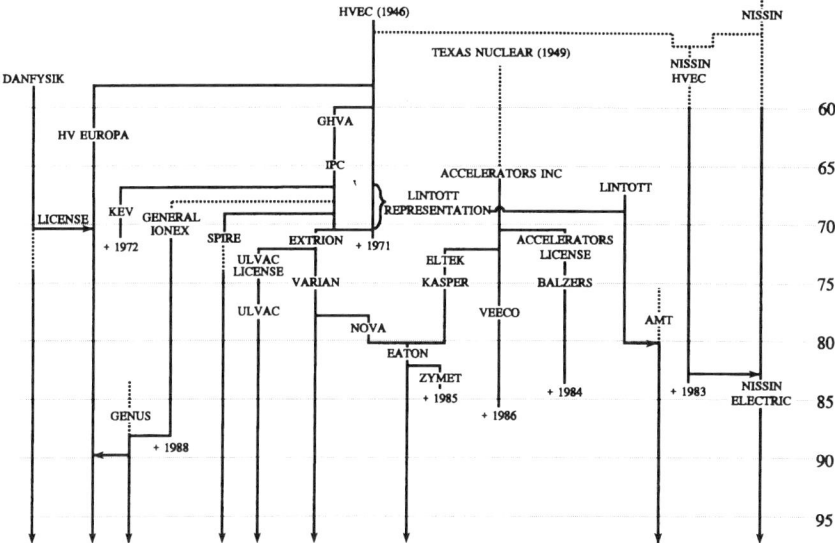

FIG. 3. Genealogy of implanter companies.

Some excellent reviews on the history of ion implantation equipment have been given in recent years. Lienhard Wegmann (1981) gave a remarkable talk at the 3rd International Conference on Ion Implantation Equipment and Technique in Kingston. Four years later, at the 5th conference in this series at Smuggler's Notch, Peter Rose gave a very personal view of the development of ion implantation equipment. He has been involved in this business since the early 1960s, a time when the term ion implantation was not yet known. At that time, ion implanters were called accelerators and were used only for basic research in solid state physics or in nuclear research Rose (1985). Also very interesting was a talk on ion beams in retrospect by Freeman in 1990, again in this series of conferences, which was similar to a paper published in 1986 (Freeman, 1986).

Figure 3 gives a genealogy of implanter companies, mainly originating from High Voltage Engineering Corporation. Not included in this genealogy are the implanters of the former Soviet Union. In total, 19 different models of Vesuvius implanters have been developed (Vyatkin, Simonov, and Kholopkin, 1991).

The special demands for ion implantation into semiconductors are homogeneity and dose control of the implanted dopant concentration, which is achieved by scanning and current integration, angular control, low contamination, cooling, and the avoidance of charging. These topics will be treated in the following sections.

2. ACCELERATION AND MASS SEPARATION

After extraction of the ions from the source (for a review of ion sources see, e.g., Stephens (1992)), the ions must be accelerated and separated according to their mass. This is usually done using electrostatic acceleration and a magnetic field for mass separation. The first systems used preacceleration with a fairly large magnet at ground potential (Fig. 4(a)). An exception was the Lintott machine, which used postacceleration with a single gap acceleration and, in case the energy was not high enough, a second acceleration with the target on high potential (Fig. 4(b)). Today, most implanters use the preanalysis concept with postacceleration as shown in Figure 4(c).

Exceptions to these three concepts were the first Accelerators, Inc. implanters, which used an E × B filter for mass separation. This worked very well in the early 1970s, with currents around one microampere; later, however, all these machines had to be retrofitted with magnetic mass separation due to space charge problems.

The current prevailing concept of postacceleration holds up very well to energies of 200 keV. In the late 1970s, machines with energies up to 400 keV were built. In the 1980s, interest arose in high-energy implantation. Varian and Eaton introduced high-energy machines in the late 1980s. These machines used a high-frequency scheme instead of an electrostatic acceleration principle. This is the basic principle of a LINAC (linear accelerator) (Lawrence and Sloan, 1931). When a standard dc beam is injected into a LINAC, ions that do not have the correct phase relation relative to the rf accelerating field are lost. Particles with the correct phase become grouped or bunched and emerge as a beam consisting of micropulses separated by a distance, $d = v/f$, where v is the final particle velocity and f is the frequency of the applied voltage. Positive ions are accelerated across the gaps when the rf voltage is negative into the next field-free drift tube. The length of this drift tube has to be adjusted so that the rf phase changes by 180° in the time it takes the ion to cross the gap. Of course, with increasing ion energies, these drift tubes are becoming longer. A schematic description of such a system is given in Figure 5. The advantage of this system is that no high voltage is required; the disadvantage is that the length of the system depends on the beam energy. Therefore, fairly large machines requiring a lot of clean room space resulted. Another disadvantage is that the system works only for a specific ion charge-to-mass ratio, which means that much flexibility is lost. Systems with energies up to 4 MeV have been built.

A different approach was pursued by several other companies using the tandem principle (Rose, 1970). Tandem accelerators were well known for many years in nuclear physics. They were used for energies of several tens

1 ION IMPLANTATION INTO SEMICONDUCTORS

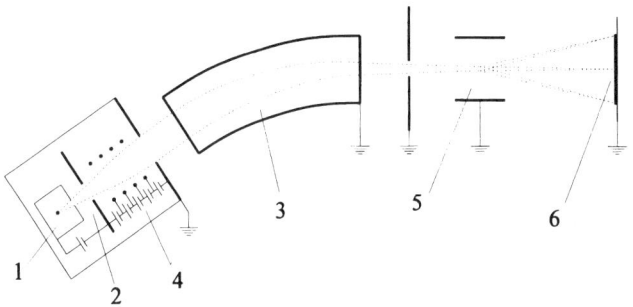

Fig. 4a

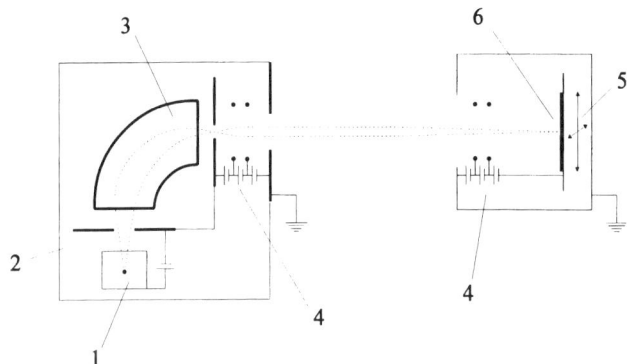

Fig. 4b

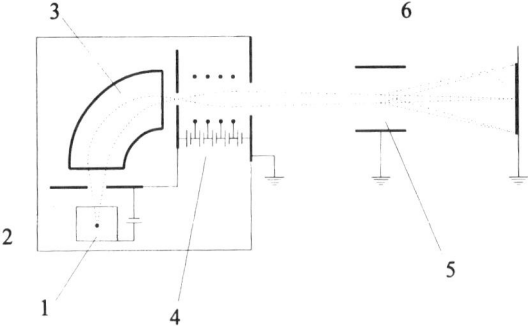

Fig. 4c

FIG. 4. Different implanter concepts; (a) preacceleration system, (b) postacceleration and additional second acceleration, and (c) postacceleration. 1 ion source, 2 extraction, 3 mass separation, 4 acceleration, 5 scanning, 6 target.

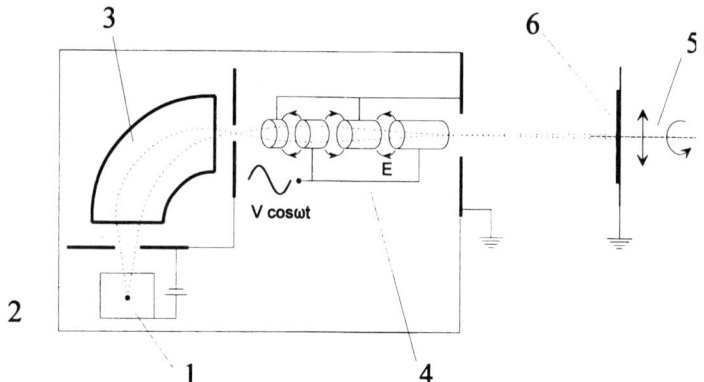

FIG. 5. Basic principle of rf-implanters. E, electric field caused by the applied voltage $V \cos \omega t$.
1 ion source, 2 extraction, 3 mass separation, 4 acceleration, 5 scanning, 6 target.

of MeV. By down-scaling and combination with a disc-type end station it was possible to develop production equipment (Fig. 6). In such a system, the positive ions extracted from the ion source are converted to negative ions by charge exchange, mass separated, accelerated, converted again to positive ions in an electron stripper and then accelerated with the same voltage to ground potential. The different charge states are separated using a second magnet. With a terminal voltage of 750 kV, ions with 1.5 MeV (singly charged) or 2.25 MeV (doubly charged) can be obtained.

In the early 1990s, device dimensions in vertical and lateral dimensions decreased so much that high-energy implants are also possible with stan-

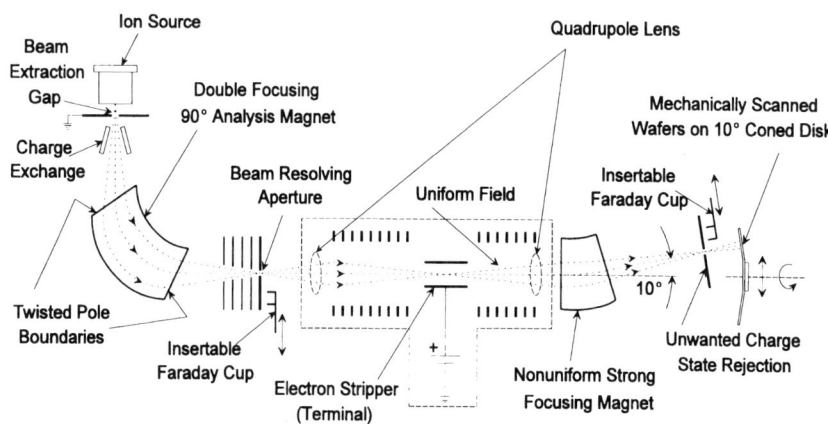

FIG. 6. Genus tandem high-energy implanter.

dard ion implantation equipment by using multiply charged ions. It remains to be seen if there will be a big demand for dedicated high-energy machines.

3. BEAM SCANNING AND DOSIMETRY

Beam scanning was done either by using electrostatic xy scanners or mechanical target scanning. In the early days, a tight implant angle control was not a special issue because wafer sizes were small. Medium current implanters used electrostatic scanners because of better access to the target, the possibility to change the wafer tilt and rotation, and short implantation times because of a much faster scan frequency. At high currents, space charge problems in the beam and wafer heating lead to batch scanning.

Electrostatic beam scanning is performed using two or three sets of capacitor electrodes (Aitken, 1989), including an extra deflection to avoid an

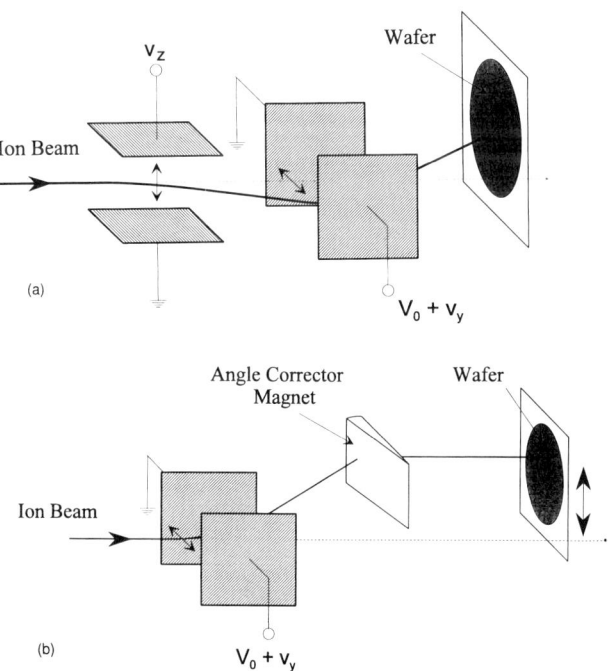

FIG. 7. Beam scanning in medium current implanters: (a) electrostatic and (b) mixed electrostatic-magnetic-mechanical scanning. v_z, v_y deflection voltage, V_0 offset voltage.

inhomogeneous implantation by neutralized atoms (neutral trap) (Fig. 7(a)). Usually, a fast scan was used for one axis and a slow scan for the other axis. With increasing wafer size, the angle variation across the wafer and the resulting channeling became unacceptable. Since electrostatic parallel scanning for large wafers is very difficult, a combination of electrostatic, magnetic, and mechanical scanning (Fig. 7(b)) can be used to obtain a parallel scan.

Mechanical batch scanning was performed in several different ways. The three most important methods are shown in Figure 8. The oldest system is the Lintott chain wheel, carousel, which was introduced around 1970. The first Varian-Extrion high-current implanter (about 1975) used a Ferris

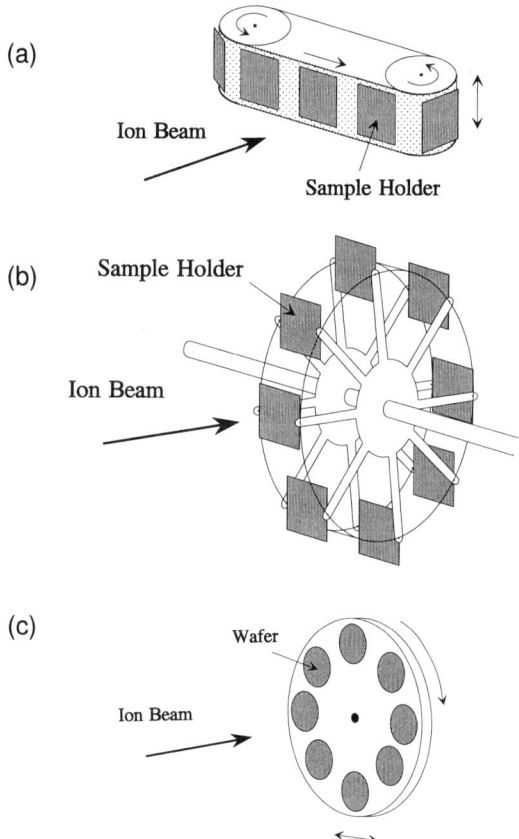

FIG. 8. Beam scanning: (a) Lintott chain wheel carousel, (b) Varian–Extrion Ferris wheel, and (c) spinning disc.

wheel-type carousel with constant wafer angle. In both designs, the wafers were mounted on special carriers. Depending on the wafer size, either several wafers or only one wafer was mounted on a carrier. It was never possible to design a cassette-to-cassette wafer loading system in both cases.

The only method still used today is the spinning disc. The main advantage is the high spinning speed without any mechanical oscillations, which leads to inhomogeneous doping in the older designs. To account for the different radial velocity of the wheel, the translation speed is changed using an electronic speed control of the drive.

Dosimetry always has been and still is performed using different designs of Faraday cups. To suppress secondary electrons, which would result in a higher current reaction, negatively charged electrodes are used in front of the cup.

4. CONTAMINATION

When implantation was introduced into semiconductor fabrication in the early 1970s, it was thought that ion implantation would be a very clean, if not the cleanest, manufacturing step. The reason for this assumption was the mass resolution provided by the separator magnet.

In the mid-1970s, it became known that particles were inadvertently being added to wafers during ion implantation. Particles may be generated by wear from moving mechanisms and may be process residues. By choosing the proper material and proper design of wafer handling components, it was possible to improve the situation considerably. Special cleaning cycles (e.g., introducing dry gas into the target chamber while spinning the disc, followed by pump downs) and improved wafer handling solved this problem.

A more severe problem arose in the early 1980s when it was realized that sputtering from the beam-line and from apertures causes heavy-metal contamination. In the course of 15 years, apertures were changed from molybdenum or stainless steel to aluminum, carbon, and silicon. Parts of the beam line also were covered with protective layers. Especially with oxygen implanters for SIMOX (separation by implanted oxygen) (Ruffell et al., 1987), contamination was a very severe issue and, therefore, after mass separation all parts of the accelerator that might be hit by the beam are covered with silicon.

However, contamination is not carried only through sputtering from the beam line and apertures onto the wafer but also by mass interferences that cannot be separated. The most prominent example is the interference between doubly charged molybdenum from the ion source material and boron fluoride ($^{11}B^{19}F_2$) (Cubina and Frost, 1991).

Further sources of contamination are the direct contact between wafer and sample holder, nonideal mass separation, charge exchange, and outgassing from implanter components. A review of these effects can be found in an article by Ryssel and Frey (1992).

5. IMPLANTER TRENDS

Ion implanters are fairly sophisticated and therefore very expensive pieces of equipment. For future machines, a parallel beam implantation is also mandatory for medium-current, single-wafer implanters. For advanced device types, large tilt angle implants will be increasingly important. An example are LATID (large angle tilt implanted drain) devices and the doping of deep trenches. This will be a domain for medium-current implanters because with high-current machines this technique seems impractical.

Further reductions of contamination levels are definitely necessary as are better solutions for reducing wafer changing. Wafer charging is an effect that became more and more important when gate oxides became thinner and ion currents higher. The gate metallization on an integrated circuit works like an antenna, conducting charge to the gate areas and forcing a current through the oxide. Depending on the gate oxide quality, it is damaged or destroyed. To avoid this effect, the ion beam in front of the wafers is flooded with electrons to compensate for this charging effect. This has been a very hot issue since the mid-1980s, but a final solution is not yet available. The big problem is that it is very difficult to compensate exactly for the ion charge and overcompensation will kill the devices. Many papers concerning this problem can be found in the IIT proceedings.[1]

To be cost-effective, ion implantation tries to use as high currents as possible. Silicon is not especially sensitive against a temperature increase during implantation; however, the photoresist frequently used as a masking layer is very sensitive. All modern implanters use gas cooling and this will continue in the future.

With shrinking device dimensions, high-energy implantation will be introduced in manufacturing. The implantation of retrograde wells (Wordeman, Demmard, Sai-Halasz, 1981; Terrill *et al.*, 1984) saves a lot of the time necessary to produce the deep junctions required and will result in better device properties. Moreover, it is expected that one can save many processing steps, which means cost savings. At the same time, p-n junction depth is becoming smaller and smaller, which means that the demand for energies will be decreasing to the keV range. This is probably not possible with present-day implanters, and low-energy machines will have to be developed.

[1] References IIT 1981, 1982, 1985, 1987, 1989, 1991, 1993, 1995.

A technique that was invented in the mid-1970s is the formation of buried SiO_2 (silicon dioxide) and Si_3N_4 (silicon nitride) layers by implanting oxygen or nitrogen into silicon. After annealing, the implanted layers are converted to SiO_2 and Si_3N_4, respectively, layers which are covered by a single crystalline silicon layer. The doses required for this technique are between 10^{17} and several times $10^{18}\,cm^{-2}$. In combination with local oxidation, fully isolated devices can be manufactured. These devices are radiation hard, faster than comparable devices in conventional techniques, and simpler to manufacture. The drawbacks at present are the high manufacturing costs of such layers and crystal defects in the silicon layer.

IV. Metal-Oxide Semiconductor Devices

The first successfully implanted devices were metal-oxide semiconductor (MOS) transistors. The introduction of ion implantation into semiconductor manufacturing would not have been possible without some unique properties offered by the technique. The first of these properties was the possibility to self-align the gate. The second was the possibility for a superior process control by current integration, which allowed adjustment of the threshold voltage for MOS devices and precise well doping for complementary MOS (CMOS) transistors. Very important for the introduction into the semiconductor fabrications was that the necessary ion dose was fairly low, from about $10^{11}\,cm^{-2}$ to $10^{14}\,cm^{-2}$. This means that the required current for a sufficient throughput was only in the microampere region, considering the small wafer diameter of 1.25 inches at that time. The pioneering work was performed at Hughes by Bower et al., 1968; Bower and Dill, 1966; and Dill, Bower, and Toombs, 1971. In the following sections, these early applications of implantation will be described.

1. Self-Aligned Gate Process

A major problem in the late 1960s and early 1970s was the mask alignment errors in aluminum gate MOS devices. The gate had to be manufactured after source and drain doping because of the low eutectic temperature of Al–Si (aluminum–silicon). In order to ensure functioning of the transistors despite the unavoidable alignment errors, the gate electrode overlapped the diffused source and drain areas by 2 to 6 μm on each side, reducing the switching speed of the transistors through the Miller feedback capacitance (Dill et al. 1971).

With ion implantation, it was possible to self-align the gate to the source-drain areas by using the gate electrodes as an implantation mask, thus producing exactly the desired channel length. With channel lengths of 5 to 7 µm, this meant a perfect registration, avoiding completely the unwanted overlap. Figure 9 shows a comparison of the conventional and the self-aligned techniques. Usually a low-dose implantation below $10^{14}\,\mathrm{cm}^{-2}$ at 60 to 100 keV boron or 100 to 150 keV phosphorus was used with a subsequent annealing at 550°C, resulting in a typical sheet resistivity 2.5 Ω/□. The low annealing temperature was necessary to avoid alloy formation between the aluminum and silicon.

At that time, everyone assumed a perfect alignment because lateral straggling of implanted ions was not yet known. Today, ion implanters deliver sufficient current, so that source and drain areas are fully implanted using the polysilicon (or polycide) gate as a mask, and annealing at higher temperature results in full activation of the implanted ions (Fig. 9). Moreover, device dimensions have been decreasing over the last 25 years and the effect of lateral straggling is well known. To avoid the overlap caused by this straggling, so-called spacers are used (Part IV, Section 4).

2. THRESHOLD ADJUST

For the general introduction and acceptance of ion implantation in semiconductor manufacturing, a second innovation by Hughes researchers was the key. By a shallow implantation through the gate oxide into the channel region of MOS devices, it was found that the threshold could be adjusted, that is, set to a desired value (Aubuchon, 1969). The threshold voltage of a MOS transistor is given by

$$V_T = \Phi_{MS} + 2\Phi_F - \frac{Q_{SS} + Q_B}{C_{OX}} \quad (1)$$

where Φ_{MS} is the difference between the work functions of the gate material and silicon, Φ_F the Fermi potential, Q_{SS} the sheet charge at the interface ($Q_{SS} = qN_{SS}$; N_{SS} is the surface state density), Q_B the space charge, and C_{OX} the gate capacitance per unit area, ($C_{OX} = \varepsilon_0 \varepsilon_{r,ox}/t_{ox}$; ε_0 the permittivity of free space, $\varepsilon_{r,ox}$ the dielectric constant of the oxide, t_{ox} oxide thickness). By implanting in the channel region, an effective charge is deposited that changes V_T to

$$V'_T = V_T \pm \frac{qK}{C_{OX}} N_\square \quad (2)$$

where $N_\square$ is the implanted dose. The constant K indicates which portion of

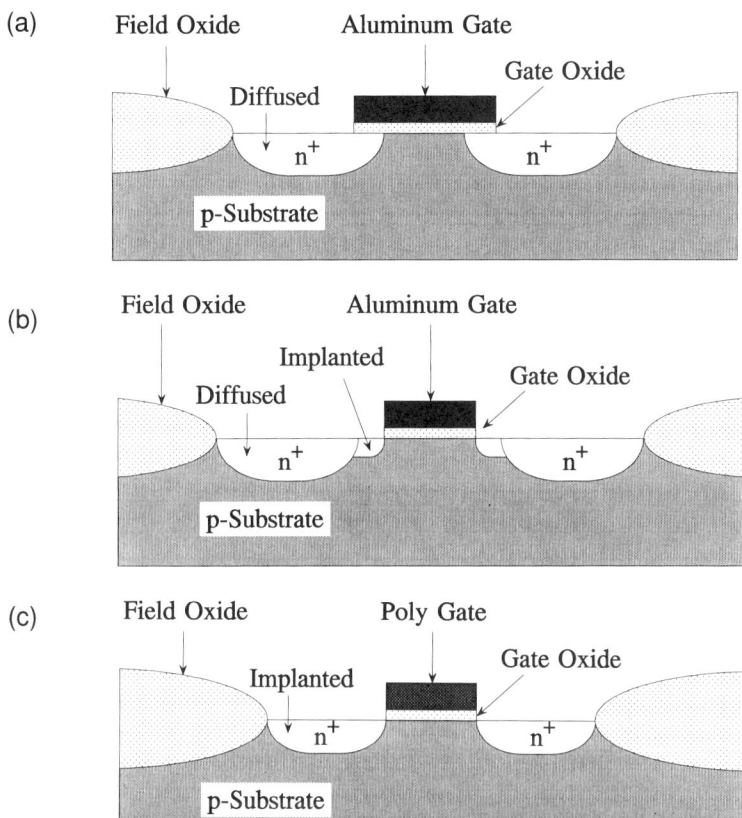

FIG. 9. Source-drain doping: (a) standard metal gate process, (b) self-aligned gate metal process, and (c) fully implanted self-aligned source drain.

the implanted ions influences the threshold voltage, that is, which portion is in the channel and is electrically active. The positive sign holds for p-type dopants and the negative sign for n-type dopants.

Before introducing ion implantation, the threshold voltage can be changed only by changing the gate metal, the dielectric layer (thickness, material, or both), the wafer orientation, or the bulk doping.

Using this method, not only can the threshold of enhancement transistors, which are normally off, be adjusted (usually $|V_T|$ is reduced), but it is also possible to manufacture depletion transistors on the same chip by selective implantation.

These two applications were decisive for large scale application of MOS circuits starting in the mid-1970s. It was soon found that it is also

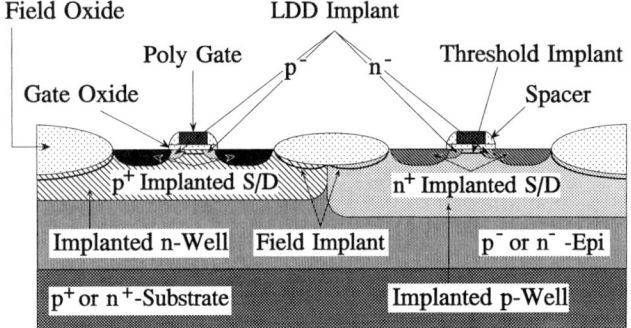

FIG. 10. Implanted basic complementary metal-oxide semiconductor (CMOS) inverter structure.

worthwhile to adjust the field threshold under the thick oxide to avoid parasitic channel formation and possible short circuits between adjacent transistors caused by the bias applied to a metal line.

Field threshold adjust can be performed before the thick oxide layer is grown or afterward by implanting through the thick oxide. The latter technique is easier to control, especially in the case of boron, because segregation effects are avoided at the interface between oxide and silicon.

3. WELL DOPING

Shortly after the introduction of self-aligned gates and threshold adjust by ion implantation, the third innovation was made. Both the p-channel MOS and the n-channel MOS techniques have advantages and disadvantages. When in the early 1970s, CMOS devices were proposed, the chances seemed to be not very good for this area-consuming technique which, in addition, required approximately two times the processing steps of simple p and n-channel devices. There was, however, one big advantage: low power consumption, which was especially important at that time for wristwatches.

For CMOS devices, it is necessary to manufacture a p-well or an n-well, depending on whether one is starting with n- or p-doped silicon wafers. It would be a very interesting subject in itself to consider all the dispute as to whether p-channel or n-channel devices or p-well-based or n-well-based CMOS devices are the right technical solutions. To manufacture a well, boron or phosphorus is implanted into a dose of around $10^{13}\,cm^{-2}$ and subsequently driven in at temperatures around 1200°C for 10 to 20 h to give a well about 4 μm deep (Swanson and Meindl, 1972). In Figure 10, an example for a CMOS inverter is shown. The advantage of ion implantation is the exact control of the dopant concentration, which is much superior to standard diffusion processes. By choosing the right well implantation, the threshold voltage of the transistor in this well can be adjusted without additional implantation (Fig. 11). Nevertheless, for higher flexibility most companies set the threshold voltage by an additional implantation step (Dill et al., 1972). Phosphorus n-well implants have been used since the mid-1980s to about 90% of total CMOS because of the superior performance of the n-channel transistor in lightly doped substrates.

The next step in the development of CMOS transistors was the use of twin wells, which means that irrespective of the substrate doping, n-wells and p-wells are implanted in order to obtain better control of the doping concentration in the well. In designing a well process, several issues must be taken into account. The surface concentration must be high enough so that sufficient isolation for active and passive devices is obtained. This, however, can be adjusted by channel implants and field implants for threshold adjust. Vertical punch-through between the substrate and the devices in a well with opposite doping must be avoided. Dynamic random access memories (DRAMs) require especially high p-well doping to suppress punch-through between adjacent trenches and to reduce the soft error rate (Küster, Müklhoff, and Cerra, 1991).

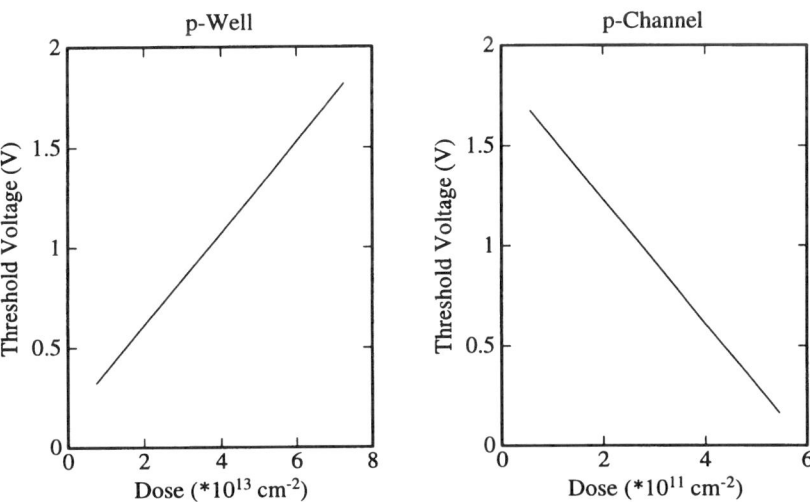

FIG. 11. Dose-dependence of threshold voltage for 60 keV boron through 100 nm silicon dioxide in 3 Ω cm ⟨100⟩ Si with drive in at 1200°C for 16 h (n-channel transistor in p-well) and 50 keV boron through 100 nm silicon dioxide with annealing at 525°C for 10 min (p-channel transistor). (Data from Dill et al., 1971.)

One big drawback of CMOS is the latch-up effect (parasitic npnp thyristor structures, which can lead to destruction of the device (Terrill *et al.*, 1984; Wordeman, Demmard, and Sai-Walasz, 1981)). This effect can be minimized by designing wells with high doping concentration and minimum current gain for the parasitic bipolar transistors.

The manufacture of p and n-wells can be done using two masking steps and a subsequent high-temperature drive-in step, as is the case of standard well formation. However, it also can be done in a self-aligned technique with local oxidation (Parillis *et al.*, 1980) or using an unmasked p-well implantation followed by overcompensation of the p-well doping by n-dopants (Muhlhoff *et al.*, 1989).

The formation of wells by high-energy implantation, thus avoiding high-temperature time-consuming drive-in steps, already had been proposed in the early 1980s (Rung, Dell Oca, and Walker, 1981). In the mid 1990s, it is slowly making its way into manufacturing. Not only is the thermal budget much lower than in conventional processing, but the inherent property of ion implantation that the maximum of the doping concentration is below the surface (which was sometimes detrimental for device applications) is a big advantage: the so-called retrograde well with a much higher doping concentration in the bulk reduces latch-up considerably (Terrill *et al.*, 1984).

4. DRAIN ENGINEERING

The properties of MOS transistors depend strongly on the impurity profile in the drain region. Lightly doped drains (LDD) are very important for n-channel devices because the hot carrier injection for electrons is much higher than it is for holes. Probably more than 50% of submicron n-channel devices have a lightly doped drain, which is achieved by a low-dose phosphorus implant in addition to the standard arsenic implant (Fig. 12(a)). The effect of the low-doped drain is to reduce the peak electric field in the channel and thus to reduce hot carrier injection into the gate. In the case of p-channel devices, LDD transistors are not yet necessary; however, they might become important in the future with further shrinking of device dimensions.

For submicron transistors, the standard 7-degree tilt during implantation leads to asymmetric transistors. This can be overcome by rotation during ion implantation. Depending also on the design, implantation from two to four different directions may be used. A further development is the use of LATID devices. To manufacture low-doped drains, angles of up to 60 degrees are used (Fig. 12(b)).

Presently, drain engineering is performed mostly on an experimental basis. To optimize processes, process modeling must be used with optimized

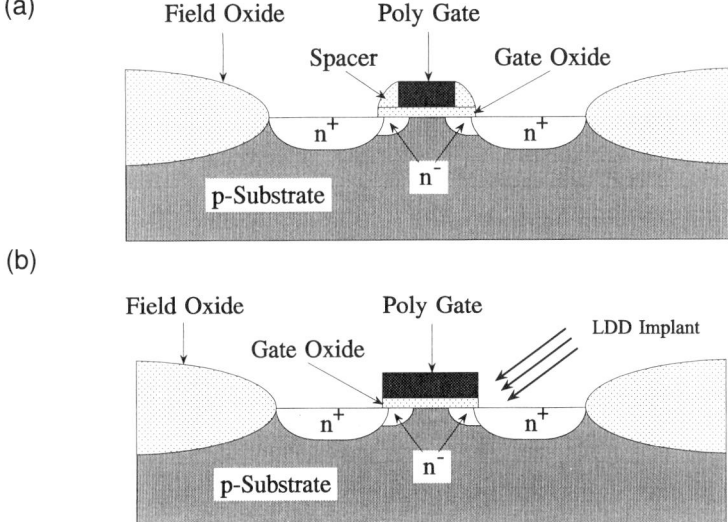

FIG. 12. Advanced transistor structures: (a) lightly doped drains (LDD) transistor and (b) large angle tilt implanted drain (LATID) transistor.

models describing the channeling effect as a function of implant angle and describing transient diffusion during annealing. Figure 13 shows an example of such a calculation.

Rotation during implantation is also necessary to obtain a homogeneous doping of deep trenches, which are used in DRAMS.

5. METAL-OXIDE SEMICONDUCTOR TRENDS

In addition to drain engineering to optimize device performance mentioned previously, research is going in the direction of elevated drains in which the dopant is implanted into polysilicon or silicides, and subsequently the junction is formed by out diffusion. Especially for p-channel devices, the formation of shallow junctions is difficult because of the light mass of boron. BF_2^+ implantation has been used for quite some time. Difficulties controlling the dissociation of BF_2^+ have led to the experimental use of germanium or silicon preamorphization, prior to boron implantation. An alternative is to use very low energy implantation (Part III, Section 5). Out-diffusion from silicides, possibly using rapid thermal annealing (RTA), seems promising and is being widely investigated but has not yet made its way into manufacturing.

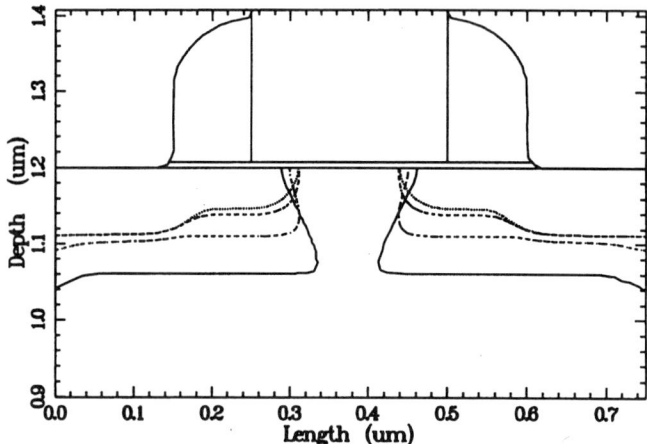

FIG. 13. Simulated equiconcentration lines of donor distributions at the level of $5 \times 10^{17}\,\text{cm}^{-3}$ in a large angle tilt implanted drain (LATID) device using different model assumptions when simulating the source–drain extensions. Dotted line is the amorphous target for $\pm 45°$ tilt and $0°$ rotation. The dashed line is the crystalline target for $\pm 45°$ tilt and $0°$ rotation. The dashed-dotted line is the crystalline target for $\pm 45°$ and $36°$ rotation. The drawn line is the crystalline target for $\pm 45°$ tilt and $45°$ rotation.

It is definite that high-tilt implants—rotation during implantation for improved homogeneity and compensation of the shaddowing effect caused by the 7-degree tilt—will be introduced on a larger scale.

The high-energy technique is very promising for well formation and could possibly reduce manufacturing costs considerably. A strong competitor in this respect is the formation of SOI (silicon on insulator) structures by high-energy oxygen implantation (silicon separation by implanted oxygen, or SIMOX) or, more recently, by bonding of oxidized wafers.

V. Bipolar Devices

The use of ion implantation of bipolar transistors was introduced much later into manufacturing than was that of MOS transistors. There were several reasons for this delay. First, most ion implanters in the late 1960s and early 1970s supplied very low current. One was fighting for microamps and implantation times for emitters that were much too long for a cost-effective process. Second, diffusion techniques for bipolar transistors (usually phosphorus for the emitter and boron for the base) were well developed.

This is still the approach used in the case in which device dimensions are large (≥ 3 μm) and the base width is wide enough (≥ 0.5 μm) (Hill and Hunt, 1991). Third, and maybe most important, the early experiments to manufacture bipolar transistors by phosphorus implantation were not really successful due to channeling tails that compensated the base.

When implantation was installed in process lines because of the manufacturing of MOS devices and annealing processes were well established, the benefits from using ion implantation for other devices was fast realized.

Table I gives a summary of the incorporation of ion implantation into bipolar integrated circuit technology. Table I shows each region of the transistor, the dopant used, the approximate implant energy and dose range, and the frequency of use.

1. Arsenic Emitter

Increasing the transit frequency above by approximately 1 GHz, by narrowing the boron base, resulted in irreproducible results because of poor process control in the case of diffusion (Tokuyama, Skeda, and Tsuchimoto,

TABLE I
Use of Ion Implantation in Bipolar Processing

Bipolar Device Region	Dopant	Energy (keV)	Dose (cm^{-2})	Frequency of Use
Buried collector	As, Sb	30–50	$>10^{15}$	Medium
Channel stop	B	30–50	$<10^{14}$–10^{15}	Medium
Collector contact	P	30–>70	$>10^{15}$	Low
Isolation	B	30–50	$>10^{15}$	Low
Base	B	<30	$<10^{14}$	High
Base contact				
Single	B	<30	$>10^{15}$	High
Crystal		<30	$>10^{15}$	High
Polysilicon	B	<30	$>10^{15}$	High
Emitter				
Single	As	30–50	$>10^{15}$	High
Crystal		30–50	$>10^{15}$	High
Polysilicon	As	30–50	$>10^{15}$	High
Resistor				
Single	B	<30	$<10^{14}$	High
Crystal		<30	$<10^{14}$	High
Polysilicon	B, P	<30	$<10^{14}$–10^{15}	High

Data from Hill and Hunt (1991).

1970). A breakthrough occurred when, for the first time, arsenic instead of phosphorus was used by Payne and Scavuzzo (1971) and Payne et al. (1974) to implant the emitter. After annealing, a very steep profile results, steeper than the original implantation profile due to concentration-dependent diffusion of arsenic. Conventional doping with arsenic is only possible in a closed ampoule system using arsenic vapor or an arsenic-doped solid germanosilicate source. Neither method was ever used in the production of microelectronic devices.

A few years later, it was found that it is possible to make an even shallower emitter when the arsenic is implanted into the polysilicon with subsequent out-diffusion into the single crystalline base region, thus completely avoiding damage in the semiconductor. This became the mainstream bipolar technique in the early 1980s, together with an implanted boron base. Figure 14 shows the schematic cross section of such a bipolar transistor. In addition to the polysilicon emitter and the base, the channel stopper also has been implanted.

2. BORON BASE

In parallel to the implantation of the emitter, Payne and Scavuzzo (1971) also implanted the base using boron. This resulted in a much better homogeneity than had been known from standard diffusion, and thus, transistors with gains of up to several thousand easily can be manufactured. Moreover, implanting base and emitter gave some additional flexibility in manufacturing of devices. With diffusion it was always the base that had to be doped first. With ion implantation this was no longer necessary. By

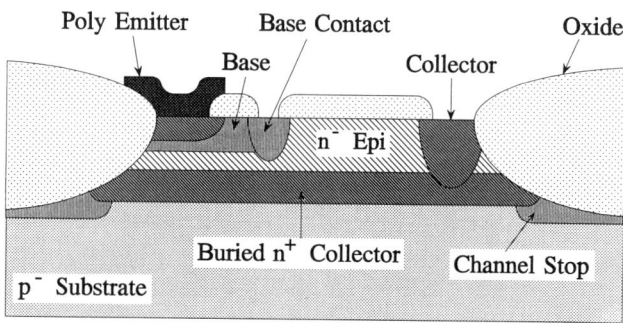

FIG. 14. Schematic cross-section through a bipolar transistor with implanted polysilicon emitter, implanted base, implanted channel stop, and implanted base collector.

changing the implantation dose and energy of the base, it is possible to change current gain and cut-off frequency. In Figure 15, typical results of the early work of Payne et al. (1974) are given, showing the independent control of cut-off frequency and current gain. To reduce contact resistance of the base, a base contact implantation usually is used (Fig. 14).

Since the late 1980s, the base contact is frequently made by out-diffusion from the implanted poly-I layer together with the arsenic emitter, which is diffused from the poly-II layer.

3. BIPOLAR TRENDS

It is difficult to see trends in bipolar implantation independent from the further development of MOS devices. The main topics probably will be to bring together bipolar and MOS circuits in bipolar CMOS (BICMOS) devices. The task here is to use as many doping steps from CMOS technology for the bipolar device as possible and, of course, vice versa.

As a stand-alone technique in bipolar circuits for extremely high frequencies, a germanium-doped hetero-base could be a new application for ion implantation. It is not certain, however, if it can compete with molecular beam epitaxy. High-energy implantation has already been investigated for

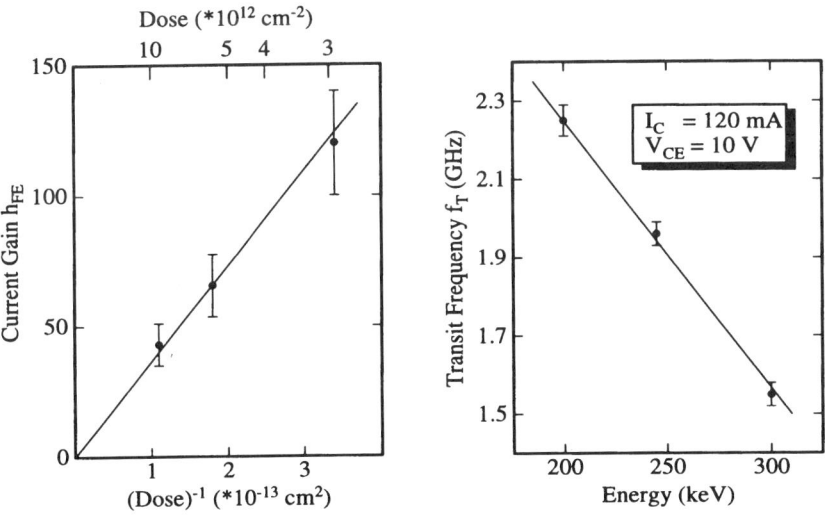

FIG. 15. Dependence of current gain and cut-off frequency on implantation dose. I_C, collector current; V_{CE}, collector-emitter voltage. (Data from Payne et al., 1974.)

the formation of buried subcollectors; in the early 1970s however, at that time, subcollectors were so deep that only very high energy implantation could be used, which resulted in unacceptably long implantation time. With the advance of high-energy implanters and additional decreasing device dimensions, there might be a new chance for this application. Theoretical calculations show that instead of 5 MeV phosphorus and a dose of 5×10^{15} cm^{-2} for half-micron devices, an energy of 600 keV and a dose of 7×10^{13} cm^{-2} would be sufficient (Hill and Hunt, 1991). Ideas of buried silicide collectors and SIMOX applications (Hill and Hunt, 1991) are purely speculative.

VI. Conclusions

This short review highlighted the development of ion implantation from a tool for nuclear physicists to a reliable production tool in semiconductor fabrication. It was not possible to cover all the aspects of devices manufactured by ion implantation, and only highlights of equipment development and MOS devices and bipolar devices could be presented.

ACKNOWLEDGMENTS

I thank Ch. McKenna from Varian, J. Gyulai from KFKI RIMS and V. Häublein, J. Pelka, M. Rommel, and J. Lorenz from my institute who assisted me through discussions, collection of material, and drawing figures.

REFERENCES

Agajanian, A. H. (1974). *Radiation Effects* **23**, 73.
Aitken, D. (1989). *Vacuum* **39**, 1025.
Alväger, T., and Hansen, N. J. (1962). *Rev. Sci. Inst.* **33**, 367.
Aubuchon, K. G. (1969). International Conference on the Properties and Use of MIS Structures, Grenoble.
Bohr, N. (1913). *Philos. Mag.* **25**, 10.
Bower, R. W., and Dill, H. G. (1966). *IEEE Electron Device Meeting*, paper 16.6. Washington, D.C.
Bower, R. W., Dill, H. G., Aubuchon, K. G., and Thompson, S. A. (1968). *IEEE Trans. Electron Devices* **ED-15**, 757.
Carter, C., and Grant, W. A. (1976). *Ion Implantation of Semiconductors*. Edward Arnold Ltd., London.
Cubina, A., and Frost, M. (1991). *Nucl. Instrum. Meth.* **B55**, 160.

Cussins, W. D. (1955). *Proc. Phys. Soc.* **368**, 213.
Dearnaley, G., Freeman, J. H., Nelson, R. S., and Stephen, J. (1973). *Ion Implantation.* North-Holland, Amsterdam.
Degen, P. L. (1973). *Phys. Stat. Sol.* **16**, 9.
Dill, H. G., Bower, R. W., and Toombs, T. N. (1971). In: *Ion Implantation* (F. H. Eisen and L. T. Chadderton, eds.), Gordon and Breach, London, p. 349.
Dill, H. G., Finnila, R. M., Leupp, A. M., and Toombs, T. N. (1972). *Solid State Technol.* **15**, 27.
Douglas, E. C., and Dingwall, A. G. F. (1975). *IEEE Trans. Electron Devices* **ED22**, 849.
Forbes, L. (1973). *IEEE J. Solid-State Circuits* **SC-8**, 226.
Freeman, J. H. (1963). *Nucl. Instrum. Meth.* **22**, 306.
Freeman, J. H. (1986). *Radiation Effects* **100**, 161.
Gibbons, J. F. (1968). *Proc. IEEE* **56**, 295.
Gibbons, J. F., Moll, J. L., and Meyer, N. I. (1965). *Nucl. Instrum. Meth.* **38**, 165.
Goode, P. D. (1970). *Report AERE-R*, 6401, Harwell.
Graul, J., Kaiser, H, Wilhelm, J. W. and Ryssel, H. (1975). *IEEE J. Solid-State Circuits* **SC-10**, 201.
Hill, C., and Hunt, P. (1991). *Nucl. Instrum. Meth.* **B55**, 1.
IBMM (1971a). *Ion Implantation* (F. H. Eisen and L. T. Chadderton, eds.), Gordon and Breach, New York.
IBMM (1971b). *Ion Implantation in Semiconductors* (I. Ruge and J. Graul, eds.), Springer Verlag, Berlin and New York.
IBMM (1973). *Ion Implantation in Semiconductors and Other Materials* (B. L. Crowder, ed.), Plenum, New York.
IBMM (1975). *Ion Implantation in Semiconductors* (S. Namba, ed.), Plenum, New York.
IBMM (1977). *Ion Implantation in Semiconductors* (F. Chernow, J. A. Borders, and D. K. Brice, eds.), Plenum, New York.
IBMM (1979). *Ion Beam Modification of Materials I, II and III*, (J. Gyulai, T. Lohner, and E. Pasztor, eds.), KFKI Publishing, Budapest (partly in *Radiation Effects* 47-49 (1980)).
IBMM (1981). *Ion Beam Modification of Materials, I–II*, (R. E. Benenson, E. N. Kaufman, G. L. Miller, and W. W. Scholz, eds.), North-Holland, Amsterdam; *Nucl. Instrum. Meth.* **182/183**.
IBMM (1983). *Ion Beam Modification of Materials*, (B. Biasse, G. Destefanis, and J. P. Gaillard, eds.), North-Holland, Amsterdam; *Nucl. Instrum. Meth.* **209/210**.
IBMM (1985). *Ion Beam Modification of Materials, I–II*, (B. M. Ullrich, ed.), North-Holland, Amsterdam; *Nucl. Instrum. Meth.* **B7/8**.
IBMM (1987). *Ion Beam Modification of Materials, I–II*, (S. U. Campisano, G. Foti, P. Mazzoldi, and E. Rimini, eds.), North-Holland, Amsterdam; *Nucl. Instrum. Meth.* **B19/20**.
IBMM (1989). *Ion Beam Modification of Materials*, (S. Namba, ed.), North-Holland, Amsterdam; *Nucl. Instrum. Meth.* **B39** (1990).
IBMM (1991). *Ion Beam Modification of Materials*, (S. P. Withrow, and D. B. Poker, eds.), North-Holland, Amsterdam; *Nucl. Instrum. Meth.* **B59/60**.
IBMM (1993). *Ion Beam Modification of Materials*, (S. Kalbitzer, O. Meyer, and G. K. Wolf, eds.), North-Holland, Amsterdam; *Nucl. Instrum. Meth.* **B80/81**.
IBMM (1995). *Ion Beam Modification of Materials*, (J. S. Williams ed.), North-Holland, Amsterdam; *Nucl. Instrum. Meth.* **B106**.
IIT (1981). *Ion Implantation Equipment and Techniques*, (C. M. McKenna, P. J. Scanlon, and J. R. Winnard, eds.), North-Holland, Amsterdam; *Nucl. Instrum. Meth.* **189**.
IIT (1982). *Ion Implantation Techniques*, (H. Ryssel and H. Glawischnig, eds.), Springer Series on Electronphysics, Vols. 10 and 11, Springer, Berlin, Heidelberg, New York.

IIT (1985). *Ion Implantation Equipment and Techniques*, (J. F. Ziegler and R. L. Brown, eds.), North-Holland, Amsterdam; *Nucl. Instrum. Meth.* **B6** (1–2).
IIT (1987). *Ion Implantation Technology*, (M. I. Current, N. W. Cheung, W. Weissberger, and B. Kirby, eds.), North-Holland, Amsterdam; *Nucl. Instrum. Meth.* **B21** (2–4).
IIT (1989). *Ion Implantation Technology*, (T. Takagi, ed.), North-Holland, Amsterdam; *Nucl. Instrum. Meth.* **B37/38**.
IIT (1991). *Ion Implementation Technology*, (K. G. Stephens, P. L. F. Hemment, K. J. Reeson, B. J. Sealy, and J. S. Colligon, eds.), North-Holland, Amsterdam; *Nucl. Instrum. Meth.* **B55**.
IIT (1993). *Ion Implantation Technology*, (D. F. Downy, M. Farley, K. S. Jones, and G. Ryding, eds.), North-Holland, Amsterdam; *Nucl. Instrum. Meth.* **B74**.
IIT (1995). *Ion Implantation Technology*, (S. Coffa, G. Ferla, F. Priolo, and E. Rimini, eds.), North-Holland, Amsterdam; *Nucl. Instrum. Meth.* **B96**.
King, W. J., Burrel, J. T., Harrison, S., Martin, F., and Kellet, C. M. (1965). *Nucl. Instrum. Meth.* **38**, 178.
Küster, K. H., Mühlhoff, H.-M., and Cerra, H. (1991). *Nucl. Instrum. Meth.* **B55**, 9.
Lawrence, E. O., and Sloan, D. H. (1931). *Proc. Natl. Acad. Sci.* **17**, 64.
Lindhard, J., Scharff, M., and Schiøtt, H. E. (1963). *Kgl. Danske Videnskab. Selskab. Mat-Fys. Medd.* **33**, 14.
Makarov, V. V., and Petrov, N. N. (1966). *Soviet Phys.-Solid State* **8**, 1272.
Martin, F. W., Harrison, S., and King, W. J. (1966). *IEEE Trans. Nucl. Sci.* **NS-5-13**, 22.
Mayer, I. W., Marsh, O. J., Mankarious, and Bower, R. (1967). *J. Appl. Phys.* **38**, 1975.
Mayer, J. W., Eriksson, L., and Davies, J. A. (1970). *Ion Implantation in Semiconductors*, Academic Press, New York.
Mazzio, J. (1971). *Ion Implantation: A Selective Bibliography.* NTIS Report No.: SC-B-710148. Sandia Laboratories, Albuquerque, New Mexico.
McCaldin, J. O., and Widmer, A. E. (1963). *J. Phys. Chem. Solids* **24**, 1073.
Mühlhoff, H.-M., Law, F., Küppe, P., and Rühl, S. (1989). *ESSDERC 89*, Springer-Verlag, Berlin, **553**.
Ohl, R. (1952). *Bell Syst. Tech. J.* **31**, 104.
Parillio, L. C., Payne, S., Davis, R. E., Reutlinger, G. W., and Field, R. L. (1980). *IEEE Internat. Electron Devices Meeting*, **752**.
Payne, R. S., and Scavuzzo, R. J. (1971). *IEEE Internat. Electron Devices Meeting*, Washington, D.C.
Payne, R. S., Scavuzzo, R. J., Olson, K. H., Nacci, J. M., and Moline, R. A. (1974). *IEEE Trans. Electron. Devices* **ED-21**, 273.
Rose, P. H. (1985). *Nucl. Instrum. Meth.* **36**, 1.
Rose, P. H., and Wittkower, A. B. (1970). *Sci. Am.* **223**, 24.
Ruffell, J. P., Douglas-Hamilton, D. H., Kaim, R. E., and Izumi, K. (1987). *Nucl. Instrum. Meth.* **B21**, 229.
Rung, R. D., Dell Oca, C. J., and Walker, L. G. (1981). *IEEE Trans. Electron Devices* **ED-28**, 1115.
Ryssel, H., and Frey, L. (1992). In: *Handbook of Ion Implantation Technology*, (J. F. Ziegler, ed.), North-Holland, Amsterdam, p. 675.
Ryssel, H., and Ruge, I. (1986). *Ion Implantation*, John Wiley & Sons, Chichester.
Seager, D. K. (1973). *Ion Implantation — A Bibliography.* NTIS Report No.: SC-B-71048, Suppl. I. Sandia Laboratories, Albuquerque, New Mexico.
Shockley, W. (1957). U.S. Patent No. 2787, 564.
Stengl, G., Löschner, H., and Murray, J. J. (1986). *Solid State Technol.* 119.

Stephens, K. G. (1992). In: *Ion Implantation Technology* (J. F. Ziegler, ed.), North-Holland, Amsterdam, p. 455.
Swanson, R. S., and Meindl, S. D. (1972). *IEEE J. Solid-State Circuits* **SC-7,** 146.
Terrill, K. W., Byrne, P. F., Hu, C., and Cheung, N. W. (1984). *Appl. Phys. Lett.* **45,** 977.
Tokuyama, T., Ikeda, T., and Tsuchimoto, T. (1970). *Proc. 4th Microelectronics* Oldenburg, München, **36.**
Vyatkin, A. F., Simonov, V. V., and Kholopkin, A. I. (1991). *Solid State Technol.* **34,** 57.
Wegmann, L. (1981). *Nucl. Instrum. Meth.* **189,** 1.
Wilson, R. G., and Brewer, G. R. (1973). *Ion Beams,* John Wiley & Sons, New York.
Wordeman, M. R., Demmard, D. H., and Sai-Halasz, G. A. (1981). *IEEE Trans. Electron Devices* **ED-81,** 40.

CHAPTER 2

Electronic Stopping Power for Energetic Ions in Solids

You-Nian Wang

DEPARTMENT OF PHYSICS
DALIAN UNIVERSITY OF TECHNOLOGY
DALIAN, P.R. CHINA[1]
AND DEPARTMENT OF PHYSICS
UNIVERSITY OF WATERLOO
WATERLOO, ONTARIO, CANADA

Teng-Cai Ma

DEPARTMENT OF PHYSICS
DALIAN UNIVERSITY OF TECHNOLOGY
DALIAN, P.R. CHINA

I. INTRODUCTION 32
II. GENERAL THEORY 33
   1. *Electron Gas Model* 33
   2. *Dielectric Function* 34
   3. *Local Density Approximation* 35
III. ELECTRONIC STOPPING POWER FOR PROTONS 36
   1. *Method of Numerical Calculation* 36
   2. *Low- and High-Velocity Approximations* 37
   3. *Fitted Formula and Comparison with Experimental Data* 38
IV. ELECTRONIC STOPPING POWER FOR HEAVY IONS 42
   1. *The Brandt-Kitagawa Model for Charge Distribution of a Projectile* . . 42
   2. *Effective Stopping Charges* 44
   3. *Comparison with Experimental Data* 45
V. ELECTRONIC STOPPING POWER FOR MOLECULAR IONS 47
   1. *Coulomb Explosion* 47
   2. *The Vicinage Effect in the Stopping Power* 49
VI. SUMMARY 52
   References 53

[1] Permanent address.

I. Introduction

Accurate values of electronic stopping power for energetic ions in solids are extremely valuable in various fields of science and technology. In particular, determination of the electronic stopping power becomes crucial for studies of ion ranges and radiation damage in some semiconductors, such as silicon (Si) and gallium arsenide (GaAS), and also as a test of up-to-date experimental data.

Over the past decades many theoretical models have been presented to describe the electronic stopping power, such as the quantum perturbance theory (Bethe, 1930), the dielectric theory (Lindhard, 1954; Lindhard and Winther, 1964), the kinetic theory (Sigmund, 1982), and the density functional theory (Echenique et al., 1986; Echenique, 1987). In this chapter we focus our attention on the linear-response dielectric theory. The dielectric theory is a many-body, self-consistent treatment of an electron gas responding to a perturbation by a charged particle. It includes the polarization of the electrons by charged particles and the resultant charge screening and the electron density fluctuations. With the local density approximation (LDA), it can be directly applied to any target and, for example, the solid effects can be treated.

Following the pioneering work of Lindhard (1954), several theoretical investigations into the dielectric theory have been made concerning calculations of the stopping power (Lindhard and Winther, 1964; Iafrate and Ziegler, 1979; Gertner, Meron, and Rosner, 1978; Ziegler, Biersack, and Littmark, 1985). Theoretical predictions for the stopping power have been found to be in good agreement with experimental data in the high-velocity region. However, at lower velocities the agreement becomes poor, especially below the maximum of the stopping-power curve. The discrepancies between theory and experiment in this region are due mainly to the fact that the random-phase approximation (RPA) dielectric function was used in all of the previous theoretical investigations. It is well known that the RPA dielectric theory is valid only in the weakly coupling limit, that is, when $r_s = [3/(4\pi n_0 a_0^3)]^{1/3} \leqslant 1$, where n_0 is the electron gas density and a_0 is the Bohr radius. In metals and semiconductors, however, the values of r_s range from about 1.5 (gold, tungsten) to 5.88 (cesium). It is evident that the RPA theory cannot provide accurate values of the stopping power for such materials. With a local-field correction (LFC) dielectric function (Utsumi and Ichimaru, 1982), in which the exchange and correlation interaction of the electron gas is included, we calculated the stopping power for energetic ions in a strongly coupled electron gas (Wang and Ma, 1990, 1991; Ma, Wang, and Cui, 1992). It has been shown that our results differ significantly from the predictions given by the RPA dielectric theory.

2 ELECTRONIC STOPPING POWER FOR ENERGETIC IONS IN SOLIDS

In fact, when a charged particle moves through a solid, it interacts not only with the conduction electrons, but also with the inner-shell electrons of atoms, especially in the case of high projectile velocity. In recent years, the LDA method has been applied to calculate the stopping power of solids, taking into account the contributions coming from the inner-shell electrons (Gertner, Merion, and Rosner, 1978; Ziegler, Biersack, and Littmark, 1985; Wang, Cui, and Ma, 1993; Cui, Wang, and Ma, 1993).

In following sections we shall review primarily some recent developments in the calculations of the stopping power with the dielectric theory and the LDA. The projectiles may be protons, heavy ions, or molecular ions.

II. General Theory

1. ELECTRON GAS MODEL

When an energetic ion with velocity $\mathbf{v}$ moves through a homogeneous electron gas with density n_0, the equilibrium state of the electron gas is disturbed and an electric field $\mathbf{E}_{\text{ind}}(\mathbf{r}, t)$ is induced around the ion. This field then acts on the ion, causing it to lose kinetic energy. Over the length dx, the energy loss of the ion, that is, the electronic stopping power, is given by

$$S_e = e \int d\mathbf{r} \rho_{\text{ext}}(\mathbf{r}, t) \mathbf{v} \cdot \mathbf{E}_{\text{ind}}(\mathbf{r}, t)/v \tag{1}$$

where e is the element charge and $\rho_{\text{ext}}(\mathbf{r}, t)$ is the charge density distribution of the projectile.

First, we need to specify the external charge density distribution in order to determine the induced electric field. For a classical charge distribution in uniform motion, the external charge density is

$$\rho_{\text{ext}}(\mathbf{r}, t) = \rho_0(\mathbf{r} - \mathbf{v}t) \tag{2}$$

that is, in Fourier space $(\mathbf{k}, \omega)$,

$$\rho_{\text{ext}}(\mathbf{k}, \omega) = 2\pi \rho_0(\mathbf{k}) \delta(\omega - \mathbf{k} \cdot \mathbf{v}) \tag{3}$$

where $\rho_0(\mathbf{k})$ is some function determined by the shape of the charge distribution, which is of different forms for different kinds of projectiles such as protons, heavy ions, and molecular ions. Second, according to the linear-response dielectric theory (Ichimaru, 1986), the induced electric field

$E_{ind}(k, \omega)$ is proportional to the external charge density distribution and the stopping power can be written as

$$S_e = (4\pi e^4/m_e v^2) n_0 L(n_0, v) \qquad (4)$$

where $L(n_0, v)$ is the stopping number, which can be expressed in terms of the longitudinal dielectric function $\varepsilon(k, \omega)$ of the electron gas as

$$L(n_0, v) = (2/\pi \omega_p^2) \int_0^\infty \rho_0^2(k)\, dk/k \int_0^{kv} d\omega \omega Im[-1/\varepsilon(k, \omega)] \qquad (5)$$

where $\omega_p = (4\pi e^2 n_0/m_e)^{1/2}$ is the plasma frequency, and m_e is the electon mass.

2. Dielectric Function

The dielectric function $\varepsilon(k, \omega)$ to be used is determined by the properties of the electron gas and is obtained from quantum mechanical considerations. In Lindhard's earlier work (1954), he assumed that the electron–electron interaction in the electron gas is weak and that the electron gas can be regarded as a free electron gas. With the RPA, the dielectric function can be written as

$$\varepsilon_{RPA}(k, \omega) = 1 - P(k, \omega) \qquad (6)$$

where $P(k, \omega)$ is Lindhard's polarizability (Lindhard, 1954). In metals and semiconductors, however, the effect of the exchange and correlation interaction should be taken into account. In this case, with LFC the dielectric function has the form

$$\varepsilon_{LFC}(k, \omega) = 1 - P(k, \omega)/[1 + G(k)P(k, \omega)] \qquad (7)$$

where $G(k)$ is the LFC function, which includes the exchange and correlation interaction of the electron gas. A parameterized expression for $G(k)$ has been given by Utsumi and Ichimaru (1982).

For numerical computation, let us introduce the dimensionless variables $z = k/(2k_F)$ and $u = \omega/(kv_F)$, where $k_F = (3\pi^2 n_0)^{1/3}$ is the Fermi wave number, $v_F = (k_F a_0) v_0$ is the Fermi velocity, and $v_0 = 2.18 \times 10^8$ cm/s is the Bohr velocity. In terms of the expressions of z and u, the polarizability can be written as

$$P(k, \omega) = -(\chi^2/z^2)[f_1(z, u) + if_2(z, u)] \quad (8)$$

where $\chi^2 = 0.166 r_s$. Explicit expressions for the dimensionless functions $f_1(z, u)$ and $f_2(z, u)$ can be found in Wang and Ma (1990).

3. LOCAL DENSITY APPROXIMATION

The local density approximation (LDA) assumes that each volume element of the solid is an independent plasma with local electron density $n(r)$. Lindhard and Winther (1964) found that this approximation yields the same velocity dependence as the shell correction known from atomic target systems. Substituting $n(r)$ for n_0 in Eq. (4), an average electronic stopping power is obtained by integrating over the atomic volume:

$$S_e = (4\pi e^4/m_e v^2)N \int_0^{R_0} n(r) L[n(r), v] 4\pi r^2 \, dr \quad (9)$$

where N is the atomic density and $R_0 = [3/(4\pi N)]^{1/3}$ is the atomic radius.

The electron density $n(r)$ in the solid differs from that of a free atom due to the overlap of electronic wave functions. For this reason, a correction to the free-atom model has been introduced by Gertner, Meron, and Rosner (GMR) (1978). In their work, the electron density $n(r)$ in the solid is obtained by dividing the atomic volume into two parts: the inner region and outer region. In the outer region, the electrons are delocalized and a homogeneous electron gas with a constant density C_0 is formed. However, in the inner region the electrons are localized and the electron density can be expressed as $n_A(r) + C_i$, where $n_A(r)$ is the free-atom electron density fixed by several types of wave functions, and C_i is a constant electron density correction. The constants C_0 and C_i, and the radius R_c, which defines the boundary between the outer and inner regions, are fixed by conservation of the total number of electrons per atom and continuity of the density at R_c. In the GMR model the electron density C_0 in the outer region is greater than the valence-electron gas density n_0. In our works (Wang, Cui, and Ma, 1993; Cui, Wang, and Ma, 1993), a model of the electron density $n(r)$, which differs slightly from that of the GMR model, is given by

$$n(r) = \begin{cases} n_0 & (r \geq R_c) \\ n_A(r) & (r < R_c) \end{cases} \quad (10)$$

where the valence-electron gas density n_0 can be derived from measurement of the plasma frequency (Pines, 1964), and R_c is fixed by

$$n_A(R_c) = n_0 \tag{11}$$

According to the atomic independent model given by Green, Sellin, and Zachor (1969), a parameterized expression of the free-atom electron density $n_A(r)$ can be expressed by

$$n_A(r) = \frac{Z_2 H}{4\pi d^2 a_0^2 r} \frac{\exp(\eta)}{(1 + H\delta)^2} \left[1 + \frac{2(H-1)}{(1 + H\delta)} \right] \tag{12}$$

where Z_2 is the atomic number of the solid, $\eta = r/(da_0)$, $\delta = \exp(\eta) - 1$, $H = d(Z_2 - 1)^{0.4}$, and d is a parameter determined by the Hartree-Fock-Slater screening functions and eigenvalues.

III. Electronic Stopping Power for Protons

1. METHOD OF NUMERICAL CALCULATION

A proton is most often regarded as a point charge, so that $\rho_0(\mathbf{k}) = 1$. From Eqs. (5), (7), and (8) the stopping power for a proton can be expressed as

$$S_p = (4\pi e^4/m_e v^2) n_0 L_p \tag{13}$$

where L_p is the stopping number of the proton

$$L_p = \frac{6}{\pi \chi^2} \int_0^\infty dz z^3 \int_0^{v/v_F} du u F(z, u) \tag{14}$$

and

$$F(z, u) = \frac{\chi^2 f_2(z, u)}{\{z^2 + \chi^2[1 - G(z)] f_1(z, u)\}^2 + \{\chi^2[G(z) - 1] f_2(z, u)\}^2} \tag{15}$$

The integral in Eq. (14) can only extend over the regions in which $f_2(z, u) \neq 0$. In fact, the contributions to stopping power are from two regions: the single-particle excitation (SPE) region in which $f_2(z, u) \neq 0$, and the collective excitation (CE) region in the electron gas in which $f_2(z, u) = 0$.

For the contribution due to SPE, the stopping number $L_p^{(SPE)}$ can be calculated by integrating Eq. (14) over the SPE region. For the contribution due to CE, however, the double integral in Eq. (14) can be reduced to a line integral and the stopping number $L_p^{(CE)}$ is given by (Iafrate and Ziegler, 1979)

$$L_p^{CE} = \frac{6}{\chi^2} \int_{u_c}^{v/v_F} du \left[\frac{zu}{\partial \mathrm{Re}\varepsilon(z,u)/\partial z} \right]_{z = z_r(u)} \quad (16)$$

where $z_r(u)$ can be found by the dispersion equation

$$z^2 + \chi^2[1 - G(z)]f_1(z, u) = 0 \quad (17)$$

for a given value of u. The lower limit u_c of the integral in Eq. (17) is the critical point at which the CE curve merges with the SPE region in z-u plane and can be determined by the solution of the equation

$$(u_c - 1)^2 + \chi^2[1 - G(u_c - 1)]f(u_c) = 0 \quad (18)$$

in which $f(u_c) = 0.5[1 + u_c \ln(1 - 1/u_c)]$.

2. Low- and High-Velocity Approximations

For a low-velocity proton ($v \leqslant v_F$), the contribution to the stopping power comes from the SPE region only. In this case, the polarizability $P(z, u)$ and the LFC function $G(z)$ can be written approximately as

$$P(z, u) = -(\chi^2/z^2)[((1 - z^2/3) + i\pi u/2] \quad (19)$$

$$G(z) = 4\gamma_0 z^2 \quad (20)$$

where the parameter γ_0 is related to the correlation energy of the electron gas (Utsumi and Ichimaru, 1982). Substituting Eqs. (19) and (20) into Eq. (15), we obtain the stopping number for a slow proton (Wang and Ma, 1990):

$$L_p = \tfrac{1}{2}(v/v_F)^3(1 - \beta\chi^2/3)^{-2}I(t) \quad (21)$$

where

$$I(t) = \ln(1 + 1/t) - 1/(1 + t) \quad (22)$$

$\beta = 1 + 12\gamma_0$, and $t = \chi^2/(1 - \beta\chi^2/3)$. When $\beta = 1$, Eq. (22) reduces to the result of Lindhard and Winther (1964). For ordinary metals and semiconductors, however, the values of β range from about 4 to 5 (Wang and Ma, 1991).

For a high-velocity proton ($v \geqslant v_F$), the effect of the exchange and correlation interaction of the electron gas in the dielectric function can be ignored, and the RPA dielectric theory is sufficiently accurate in calculating the stopping power. In this case, the analytical expression for the stopping power has been given by Gertner, Meron, and Rosner (1978):

$$L_p = \ln\left(\frac{2m_e v^2}{\hbar\omega_p}\right) - (3/5)(v_F/v)^2 - (3/14 + \chi^2/3)(v_F/v)^4 \quad (23)$$

where $\hbar$ is the Planck constant.

3. FITTED FORMULA AND COMPARISON WITH EXPERIMENTAL DATA

Now we consider the solid effect on the stopping power for protons. According to Eq. (9), the stopping power for protons under the LDA is given by

$$S_p = (4\pi e^4/m_e v^2)N \int_0^{R_0} L_p[n(r), v] 4\pi r^2 dr \quad (24)$$

Using Eq. (24), we have calculated the electronic stopping power based on LFC dielectric theory and the LDA for protons in solids for energies E from 1 to 10^4 keV. The parameters adopted in the calculation are given in Table I.

The comparison between the results of LFC and RPA dielectric theory for protons in titanium (Ti) is shown in Figure 1. From Figure 1 we can see that the agreement between predictions of both theories is quite good in the high-energy region; below about 700 keV, however the values of S_p from LFC are larger than those from RPA. This is because a slow proton can interact with a number of electrons in the electron gas along its path, so the effect of the exchange and correlation interaction of the electron gas is rather strong. However, in the high-energy region the effect of the exchange and correlation interaction can be ignored.

In order to avoid computational difficulties in calculating the stopping power, we used the least squares method based on the numerical results of

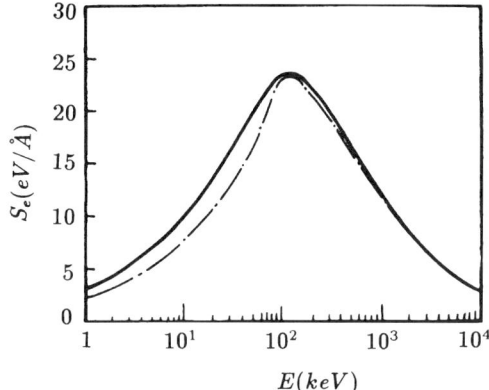

FIG. 1. The stopping power for a proton in a titanium (Ti) target as a function of the projectile energy. The solid and dashed lines, respectively, correspond to the results based on local-field correction (LFC) and random-phase approximation (RPA) dielectric theory. (Reprinted from Cui, Wang, and Ma (1993). *Nucl. Instrum. Meth.* **B73**, 123, with permission from the author and Elsevier Science, Amsterdam.)

Eq. (24) and obtained a fitted function dependent on eight parameters (Cui, Wang, and Ma, 1993)

$$S_p = \begin{cases} b_0 E^{1/2} & (E < 10^4 \text{ keV}) \\ \dfrac{b_1 E^{d_1}}{1 + b_2 E^{d_1} + b_3 E^{d_2} + b_4 E^{d_3}} & (10 \leqslant E \leqslant 10^4 \text{ keV}) \end{cases} \quad (25)$$

where the stopping power S_p is in eV/Å and the energy E is in keV. The parameters b_i and d_i for some solids are given in Table II.

Figure 2 compares the results of numerical integration of Eq. (24) (solid curve) and the fitted data (dashed curve) for gold (Au). At low and high proton energies, our numerical results and fit data agree very well. Only for energies ranging from 200 to 400 keV do they differ, but only slightly. In the following discussion, we will use the values calculated from Eq. (25) to compare with experimental data.

Figure 3 compares the predictions of our theory with some experimental data (Blume, Eckstein, and Verbeck, 1980; Gott and Tel'kovskiy, 1968; Valenzuela *et al.*, 1972; Mertens and Krist, 1980; Andersen and Nielsen, 1981) and Ziegler's results (1980) for protons in Au. The solid curve

TABLE I

PARAMETERS USED IN THE CALCULATION

Element	Z_2	d	R_c/a_0	Na_0^3	$n_0 a_0^3$
Be	4	0.585	1.544	1.766E-2	4.026E-2
B	5	0.979	1.628	1.939E-2	3.909E-2
C	6	0.880	1.561	1.688E-2	5.276E-2
Al	13	0.927	2.182	8.912E-3	2.505E-2
Si	14	0.817	2.108	7.372E-3	3.122E-2
Ti	22	1.060	2.246	8.408E-3	3.316E-2
V	23	0.966	2.034	1.067E-2	5.008E-2
Cr	24	0.837	1.917	9.374E-3	6.436E-2
Mn	25	0.866	2.085	9.104E-3	4.749E-2
Fe	26	0.807	2.391	9.539E-3	2.692E-2
Co	27	0.751	2.265	1.331E-2	3.378E-2
Ni	28	0.700	2.117	1.352E-2	4.501E-2
Cu	29	0.606	2.149	9.593E-3	3.908E-2
Zn	30	0.612	2.265	9.696E-3	3.122E-2
Ge	32	0.649	2.344	6.558E-3	2.896E-2
As	33	0.663	2.164	6.809E-3	4.376E-2
Se	34	0.675	2.246	5.403E-3	3.801E-2
Zr	40	0.866	2.395	6.321E-3	3.587E-2
Nb	41	0.831	2.401	8.260E-3	3.692E-2
Mo	42	0.825	2.079	9.481E-3	6.601E-2
Ru	44	0.803	2.259	1.075E-2	4.797E-2
Rh	45	0.788	2.306	1.075E-2	4.379E-2
Pd	46	0.737	2.068	1.022E-2	6.920E-2
Ag	47	0.754	2.102	8.660E-3	6.601E-2
Sb	51	0.870	2.637	4.841E-3	2.725E-2
Te	52	0.896	2.529	4.352E-3	3.378E-2
Ta	73	0.676	2.365	8.172E-3	4.749E-2
W	74	0.679	2.223	9.356E-3	6.436E-2
Re	75	0.680	2.472	1.007E-2	3.908E-2
Os	76	0.680	2.448	1.057E-2	4.142E-2
Ir	77	0.679	2.410	1.044E-2	4.500E-2
Pt	78	0.661	2.229	9.800E-3	6.436E-2
Au	79	0.657	2.178	8.734E-3	7.246E-2
U	92	0.880	2.677	7.135E-3	3.692E-2

(Reprinted from Cui, Wang, and Ma (1993). *Nucl. Instrum. Meth.* **B73**, 123, with permission from the author and Elsevier Science, Amsterdam.)

represents our fitted values, and the dashed curve corresponds to Ziegler's values based on many experimental data. One can see that the agreement between the predictions of our theory and experimental data is quite good, even at low proton energies.

In order to determine whether the predictions of our theory agree with

TABLE II

FITTING COEFFICIENTS IN EQS. (26) AND (27)

Element	b_0	b_1	b_2	b_3	b_4	d_1	d_2	d_3
Be	2.72230	0.735928	−0.275101	0.0895449	0.156353E-3	0.47	0.70	1.48
B	2.74319	0.640348	−0.164740	0.0441515	0.166658E-3	0.50	0.80	1.51
C	2.94009	0.716866	−0.234499	0.0602011	0.329499E-4	0.49	0.74	1.59
Al	2.62783	0.683166	−0.150977	0.0482755	0.368344E-4	0.58	0.86	1.68
Si	2.64725	0.837879	−0.160684	0.0609333	0.649054E-4	0.63	0.86	1.66
Ti	2.74994	0.672951	−0.238547	0.0892033	0.340257E-4	0.53	0.86	1.60
V	2.92506	0.722713	−0.157293	0.0520610	0.943595E-5	0.59	0.83	1.75
Cr	2.98483	0.769477	−0.284344	0.1093660	0.614877E-5	0.52	0.70	1.72
Mn	2.86099	0.868364	−0.167323	0.0722461	0.156512E-4	0.65	0.84	1.75
Fe	2.74189	0.604047	−0.241641	0.074573	0.212977E-4	0.48	0.72	1.58
Co	3.19442	0.483433	−0.215466	0.0577888	0.636177E-5	0.46	0.72	1.66
Ni	2.97447	0.541543	−0.182082	0.0528453	0.542216E-5	0.50	0.75	1.70
Cu	2.83818	0.591249	−0.269594	0.0767048	0.237160E-5	0.45	0.69	1.75
Zn	2.78841	0.552786	−0.203778	0.0623401	0.109040E-4	0.49	0.74	1.64
Ge	2.68598	0.876786	−0.101682	0.0516493	0.431769E-4	0.74	0.95	1.75
As	2.82841	0.821977	−0.430777	0.0545955	0.608200E-5	0.65	0.88	1.85
Se	2.74145	1.438484	−0.056751	0.0383625	0.316686E-4	0.91	1.13	1.91
Zr	2.77629	0.660743	−0.227789	0.0763165	0.134688E-4	0.52	0.75	1.65
Nb	2.84463	0.680576	−0.263506	0.1164784	0.143637E-4	0.49	0.74	1.64
Mo	3.04181	0.671192	−0.122544	0.0403575	0.322006E-5	0.61	0.86	1.80
Ru	2.98649	0.655094	−0.140717	0.0571504	0.444266E-5	0.61	0.86	1.80
Rh	2.97527	0.656296	−0.133360	0.0575443	0.312124E-5	0.62	0.83	1.85
Pd	3.08779	0.577753	−0.261893	0.0783717	0.583941E-6	0.46	0.83	1.86
Ag	3.03612	0.631896	−0.230248	0.0598953	0.831050E-7	0.48	0.73	2.08
Sb	2.66998	0.930059	−0.081161	0.0403400	0.554310E-4	0.77	1.00	1.74
Te	2.80650	0.817967	−0.174794	0.0727854	0.351027E-4	0.64	0.85	1.67
Ta	2.96941	0.848046	−0.671026	0.0332715	0.538633E-5	0.76	0.98	1.89
W	3.09170	0.732550	−0.131096	0.0692404	0.839776E-5	0.68	0.84	1.77
Re	3.04791	0.624166	−0.628265	0.0299930	0.775466E-5	0.68	0.92	1.78
Os	3.04027	0.508015	−0.251244	0.0869151	0.424426E-6	0.46	0.66	1.86
Ir	3.05050	0.616627	−0.624042	0.0246726	0.247387E-5	0.66	0.93	1.89
Pt	3.11367	0.592722	−0.327653	0.1663105	0.297856E-5	0.51	0.54	1.71
Au	3.12030	0.662433	−0.202420	0.0101176	0.795602E-5	0.60	0.75	1.70
U	2.92422	0.588173	−0102614	0.0353953	0.224912E-5	0.60	0.87	1.84

(Reprinted from Cui, Wang, and Ma (1993). *Nucl. Instrum. Meth.* **B73**, 123, with permission from the author and Elsevier Science, Amsterdam.)

the experimental data for protons through many solid targets, we compare the results of our model with Ziegler's values for protons in various materials. The electronic stopping power for a 50 keV proton moving in various solids is shown in Figure 4. From Figure 4 we can see that the results of our model agree quite well with Ziegler's fits for many solids.

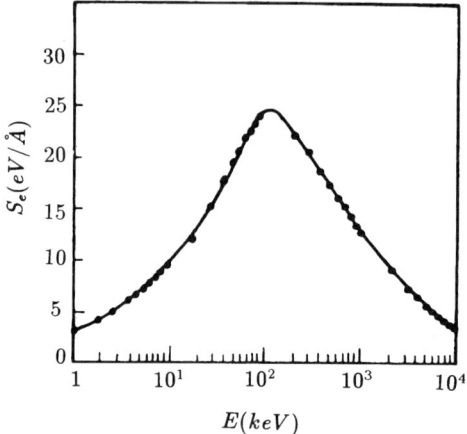

FIG. 2. The stopping power for a proton in a gold (Au) target as a function of the projectile energy. The solid and dashed lines, respectively, are calculated from Eqs. (25) and (26). (Reprinted from Cui, Wang and Ma (1993). *Nucl. Instrum. Meth.* **B73**, 23, with permission from the author and Elsevier Science, Amsterdam.)

IV. Electronic Stopping Power for Heavy Ions

1. THE BRANDT–KITAGAWA MODEL FOR CHARGE DISTRIBUTION OF A PROJECTILE

When a heavy ion with the atomic number Z_1 moves through an electron gas, some electrons in the electron gas are captured and bound to the ion. Thus, the ion cannot be regarded as a bare particle, and changes of its charge state should be considered in this case. Brandt and Kitagawa (BK) (1982) attempted to include the effects of electrons bound to the heavy ion, using a nonlocal density distribution of bound electrons.

Assuming that there are N electrons bound to the ion, its charge density distribution is given by

$$\rho_{\text{ext}}(\mathbf{r}, t) = Z_1 \delta(\mathbf{r} - \mathbf{v}t) - \rho_e(\mathbf{r} - \mathbf{v}t) \quad (26)$$

where $Z_1 \delta(\mathbf{R})$ and $\rho_e(\mathbf{R})$ denote the nuclear and electronic charge densities, respectively, subject to the condition $\int d\mathbf{R}\rho_e(\mathbf{R}) = N$. In the BK model, the

2 ELECTRONIC STOPPING POWER FOR ENERGETIC IONS IN SOLIDS   43

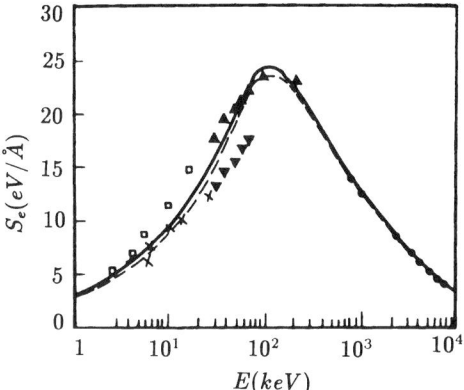

FIG. 3. The stopping power for a proton in a gold (Au) target as a function of the projectile energy. The solid line is calculated from Eq. (26), the dashed line corresponds to Ziegler's values (1980), and the experimental data are the following references: □, Blume, Eckstein, and Verbeck (1980); ×, Gott and Tel'kovskiy (1968); ▲, Valenzuela et al. (1972); ▼, Mertens and Krist (1980); and ●, Andersen and Nielsen (1981). (Reprinted from Cui, Wang, and Ma (1993). *Nucl. Instrum. Meth.* **B73**, 123, with permission from the author and Elsevier Science, Amsterdam.)

form of $\rho_e(\mathbf{R})$ is chosen to be

$$\rho_e(R) = N\exp(-R/\Lambda)/(4\pi R\Lambda^2) \quad (27)$$

Here Λ is the screening length determined by a variational statistical approximation

$$\Lambda = 0.48(1-q)^{2/3}a_0/\{Z_1^{1/3}[1-(1-q)/7]\} \quad (28)$$

where $q = 1 - N/Z_1$ is the ionization degree of the projectile. An empirical formula of q has been given by Ziegler, Biersack, and Littmark (ZBL) (1985).

From Eqs. (27) and (28), the function $\rho_0(k)$ appearing in the stopping number (see Eq. (5)) is given by

$$\rho_0(k) = Z_1[q + (k\Lambda)^2]/[1 + (k\Lambda)^2] \quad (29)$$

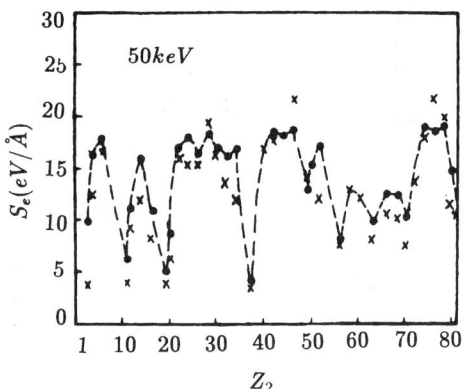

FIG. 4. Z_2-dependence of the electronic stopping power for 50 keV protons. Our results are denoted by (●---●), and the ×'s are Ziegler's fits (1980). (Reprinted from Cui, Wang and Ma (1993). *Nucl. Instrum. Meth.* **B73**, 123, with permission from the author and Elsevier Science, Amsterdam.)

2. EFFECTIVE STOPPING CHARGES

In the BK theory the effective stopping charge Z_{eff} can be defined as

$$Z_{\text{eff}} = (L_{Z_1}/L_p)^{1/2} \tag{30}$$

where L_{Z_1} is the stopping number for the heavy ion

$$L_{Z_1} = \frac{6}{\pi \chi^2} \int_0^\infty dz z^3 \rho_0^2(z) \int_0^{v/v_F} du u F(z, u) \tag{31}$$

Thus, from Eqs. (5) and (30) the stopping power for the heavy ion can be written as

$$S_{Z_1} = (4\pi e^4/m_e v^2) Z_{\text{eff}}^2 n_0 L_p \tag{32}$$

In the case of low velocity ($v \leqslant v_F$), substituting Eqs. (19), (20), and (29) into Eq. (30), we obtain the effective charge fraction $\zeta = Z_{\text{eff}}/Z_1$ as (Wang and Ma, 1993a)

$$\zeta^2 = [(q - tt_0)^2 + (1 - q)^2 I(1/t_0)/I(t) \\ + 2(1 - q)(q - tt_0) K(t, t_0)/I(t)]/(1 - tt_0)^2 \tag{33}$$

where

$$K(t, t_0) = [\ln(1 + t_0) - tt_0 \ln(1 + 1/t)]/(1 - tt_0) \qquad (34)$$

and $t_0 = (2k_F \Lambda)^2$. When $\beta = 0$, the expression for Z_{eff} reduces to the BK results (Brandt and Kitagawa, 1982) based on the RPA dielectric theory.

At high velocities ($v \geqslant v_F$), the LFC effect in the dielectric function can be ignored, and an approximate expression of the dielectric function is given by (Lindhard, 1954)

$$\varepsilon(k, \omega) = 1 + \omega_p^2/[(\hbar k^2/2m_e)^2 - (\omega + i\delta)^2] \qquad (35)$$

where δ is an infinitesimally small positive quantity. This approximation corresponds to the case in which $\omega/k \gg v_F$, but arbitrary k, and describes in a very simple way the collective and individual excitations of the electron gas. Using this dielectric function, we obtain the effective charge fraction ζ (Wang and Ma, 1993a):

$$\zeta^2 = q^2 + (1 - q)\{(1 - q)(x_1 - x_2)/[(1 + x_1)(1 + x_2)]$$
$$+ (1 + q) \ln[(1 + x_2)/(1 + x_1)]\}/(2L) \qquad (36)$$

where $L = \ln(2m_e v^2/\hbar\omega_p)$. In Eq. (36) x_1 and x_2 are defined by $x_1 = (\Lambda\omega_p/v)^2$ and $x_2 = (2\Lambda v/a_0 v_0)^2$.

3. COMPARISON WITH EXPERIMENTAL DATA

Using the inverted Doppler-shift-attenuation analysis method, Arstila, Keinonen, and Tikkanen (1990) measured the electronic stopping power for magnesium (Mg) ions in 17 elemental solids in the velocity region of v from $0.2v_0$ to $5v_0$. Figures 5 and 6 show the velocity dependence of the stopping power for Mg ions in a silver (Ag) target and a tantalum (Ta) target, respectively. The solid and dashed lines correspond to our calculations predicted by Eq. (32) and the results of the ZBL empirical formula (Ziegler, Biersack, and Littmark, 1985), respectively; the dots represent the experimental data. At low velocities ($v < 2v_0$), our results agree rather well with the data, but the ZBL predictions are obviously lower than are the data. At high velocities ($2v_0 < v < 5v_0$), the ZBL results are closer to the experimental data than are our results.

Figures 7 and 8 compare our calculations with experimental data (Porat and Ramavataram, 1961; Santry and Werner, 1991; Abdesselam et al., 1991) for carbon (C) ions in an Au target and aluminium (Al) ions in a copper

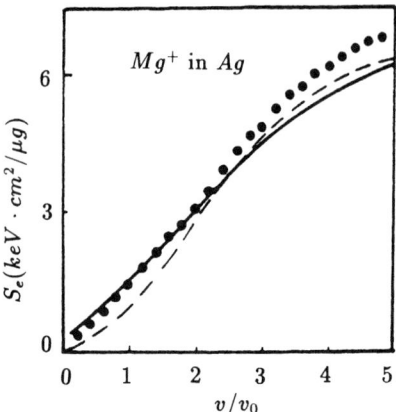

FIG. 5. The electronic stopping power for magnesium (Mg) ions in a silver (Ag) target. The solid line corresponds to our calculations, the dashed line is the result of the Ziegler-Biersack-Littmark (ZBL) empirical formula (1985), and the dots represent experimental data (Arstila, Keinonen, and Tikkanen, 1990). (Reprinted from Wang and Ma (1993a). *Nucl. Instrum. Meth.* **B80/81**, 16, with permission from the author and Elsevier Science, Amsterdam.)

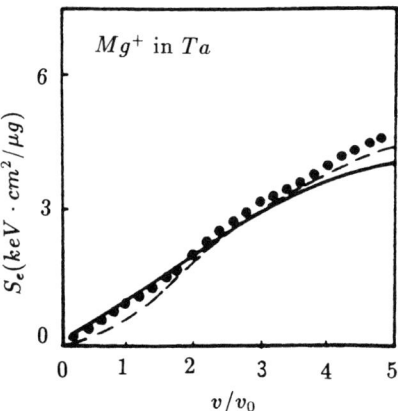

FIG. 6. The electronic stopping power for magnesium (Mg) ions in a tantalum (Ta) target. The solid line corresponds to our calculations, the dashed line is the result of the Ziegler-Biersack-Littmark (ZBL) empirical formula (1985), and the dots represent experimental data (Arstila, Keinonen, and Tikkanen, 1990). (Reprinted from Wang and Ma (1993a). *Nucl. Instrum. Meth.* **B80/81**, 16, with permission from the author and Elsevier Science, Amsterdam.)

(Cu) target, respectively. We can see from Figures 7 and 8 that our calculations are very close to the data for a wide range of projectile-velocity regions (v from v_0 to $12v_0$), especially for C ions in the Au target. Abdesselam et al. (1991) have calculated the stopping power for C and Al ions in solids using the RPA dielectric function and the model of solid-state electron density given by Gertner, Meron, and Rosner (1978). Their work reproduces the experimental data above the maximum of the stopping curve rather well. At low energy, however, a discrepancy is observed and the magnitude of the stopping power varies with different targets. These deviations might be attributable to inaccuracies of the RPA dielectric function.

V. Electronic Stopping Power for Molecular Ions

1. COULOMB EXPLOSION

A molecular ion is an aggregate of two or more atomic ions held together by the binding valence electrons. When the ion impinges on a target, the binding valence electrons are stripped away in the first few atomic layers. The residual ionic fragments then immediately begin to recede from one

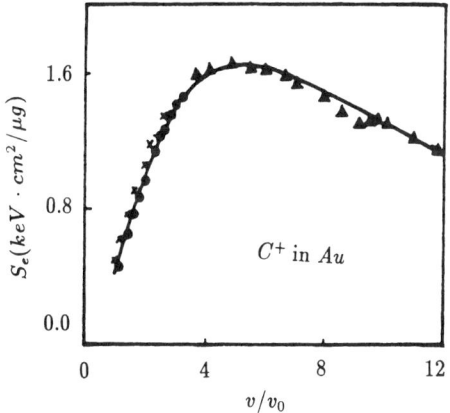

FIG. 7. The electronic stopping power for carbon (C) ions in a gold (Au) target. The solid line corresponds to our calculations, and the experimental data are from the following references: ●, Porat and Ramavataram (1961); ×, Santry and Werner (1991); and ▲, Abdesselam et al. (1991). (Reprinted from Wang and Ma (1993a). *Nucl. Instrum. Meth.* **B80/81**, 16, with permission from the author and Elsevier Science, Amsterdam.)

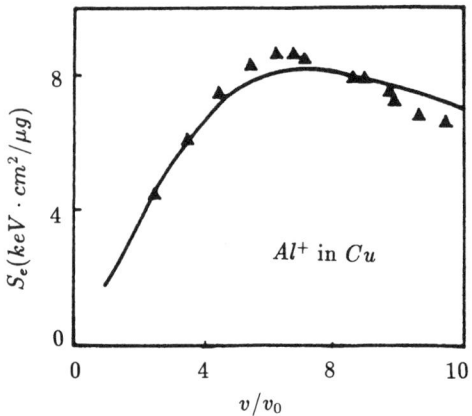

FIG. 8. The electronic stopping power for aluminum (Al) ions in a copper (Cu) target. The solid line corresponds to our calculations, and the experimental data are from the following references: ●, Porat and Ramavataram (1961); ×, Santry and Werner (1991); and ▲, Abdesselam et al. (1991). (Reprinted from Wang and Ma (1993a). *Nucl. Instrum. Meth.* **B80/81**, 16, with permission from the author and Elsevier Science, Amsterdam.)

another under the influence of interionic Coulomb forces. Thus, the distance between the fragments increases and the fragments finally separate in a few femtoseconds. This is called "Coulomb explosion" of the molecular ion.

With an unscreened Coulomb potential $V(r) = Z_{1\text{eff}} Z_{2\text{eff}} e^2/r$, the relation between the internuclear distance r of a diatomic molecular ion and the penetrating depth D in the solid can be expressed as

$$D = D_0 [\sqrt{r/r_0}\sqrt{r/r_0 - 1} + \ln(\sqrt{r/r_0} + \sqrt{r/r_0 - 1})] \quad (37)$$

where $D_0 = v(\mu r_0^3/2 Z_{1\text{eff}} Z_{2\text{eff}} e^2)^{1/2}$, $Z_{i\text{eff}} (i = 1, 2)$ are the effective charges predicted by Eq. (30), $\mu = M_1 M_2/(M_1 + M_2)$ is the reduced mass, r_0 is the initial internuclear distance (the bond length), and M_1 and M_2 are the masses of the corresponding atoms. For high velocity N_2^+, Figure 9 shows the results for the internuclear distance r as a function of the penetrating depth D. We can observe that the values of the internuclear distance r increase as the projectile velocity decreases. It is very easy to understand that for high-velocity molecular ions, it is difficult to yield a Coulomb explosion.

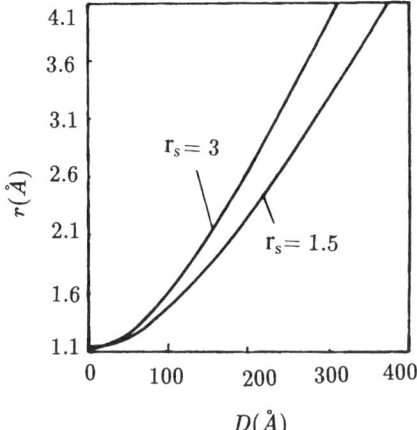

FIG. 9. Variation of the internuclear distance with target thickness for a high-velocity nitrogen ion (N_2^+) at $v = 2v_F$ and various values of r_s. D, penetrating depth. (Reprinted from Wang and Ma (1993b). *Phys. Lett.* **A178**, 209, with permission from the author and Elsevier Science, Amsterdam.)

2. THE VICINAGE EFFECT IN THE STOPPING POWER

The interference of electronic excitation in the solid due to the correlated motion of the ions in the cluster makes the energy loss of the molecular ion differ from that of the separated ions. This is usually called the "vicinage effect."

For hydrogen molecular ions (H_2^+) the vicinage effect was first shown by Brandt, Ratkowski, and Ritchie (1974). They reported measurements of the energy loss of a hydrogen molecular ion beam through films of C and Au at energies of 75 and 150 keV per proton, in which the molecular stopping power per nucleus exceeded the atomic stopping power by as much as 20%. Following the work of Brandt *et al.*, similar experimental measurements for the stopping power of the molecular ions hydrogen, oxygen, and nitrogen (H_2^+, O_2^+, and N_2^+) have been made by several authors (Eckardt *et al.*, 1978; Tape *et al.*, 1976; Steuer *et al.*, 1983). In particular, Ray *et al.* (1992) reported experimental results for the vicinage effect of large hydrogen clusters H_n^+ (up to $n = 25$) in the energy range of from 10 to 120 keV per proton.

Although this general feature has been known for hydrogen molecular ions (H_2^+) (Arista, 1978; Babas and Ritchie, 1982), few detailed investigations have been made for the molecular ions of heavy atoms, such as nitrogen (N_2^+) and carbon monoxide (CO^+). Hydrogen molecular ions can be regarded as a point-charge pair. For molecular ions constituted of heavy atoms, however, each atomic ion can be represented as a nuclear charge and a statistically equilibrated distribution of electrons that move together through the medium. Here, we attempt to extend the BK effective charge model for atomic ions (Section IV) to describe molecular ions and investigate the vicinage effect in the stopping power with the linear-response dielectric theory.

For a diatomic molecular ion, the function $\rho_0(k)$ appearing in Eq. (5) can be written as

$$\rho_0(\mathbf{k}) = \rho_1(k) + \rho_2(k)e^{i\mathbf{k}\cdot\mathbf{r}} \tag{38}$$

where $\rho_i(k)$ are given by Eq. (29). In many experimental situations, the orientations of $\mathbf{r}$ are randomly distributed. Substituting Eq. (38) into Eq. (5) and carrying out a spherical average over $\mathbf{r}$, the stopping power for the molecular ion can be expressed as (Wang and Ma, 1993b)

$$S_{\text{mol}} = [Z_{1\text{eff}}^2 + Z_{2\text{eff}}^2 + 2Z_{1\text{eff}}Z_{2\text{eff}}g(r)]S_p \tag{39}$$

where S_p is the proton stopping power (see Eq. (13)) and $g(r)$ is the vicinage function given by

$$g(r) = (Z_{1\text{eff}}Z_{2\text{eff}}L_p)^{-1} \int_0^\infty dk \rho_1(k)\rho_2(k)H(k,v)\frac{\sin(kr)}{kr} \tag{40}$$

where

$$H(k,v) = \int_0^{kv} d\omega\omega \text{Im}[-1/\varepsilon(k,\omega)] \tag{41}$$

The vicinage function $g(r)$ measures the interference effects on the stopping power of the cluster. For large internuclear distances ($r \to \infty$), the vicinage function vanishes and Eq. (39) becomes the uncorrelated stopping power of two independent charged particles

$$S_{\text{in}} = (Z_{1\text{eff}}^2 + Z_{2\text{eff}}^2)S_p \tag{42}$$

To further show the interference effects, we introduce a factor R, which is the ratio of the stopping power of the molecular ion, to that of the sum of its separate atomic constituents; that is,

$$R = \frac{S_{\text{mol}}}{S_{\text{in}}} = 1 + \frac{2Z_{1\text{eff}}Z_{2\text{eff}}}{Z_{1\text{eff}}^2 + Z_{2\text{eff}}^2}g(r) \tag{43}$$

For a low-velocity molecular ion ($v \leqslant v_F$), the vicinage function can be written as (Wang and Ma, 1993b)

$$g(r) = (C_1 C_2)^{-1/2} \int_0^1 dz z^3 \rho_1(z) \rho_2(z) f(z) \frac{\sin(2k_F rz)}{2k_F rz} \quad (44)$$

where

$$f(z) = [(1 - \beta\chi^2/3)z^2 + \chi^2]^{-2} \quad (45)$$

$$C_i = \int_0^1 dz z^3 \rho_i^2(k) f(z) \quad (i = 1, 2) \quad (46)$$

At high velocities ($v \gg v_F$), substituting Eq. (35) into Eq. (40), we obtain (Wang and Ma, 1993b)

$$g(r) = (L_1 L_2)^{-1/2} \int_{z_{\min}}^{z_{\max}} dz \rho_1(z) \rho_2(z) \frac{\sin(2k_F rz)}{2k_F rz^2} \quad (47)$$

where

$$L_i = \int_{n_{\min}}^{z_{\max}} \rho_i^2(z)/z \quad (i = 1, 2) \quad (48)$$

are the stopping numbers of the heavy ions. The upper and lower limits in the integral are given, respectively, as $z_{\max} = m_e v/\hbar k_F$ and $z_{\max} = \omega_p/2k_F v$. For high velocity N_2^+, we show the dependence of the stopping power ratio R with the depth D in Figure 10. Note from Figure 10 that the interference effects in the stopping power are more obvious for a higher projectile velocity ($v = 4v_F$). This is because the electronic excitations yielded by the motion of the ions in the cluster are very strong in the case of high velocity.

The measurement of the energy loss for nitrogen molecular ions (N_2^+) in carbon was made by Steuer et al. (1983). Figure 11 compares our calculations based on Eq. (43) with the experimental data for the stopping power ratio. At large dwell time, $t_D = D/v$, our calculations are very close to the data; however, our calculations are lower at short dwell times.

We investigated (Wang and Ma, 1994) the influence of the inner-shell electrons of target atoms on the vicinage effect for H_2^+, based on the LDA theory. It has been found that the inner-shell electrons of the target atoms

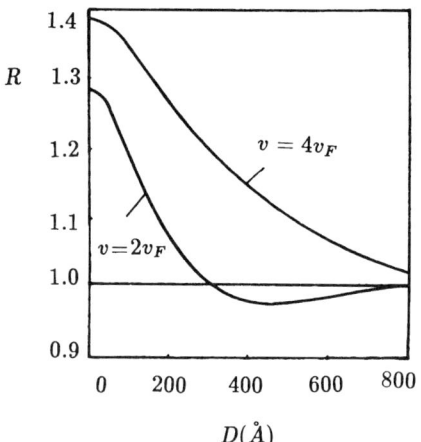

FIG. 10. Variation of the stopping power ratio (R) with target thickness for a high-velocity nitrogen ion (N_2^+) at $r_s = 2$ and various values of the velocity v. D, depth. (Reprinted from Wang and Ma (1993b). *Phys. Lett.* **A178**, 209, with permission from the author and Elsevier Science, Amsterdam.)

act by dropping the vicinage effect at high velocities. Comparison with experimental data has shown that our investigation is reasonable.

VI. Summary

With the linear-response dielectric theory and the LDA method, we have investigated the electronic stopping power for energetic ions moving through solids. First, for the stopping power for protons we obtained a fitted expression dependent on eight parameters, which is based on the results of numerical intergration. It has been shown that our theory agrees with some of the experimental data and with Ziegler's results (1980). Second, we use the BK theoretical model to determine the effective charges of heavy ions. Some analytical expressions of the effective charges are obtained in low- and high-velocity cases, respectively. Finally, we discussed the "Coulomb explosion" course and the "vicinage effect" when a diatomic molecular ion is moving through solids.

Although much theoretical work has been done concerning the calculations of the electronic stopping power over the past decades, there are still some attractive problems that remain to be investigated, such as the stopping powers for energetic ions in two-dimensional systems and in hot plasmas, and the stopping power for ion clusters (for example, C_{60}^+).

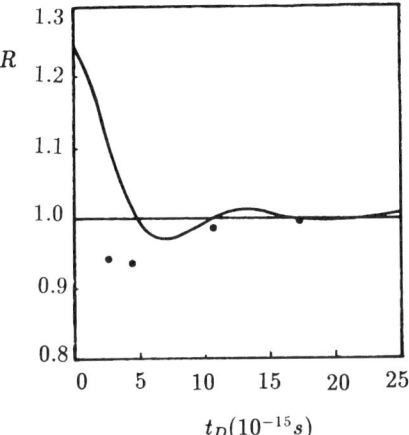

FIG. 11. Stopping power ratio (R) as a function of the dwell time t_0 for a nitrogen ion (N_2^+) in a carbon (C) target at a projectile energy $E = 1.8$ MeV/atom. The solid line corresponds to our calculations, and the dots indicate experimental data (Steuer et al., 1983). (Reprinted from Wang and Ma (1993b). *Phys. Lett A* **178**, 209, with permission from the author and Elsevier Science, Amsterdam.)

ACKNOWLEGMENTS

This work of You-Nian Wang has been partially supported by the Ion-Beam Laboratory of the Shanghai Metallurgy, Chinese Academy of Sciences. YNW thanks the Department of Physics, University of Waterloo, Waterloo, Ontario, Canada, for its hospitality when this review is written.

REFERENCES

Abdesselam, M., Stoquert, J. P., Guillaume, M., Grob, J. J., and Siffert, P. (1991). *Nucl. Instrum. Meth.* **B61**, 385.
Andersen, H. H., and Nielsen, B. R. (1981). *Nucl. Instrum. Meth.* **191**, 475.
Arista, N. R. (1978). *Phys. Rev. B* **18**, 1.
Arstila, K., Keinonen, J., and Tikkanen, P. (1990). *Phys. Rev. B* **41**, 6117.
Babas, G., and Ritchie, R. H. (1982). *Phys. Rev. A* **25**, 1943.
Bethe, H. (1930). *Ann. Phys.* **5**, 325.
Blume, R., Eckstein, W., and Verbeck, H. (1980). *Nucl. Instrum. Meth.* **168**, 57.
Brandt, W., and Kitagawa, M. (1982). *Phys. Rev.* **B25**, 5631.
Brandt, W., Ratkowski, A., and Ritchie, R. H. (1974). *Phys. Rev. Lett.* **33**, 1325.
Cui, T., Wang, Y. N., and Ma, T. C. (1993). *Nucl. Instrum. Meth.* **B73**, 123.

Echenique, P. M. (1987). *Nucl. Instrum. Meth.* **B27**, 256.
Echenique, P. M., Nieminen, R. M., Ashely, J. C., and Ritchie, R. H. (1986). *Phys. Rev. A* **33**, 897.
Eckardt, J. C., Lantschner, G., Arista, N. R., and Baragiola, R. A. (1978). *J. Phys. C* **11**, L851.
Gertner, I., Meron, M., and Rosner, B. (1978). *Phys. Rev. A* **18**, 2022.
Gott, Yu., and Tel'kovskiy, V. G. (1968). *Soviet Phys.-Solid State* **9**, 1741.
Green, A. E. S., Sellin, D. D., and Zachor, A. S. (1969). *Phys. Rev.* **184**, 1.
Iafrate, G. J., and Ziegler, J. F. (1979). *J. Appl. Phys.* **50**, 5579.
Ichimaru, S. (1986). *An Introduction to Statistical Physics of Charged Particles*, Benjamin, Menlo Park, p. 98.
Lindhard, J. (1954). *Kgl. Danske Videnskab. Selskab. Mat.-Fys. Medd.* **28**(8).
Lindhard, J., and Winther, A. (1964). *Kgl. Danske Videnskab. Selskab. Mat.-Fys. Medd.* **34**(4).
Ma, T. C., Wang, Y. N., and Cui, T. (1992). *J. Appl. Phys.* **72**(8), 3838.
Mertens, P., and Krist, Th. (1980). *Nucl. Instrum. Meth.* **168**, 33.
Pines, D. (1964). In *Elementary Excitations in Solids*. Benjamin, New York.
Porat, D. I., and Ramavataram, K. (1961). *Proc. Phys. Soc.* **77**, 97.
Ray, E., Kirsch, R., Mikkelsen, H. H., Poizat, J. C., and Remillieux, J. (1992). *Nucl. Instrum. Meth.* **B69**, 133.
Santry, D. C., and Werner, R. D. (1991). *Nucl. Instrum. Meth.* **B53**, 7.
Sigmund, P. (1982). *Phys. Rev.* **A26**, 2497.
Steuer, M. F., Gemmell, D. S., Kanter, E. P., Jojnson E., and Zabransky, B. (1983). *IEEE Trans. Nucl. Sci.* **NS-30**, 1069.
Tape, J. W., Gibson, W. M., Remillieux, J., Laubert, R., and Wegner, H. E. (1976). *Nucl. Instrum. Meth.* **132**, 75.
Wang, Y. N., Cui, T., and Ma, T. C. (1993). *J. Appl. Phys.* **73**(9), 4257.
Wang, Y. N., and Ma, T. C. (1990). *Nucl. Instrum. Meth.* **B51**, 216.
Wang, Y. N., and Ma, T. C. (1991). *Phys. Rev. A* **44**, 1768.
Wang, Y. N., and Ma, T. C. (1993a). *Nucl. Instrum. Meth.* **B80/81**, 16.
Wang, Y. N., and Ma, T. C. (1993b). *Phys. Lett. A* **178**, 209.
Wang, Y. N., and Ma, T. C. (1994). *Phys. Rev. A* **50**, 3192.
Utsumi, K., and Ichimaru, S. (1982). *Phys. Rev. A* **26**, 603.
Valenzuela, A., Meckbach, W., Kestelman, A. J., and Eckardt, J. C. (1972). *Phys. Rev. B* **6**, 95.
Ziegler, J. F. (1980). *Hydrogen Stopping Powers and Ranges in All Elements*, Pergamon, New York.
Ziegler, J. F., Biersack, J. P., and Littmark, U. (1985). *The Stopping Power and Ranges of Ions in Matter*, Vol. 1. Pergamon, New York, p. 66.

CHAPTER 3

Solid Effect on the Electronic Stopping of Crystalline Target and Application to Range Estimation

Sachiko T. Nakagawa

DEPARTMENT OF APPLIED PHYSICS
OKAYAMA UNIVERSITY OF SCIENCE
OKAYAMA, JAPAN

I. INTRODUCTION 55
II. LOCAL DENSITY APPROXIMATION FOR BINARY COLLISION 58
    1. Electron Density of Solid-State Target Atoms 58
    2. Local Density Approximation for the Nuclear Stopping Power 59
    3. Local Density Approximation for the Electronic Stopping Power . . 61
III. IMPACT PARAMETER-DEPENDENCE OF THE ELECTRONIC STOPPING POWER IN
     CRYSTALLINE SOLIDS 64
    1. Original Oen-Robinson Model and Innovation 64
    2. Determination of the Impact Parameter-Dependence of the Electronic Stopping
       Power in a Cluster 65
    3. Solid Effects on the Electronic Stopping Power from Cluster Calculation . . . 68
IV. ELECTRONIC STOPPING POWER OF CHEMICAL COMPOUNDS 69
    1. Axial Channeling in Zinc-Blende 69
    2. Combination Rule as an Alternative to the Bragg Rule for the Electronic
       Stopping Power of Compounds 71
V. ELECTRONIC STOPPING POWER AND RANGE PROFILES VIA COMPUTER
    SIMULATIONS 72
    1. Influence of the Electronic Stopping Power on Range Profiles . . . 73
    2. Solid-Effects on the Electronic Stopping Power from Computer
       Simulations 76
VI. CONCLUDING REMARKS 78
    References 81

I. Introduction

The possibility of using an ion beam to introduce impurity atoms into the surface layer of a solid target, particularly for the doping of semiconductors, has led to extensive research activity in the field of atomic collisions in solids and has accelerated the development of ion beam technology. The key point of the ion implantation process is the control of the concentration profile of implanted species. The first theoretical approach for a quantitative prediction of implantation profiles was established by Lindhard, Scharff, and Schiøtt (1963), hereafter referred to as the LSS theory. Basic data for elemental solids, such as silicon and germanium, have been compiled and reveal a reasonable agreement with measurements. The LSS theory uses the linear cascade approximation, or low-fluence approximation (LFA), in which it is assumed that the probability of collision of a given projectile is independent of the damage introduced by previous ion impacts. It must be noted that LFA is basically valid, as far as the dose rate being low. [1] In fact, LFA breaks down when the energy deposited in a spike exceeds a given threshold energy during ion bombardment at a high dose rate; in such a situation nonlinear effects occur (Sigmund, 1969). Recently, quantitative predictions have been required for more complicated cases: high-current implantation, cluster implantation, and bombardment of semiconducting compounds and organic matter. Such tasks can be achieved by using computer simulations provided that the number of free parameters remains reasonable.

An incoming projectile excites the electronic or atomic system, or both, of the target, with a possible perturbation of the whole lattice, as long as the transferred energy exceeds the relevant threshold energies. The former process is called the electronic stopping power Se; the latter process, which occurs without any changes in inner states of both the projectile and target atoms, is called the nuclear stopping power Sn.[2] This classification, due to Bohr (1948), is rigorous only at the colliding moment, which is regarded as the primary radiation effect. A relaxation step then follows, which is due to the necessity for the target to accommodate the energy deposited locally and which can be regarded as the secondary radiation effect. This process leads to more or less important changes in both the surface and the bulk of the irradiated solid. LFA is based on the hypothesis that the primary radiation effect is disconnected from the secondary effect. In that sense this approximation is essential for a discussion of the radiation effect by means of the stopping power. In fact, the stopping power is a conceptual quantity, except

[1] For example, as suggested by Wilson and Baskes (1978), if the sequential collisions are described as the recurrent processes of a violent collision followed by relaxation, LFA is possible for each collision period. That is, the kinematic and diffusion equations are solved alternately. This model successfully reproduced ion beam mixing experiments (Nakagawa and Yamamura, 1988a).

[2] Usually either Se or Sn is predominant, depending on the ion energy E. Roughly speaking, Se is predominant at $\varepsilon \gg 0.3$ in the LSS unit, where $\varepsilon = E\ M_2/(M_1 + M_2)/(Z_1 Z_2/a_{12})$ for a projectile with atomic number Z_1 and mass M_1, when it collides with a target atom with atomic number Z_2 and mass M_2. The relevant screening length is defined using the Bohr radius: $a_{12} = 0.8853\,a_0/\,(Z_1^{2/3} + Z_2^{2/3})^{1/2}$ ($a_0 = 0.052917\,\text{nm}$).

for the case of ion transmission in very thin films. Nevertheless, the concept of stopping power manifests a reality when computer simulations are used, since in such cases the target can be artificially decomposed in an ensemble of very thin films.

This article does not discuss the secondary radiation effect. The basic idea is that the atomic rearrangements due to ion irradiation are generally induced by Sn. However, transfer of energy to target atoms can involve Se in some cases. For example, the giant deformation of amorphous solids irradiated with very high energy heavy ions was demonstrated to be due to a Coulomb explosion phenomenon generated by strong electronic excitation (Thomé and Garrido, 1995). Strong electronic excitation is the primary effect described by Se; however, it is the trigger for atomic displacements as the secondary effect. The reverse process also exists, namely electron excitation triggered by Sn. One example is provided by the Fano-Lichten process (1965) caused by the primary knocked-on atom. In this case the energy transferred via Sn can promote electrons along newly made paths in an energy scheme of a quasi-molecule formed during the collision.

The material dependence of the radiation effect can be examined in two different ways. The first takes into account the specific properties of a given solid. The second considers more general quantities, such as atomic and mass numbers of the elements, that comprise the target. In this essay we use the second description, which better fits the local density approximation (LDA) employed for both Se and Sn. We consider two solid effects to account for the material dependence of the radiation effect: the phase effect, which expresses the nature of the irradiated target as a solid, and the directional effect, which is connected to the crystallographic character of the solid. The phase effect was observed as the difference in Se between a solid target and a gaseous target (Arnau et al., 1994), due to the spatial redistribution of valence electrons when atoms cohere to form an amorphous solid. This is interpreted by the difference in the electronic states, that is, a band structure for condensed matter and a series of individual states for each free atom. The directional effect can be observed through the different implantation profiles obtained when the incident direction of the projectiles is varied with respect to the main crystallographic axes or planes in crystalline solids. Such a topic was discussed extensively by Rimini (1994) from the experimental viewpoint.

Similar to the case of electron excitation due to photon irradiation, according to the "Wigner policy," a collision event of a projectile with a target atom is a sequence of three stages: (1) long-range interaction on approaching the collision spot; (2) short-range interaction at the vicinity of the lower apsis to the target atom; and (3) long-range interaction on leaving the collision region. Inokuti (1991) discussed the difference that exists in Se between the two cases of solid and gaseous targets. This difference concerns only the long-range interactions due to surrounding atoms in the solid. More precisely, in the framework of LDA the difference is due to valence electrons.

Part II, Sections 1–3, of this chapter is devoted to the description of Se and Sn in the LDA. This part is necessary for the discussion of the impact parameter, p, dependence of Se. Part III, Sections 1–3, concerns the establishment of a database of Se in a solid as a function of p. The solid effects in Se are discussed in connection with the material dependence of the radiation effect. The calculations of Se for compounds as an alternative to the Bragg rule are presented in Part IV, Sections 1 and 2, regarding axial channeling. Then, in Part V, Sections 1 and 2, range profiles are estimated by making use of computer simulations. A short summary of the results is presented in Part VI.

II. Local Density Approximation for Binary Collision

In the LDA formalism, the value of Se is expressed by the electron density $\rho(r)$ (Nakagawa and Biersack, 1991), according to the Lindhard and Winther (LW) model (1964). However, Sn is derived from $\rho(r)$ by means of an adiabatic interatomic potential $V(R)$ between atoms separated by a distance R, according to Gordon and Kim (1972). On describing atomic collisions in solids, the value of $\rho(r)$, which defines both Se and Sn should be that of a solid-state phase (ρ_{sol}), different from that of a gas-state phase (ρ_{gas}). This detail about $\rho(r)$ is very important when discussing the effects on collision events due to the solid.

1. ELECTRON DENSITY OF SOLID-STATE TARGET ATOMS

The main difference between ρ_{sol} and ρ_{gas} lies in the valence electrons, which can be free electrons in metals. This difference is more important for outer-shell electrons than for core electrons forming closed shells around each nucleus. However, $\rho(r)$ due to inner-shell or core electrons is almost invariant, regardless of whether an atom is in the gaseous or solid state. The radius of the sphere formed by closed shells (r_{core}) is defined by the number of enclosed electrons equivalent to that of closed-shell electrons. For example, in the case of silicon, the structure of the closed shells is $1s^2 2s^2 2p^6$ (neon core), thus 10 electrons are confined in a sphere of $r_{core} = 0.05934\,nm$). Gertner, Meron, and Rosner (1978) presented a method to obtain ρ_{sol} from ρ_{gas}, in which ρ_{sol} is enclosed in an Atomic-Wigner-Seitz sphere (AWS) of radius r_{AWS} and with a volume equivalent to the Wigner-Seitz cell in a solid. In the case of silicon, $r_{AWS} = 0.1691$ nm (Part II, Section 3).

First, we obtained ρ_{sol} using the value of ρ_{gas} of the Dirac-Hartree-Fock-Slater atoms (DHFS) (Carlson et al., 1970), which hereafter is called the AWS approach,

$$\rho_{sol}(r) = \rho_{gas}(r) + \rho_o \qquad (1)$$

where

$$\rho_o = \left(\frac{4}{3}\pi r_{AWS}^3\right)^{-1} 4\pi \int_{r_{AWS}}^{\infty} r^2 \rho_{gas}(r)\, dr$$

The influence of the relativistic term, which is the difference of DHFS from the Hartree-Fock-Slater (HFS) approximation, is effective for inner-shell electrons of high-Z atoms. In fact, this difference was remarkable even for the case of indium ($Z = 49$).

2. LOCAL DENSITY APPROXIMATION FOR THE NUCLEAR STOPPING POWER

Conventionally, Sn is defined by means of $V(R)$. This potential is obtained by either self-consistent HFS calculations using an adiabatic molecular orbital (MO) approach or the density functional (DF) method. (If the collision event cannot be described adiabatically, nonadiabatic MO should be employed, as calculated by Schiwietz in 1990 using the coupled channel method). The former ab initio approach was applied to systems with a few electrons, for example, He–He, Ne–Ne, Ar–Ar (helium, neon, argon) (Gilbert and Wahl, 1967) and Al–Al (aluminum) (Sabelli, Benedek, and Gilbert, 1979), which were expected to give almost exact values of $V(R)$ for individual collision systems. The latter approach originated in the Thomas-Fermi statistical model for a single atom; it was developed for diatomic systems (Firsov, 1957, 1958), with a formalism using two terms of kinetic and electrostatic energy.

The DF method has two merits: (1) it is particularly suited to the study of many electron systems due to the characteristics of the statistical model, and (2) it can be applied for any collision systems due to the use of a scaling factor (or screening length). According to Gordon and Kim (1972), Firsov's formalism was improved by adding higher-order electron–electron interactions, with the exchange term including the difference of spin states and the correlation term beyond the Hartree-Fock approximation. The addition of these two term breaks the self-consistency between the electron density $\rho(r)$ and the minimization of the total energy of the system, although this self-consistency was achieved in the original Thomas-Fermi statistical model for a single atom. More precisely, because of the different ρ-dependence introduced by higher precision terms, such a breakdown leads to the difficulty of scaling $V(R)$ using a single screening length in order to obtain a "universal potential." Nevertheless, the more accurate $\rho(r)$ is, the better is the $V(R)$ that is obtained.

We deduced $V(R)$ values from the first principles and called them MLJ (with a numerical table), which are comparable to ab initio calculations by MO. The general expression for the screening length of a heteronuclear $V(R)$ is $a_{ij} = 0.8853\, a_0 (Z_i^p + Z_j^p)^{-1/m}$, where p and m are chosen arbitrarily. Note that $p/m = 1/3$ is exact only for the case of a free atom according to the Thomas-Fermi model. The expression of a_{ij} can be rewritten by the screening lengths a_{ii} and a_{jj} of homonuclear $V(R)$'s, which works as a combination rule:

$$a_{ij}^{-m} = \tfrac{1}{2}(a_{ii}^{-m} + a_{jj}^{-m}) \qquad (2)$$

Then a more general potential, called AMLJ, is derived (as a function) by means of this combination rule for the screening lengths. The value m is chosen in order to reproduce the measured data. For instance, $m = 1.5$ is taken for the calculation of the projected ranges of various ions into silicon. The MLJ potential includes both the shell effect and a global Z-dependence of composite solid-state atoms, whereas the AMLJ potential includes only the global Z-dependence. The expression of AMLJ or MLJ available for any colliding systems is

$$V(R) = \frac{Z_1 Z_2}{R} \exp\left(-C_1 \frac{R}{a_s} + C_2 \left(\frac{R}{a_s}\right)^{1.5} - C_3 \left(\frac{R}{a_s}\right)^2\right) \qquad (3)$$

where

$$a_s = \frac{0.8853\, a_0}{(Z_1^{0.307} + Z_2^{0.307})^{2/3}}$$

The values C_1, C_2, and C_3, for the MLJ potential are tabulated and those for the AMLJ potential are the following (Nakagawa and Yamamura,

1988a):

$$C_1 = 1.51 \tag{4a}$$

$$C_2 = 0.763 \left[\frac{Z_1^{0.169} + Z_2^{0.169}}{Z_1^{0.307} + Z_2^{0.307}} \right] \tag{4b}$$

$$C_3 = 0.191 \left[\frac{Z_1^{0.0418} + Z_2^{0.0418}}{Z_1^{0.307} + Z_2^{0.307}} \right]^{4/3} \tag{4c}$$

The particularity of MLJ or AMLJ potentials is that they follow a convex function; conversely, other universal potentials, expressed by a Dirichlet-type function, $\Sigma A_j \exp(-b_j R/a)$, present a concave profile. The clear drop-off of the potential tail is necessary to describe the phase effect observed in Sn for low-energy ions in solids. An artificial truncation was introduced for that sake (Wilson and Bisson, 1971; Latta and Scanlon, 1974). The potential tail of the AMLJ potential agrees well with the "Gibson-2" potential (Gibson et al., 1960), which was proposed empirically for molecular dynamics calculations made in order to reproduce the collision events in the case of self-irradiation of copper. The accuracy of the AMLJ potential was ascertained by comparison with ion ranges (Nakagawa and Yamamura, 1988b), stoichiometric changes in ion beam mixing (Nakagawa and Yamamura, 1988a), and other experiments involving ion beams (Nakagawa, 1991; Robinson, 1994).

3. LOCAL DENSITY APPROXIMATION FOR THE ELECTRONIC STOPPING POWER

A partially stripped projectile of atomic number Z_1 senses the repulsive forces from atoms in its range of interaction and dissipates its energy dominantly via Se. The trajectory r is determined by the Newton equation under the influences of both centrifugal forces due to the surrounding atoms (located at $\{R_j\}$) and a friction force due to target electrons or the electronic stopping power acting on an ion of energy E:

$$M_1 \ddot{r} = - \sum_{j=1} \frac{\partial V(R_j)}{\partial R_j} + \frac{dE}{dr} \tag{5a}$$

where the elastic term expresses the multicollisions, and the friction force is

$$\frac{dE}{dr} = (Z_1 \gamma)^2 \left(\frac{dE}{dr} \right)_p = (Z_1 \gamma)^2 (S_L^{-1} + S_H^{-1})^{-1} \tag{5b}$$

The value of γ (less than unity) depends on the energy E according to Brandt (1982). The first term on the right-hand side of Eq. (5a) is a sum of vectors representing internuclear forces. The second one is the product of Se for a change unit $(dE/dr)_p$ and the effective point charge $(Z_1\gamma)^2$. The components S_H and S_L are asymptotic solutions to the LW theory for the cases of high- and low-energy protons, respectively. Each is expressed as a function of the electron density $\rho(r)$ (Å^{-3}) and energy E (keV/amu) in eV/Å units (Nakagawa, 1993a):

$$S_H = 2349 S_\rho \frac{\ln(1 + (E/17.06\sqrt{\rho(r)}))}{E} \quad (6a)$$

$$S_L = \alpha 4.6\sqrt{E}[-(0.5 + 0.31 X)\ln X - 0.5 + 0.634 X] \quad (6b)$$

with

$$X = 0.1045\,\rho(r)^{-1/3}$$

The multiplying factor α in Eq. (6b) is introduced to compensate for the too small values of S_L provided by the LW theory, especially at low energy (Mann and Brandt, 1981; ENR theory by Echenique, Nieminen, and Ritchie, 1981). This factor can be derived by comparing the values of Se obtained from single collisions and those obtained experimentally for cases of random incidence. The ENR theory, which adopts the DF formalism for Se based on the dielectric approximation, is one of the most reliable ones (Mann and Brandt, 1981). This gives $\alpha = 1.63$. In the single collision approximation, Se is accounted for by averaging the energy loss on respective binary collision with various values of p, on solving Eq. (5a) numerically as a nonlinear difference equation with respect to time (or velocity). The numerical integrations were executed by means of the 40th-order Legendre-Gauss algorithm, which was checked by counting the total number of electrons of a silicon atom. The numerical error (less than 1%) was small enough to be compensated for by adjusting the radius r_{AWS} of AWS from 0.1690 nm (Ziegler, Biersack, and Littmark, 1985) to 0.1691 nm.

The factor α is defined by comparing the values of the averaged Se due to the single collision approximation with those measured for amorphous solids. This is a useful benchmark for checking the validity of the binary collision approximation (BCA) coupled with the AWS approach, as is often used in computer simulations. Since the AWS approach requires locating each AWS at a specific site in a solid, if an ion selects only one closest AWS as its colliding partner, it is likely that AWS provides ρ_{sol}. More precisely, the sequence of single collisions yields Se for random ions as long as the ion energy is high and the relevant p is less than r_{AWS}. The former condition is related to the BCA approximation and requires $E_{ion} \gg E_B$ (Robinson, 1994),

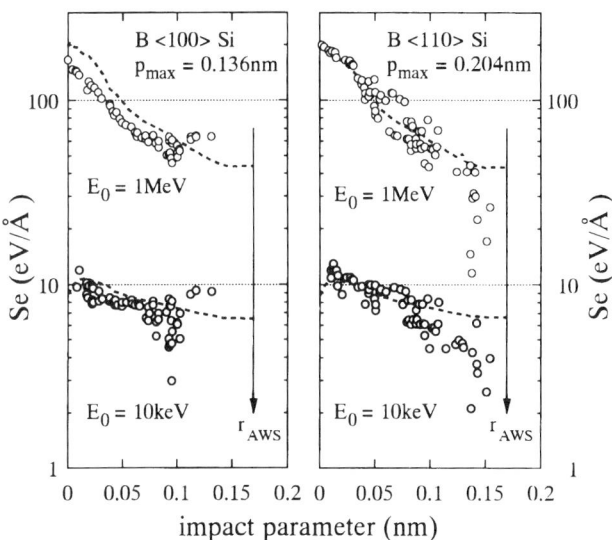

FIG. 1. Calculated Se for axially channeled boron (B) ions into silicon (Si) as a function of the distance from the closest string. Dotted curves indicate the results obtained for random ions on the basis of binary collisions. Here the ZBL (Ziegler-Biersack-Littmark) interatomic potential (1985) and the Hartree-Fock-Slater approximation is used for $\rho(r)$. (Reprinted from Nakagawa (1993a) "Impact Parameter Dependence of the Electronic Stopping Power of Channeled Sons," *Phys. Stat. Sol. (b)* **178**, 87–89, with permission from the author and Elsevier Science, Amsterdam).

where $E_B (\cong 25\,\text{eV})$ is the binding energy of bulk target atoms. Note that the recent version of the MARLOWE is beyond BCA (Robinson, 1994). The latter condition is connected to a deficiency of the AWS approach itself, as is shown in Figure 1. Figure 1, which provides a comparison of Se for channeled and random ions, suggested that a channeled ion passing through the ion core with an incident energy $E_0 > 100\,\text{keV}$ would experience the same p-dependence of Se as that of a random ion. This observation partly supports BCA.

The particular case of boron implantation into amorphous silicon in an energy range of from 10 keV to 10 MeV has been examined by Nakagawa (1993a). It must be noted that the Coulomb barrier of silicon for incident boron ions is 16 MeV, that is, beyond the energy range under consideration. Various calculations of Se were compared: the MARLOWE code (the original idea was by Robinson and Torrents, 1974), which gives the best result for the case of boron ions implanted into silicon by using a multiplying factor of 1.6 on Se (La Ferla *et al.*, 1992); theoretical predictions by Besenbacher, Andersen, and Bonderup (1980) based on the modified Bethe-Bloche theory; and the PRAL code (Biersack, 1982), which is expected to cover at least the whole energy range. The ratios of our calculated values to other values are slightly decreasing with E_0, but keep a value on the order of 2 in the energy range under consideration. Therefore, we have taken $\alpha = 2$ in Eq. (6b) for all calculations in order to compensate for the lower values of Se provided by the LW theory. There are several explanations for the too small Se values obtained, such as the additive contribution arising from the higher-order corrections, the Z_1^3 term (Lindhard, 1976), or the Barkas effect, that is, the so-called distant collision concerning plasmon excitation. The same multiplying factor α gave reasonable values of Se (Nakagawa, 1993b), in comparison with the database given by Northcliffe and Schilling (1970) and TRIM (Biersack and Haggermark, 1980), for MeV hydrogen, helium, boron, and arsenic ions implanted into carbon and germanium in a random direction.

III. Impact Parameter-Dependence of the Electronic Stopping Power in Crystalline Solids

In computer simulations the electronic stopping power Se can be described in two ways: the path-length dependent and the p-dependent energy loss approaches (note that the impact parameter description is valid when the energy of incident ions is much higher than thermal energies). The borderline between the two descriptions lies at the energy below which the ion trajectory can no longer be described using purely geometrical considerations. Originally the p-dependent approach arose from the fundamental interest and also was approved in channeling experiments (Datz et al., 1969; see also Gemmel, 1974). Practically, the pathlength-dependent approach is often adopted for very high-energy ions, which presumes nonlocal $\rho(r)$ consequently. Even for very high energy ions — for example, in the case of a silicon target, 27 MeV/u Xe^{35+} ions into $\langle 110 \rangle$ axial channeling (L'Hoir et al., 1990) or 103.5 MeV oxygen ions into (110) planar channeling (Azuma et al., 1995) — a significant p-dependence of Se was observed. In such a case the LDA should be taken. Therefore, in the following we discuss the p-dependence of Se in solids with the purpose of interpreting the phase effect for low-energy ions and the directional effect for high-energy ions.

1. ORIGINAL OEN-ROBINSON MODEL AND INNOVATION

In the LSS theory the excitation of a free electron gas with a uniform density is considered. Nevertheless, as long as the ion energy does not exceed a value that results in the excitation of all bound electrons, it is reasonable to suppose the existence of frozen inner electrons (Oen and Robinson, 1976). This assumption was introduced in the well-known MARLOWE code, which simulates atomic collisions in crystals. This code uses the expression

3 SOLID EFFECT ON THE ELECTRONIC STOPPING

$Se(E) = A_0(E)\exp(-kr_0(E)/a)$, where $A_0(E)$ is given by LSS, $r_0(E)$ is the lower apsidal distance, and $k = 0.3$. In this expression it is implicitly expected that the factor $A_0(E)$ accounts for the energy dependence of Se, and that the exponential term including $r_0(E)$ reflects the p-dependence of Se. The constant value $k = 0.3$ is related to the choice of the Molière potential for $V(R)$. In imagining a boundary for frozen core electrons embedded in a sea of valence electrons around each nucleus, the extent of the freezing zone should increase as the energy of the projectile damps during the slowing-down process.

The original Oen-Robinson model can be improved, based on the following three considerations. (1) The value $k = 0.3$ can be reconsidered from the first principles (Robinson, 1994). (2) The factor $A_0(E)$, which was originally given by the LSS theory (Lindhard, Scharff, and Schiøtt, 1963), requires a correction factor specific to each colliding system for computation (Eckstein, 1991). (3) An additional exponential term with a different value of k is necessary to account for the directional effect obtained in various depth profiles. The last point is related to the difference between the value of $\rho(r)$ for core and valence electrons and appears as a bending of Se as a function of p. This feature is coincident with the separate treatment for the fraction of Se due to core electrons from that due to valence electrons, as has been suggested by Robinson (1994) and Bulgakov, Nikolaev, and Shulga (1974). Bulgakov *et al.* described the p-dependence of Se differently for core electrons (L shell for silicon) and valence electrons (M shell for silicon) in the case of helium ions implanted into silicon at a given E_0. Ziegler (1988) also has predicted the bending of Se with respect to ρ, as a numerical result of the LW theory; this means that he obtained the same bending by converting the abscissa of ρ to the distance from a nucleus of the atom concerned.

In order to solve problems mentioned previously, the following innovations were brought to the original Oen-Robinson model. Concerning the first two points, we performed precise computations by means of the LDA coupled with nonlinear molecular dynamics in a cluster representing low-index axial channels for each projectile impinging on a crystalline target. This procedure provides an exact expression of the p-dependence of Se in the collision system without any free parameters. Concerning the last point, we propose a hybrid expression of Se, which is composed of two fractions due to core and valence electrons, valid for any crystalline target.

2. DETERMINATION OF THE IMPACT PARAMETER-DEPENDENCE OF THE ELECTRONIC STOPPING POWER IN A CLUSTER

When we examine the p-dependence of Se from first principles, the nonlinear molecular dynamics approach is the best, in order to take the energy dissipation on multicollision events into account. As the ion energy increases, the required computation time becomes too long to execute molecular dynamics practically. Instead of applying nonlinear molecular dynamics to the whole crystal, we adopt the following cluster model coupled with LDA based on the LW theory, and obtain the values of Se precisely as functions of p. This procedure allows us to predict the details of collision events in a simulation and considerably shortens the computation time.

Normally molecular dynamics are developed in real time, in accordance with the time mesh on solving the difference equation called the Newton equation. Let us recall the complementarity between energy mesh and time mesh on the slowing-down process: in the framework of LFA (i.e., when the implantation dose rate is not too high), the current depth (x) of an implanted ion can be predicted by integrating the energy loss $(dE/dx)^{-1}$ with respect to the energy E, or integrating (dx/dt) with respect to the elapsed time t. In other words, molecular dynamics can be developed basically by means of the energy mesh taking advantage of the reality of the stopping powers, in the framework of LFA. Therefore, here we prefer to discuss the slowing-down process using energy mesh where the accuracy of Se or Sn is again discussed in relation to that of $\rho(r)$.

Basically, we represent a solid as an ensemble of axial channels, and we build a cluster by selecting a bunch of these channels that form an infinitely long tube composed of a central channel and peripheral encircling subchannels. In fact, the length of such a cluster is finite, as will be mentioned subsequently. The database we need is $\{Se(p,E)\}$ for ions passing through a specific axial channel in a crystal, in cases of various sets of E and incident p (measured from the closest atomic string forming the cluster). Since p distributes in a cross section of the main channel, $Se(p,E)$ is the average of $\{Se(p,E)\}$ having the same scalar value of p.

In practice, the length of a cluster is not infinite. The minimum length of the main axial channel can be taken to be equal to one period of the crystallographic arrangement, called a central segment, and that of subchannels should be a little longer. Atoms belonging to these surrounding segments play a role similar to the so-called periodic image atoms in molecular dynamics, and can even govern the movement of an ion passing through the central segment. Each main segment has five and nine atoms in the case of the $\langle 100 \rangle$ and $\langle 110 \rangle$ channels, respectively; the total number of atoms forming the cluster to represent a zinc-blende structure is 36 and 30 for the $\langle 100 \rangle$ and $\langle 110 \rangle$ directions, respectively. The size of such a cluster is large enough for molecular dynamics calculations, and also for determining the crystallographic influence of axial channels on Se.

3 SOLID EFFECT ON THE ELECTRONIC STOPPING

The AWS approach cannot be used as it is to calculate $\rho(r)$ in a crystal or even in an amorphous solid. In fact, only the calculation of $\rho(r)$ around nuclei seems reasonable. In the peripheral region of an AWS, no electrons are located in some places, whereas excess electrons are located in other places. The most obvious example of such an unrealistic situation is the channeling of particles in open axial channels, such as the $\langle 110 \rangle$ channel in the diamond structure. Let us now consider the case of $\langle 110 \rangle$ silicon channeling. An ion entering this direction sees the target as an ensemble of hexagonal axial channels that looks like a honeycomb, each channel being made of six monatomic strings. The center of each axial channel is vacant because the length of diagonals of one hexagon is 0.4074 nm, that is, much larger than twice $r_{AWS} = 0.1691$ nm, and thus no electronic stopping is experienced by the ion. This problem, due to the deficiency of the AWS approach (see the drastic damping of Se profiles for $\langle 110 \rangle$ channeled ions in Fig. 1), should be overcome.[1]

Results obtained in the case of boron ions implanted into silicon in two channeling directions ($\langle 110 \rangle$ and $\langle 110 \rangle$) and at two energies ($E_0 = 10$ keV, $E_0 = 1$ MeV) are shown in Figure 1. Note that the data at the same p are scattered due to topological differences in electron density. The maximum p range is 0.2 nm, which roughly corresponds to $p_{max} = 0.2037$ nm for the case of the wider channel ($\langle 110 \rangle$), whereas $p_{max} = 0.1358$ nm for the narrower channel ($\langle 100 \rangle$). Dotted curves in Figure 1 correspond to values of Se for random ions in the single collision approximation; these data serve the purpose of checking the reliability of BCA. A value $r_{AWS} = 0.1691$ nm corresponds to p_{max} in the case of the single collision approximation. The most remarkable feature exhibited in Figure 1 is the bending of Se as a function of p at $p \approx r_{core}$. For hydrogen, helium, boron, silicon, and arsenic ions implanted into carbon, silicon, and germanium crystals in axially channeled directions, it is shown that the p-dependence of Se is much stronger for core electrons than it is for valence electrons (Nakagawa, 1993b). Such a bending could be attributed to the phase effect on Se and looked to be enhanced for high-energy ions. The bending point almost corresponds to the border between core and valence electrons in the case of a silicon target. For example, it appears at $p \approx 0.055$ nm (for a value of $r_{core} = 0.05934$ nm) in silicon.

---

[1] A better way to determine ρ_{sol} in a crystal is to use the muffin-tin model, which separates an inscribed sphere from the Wigner-Seitz cell, for example, by the KKR (Korringa-Kohn-Rostoker) or the APW (augmented plane wave) method. Because of its different symmetry, ρ_{sol} is spherical inside the sphere and uniform in the peripheral region. Morruzi, Janak, and Williams, (1978) obtained HFS-ρ_{sol} of elemental metals using KKR. This is a useful database, as far as we treat the case of a low-Z metal bombarded with light ions.

3. SOLID EFFECTS ON THE ELECTRONIC STOPPING POWER FROM CLUSTER CALCULATION

Two kinds of solid effects, directional and phase, are exhibited in Figure 1. The phase effect causes the bending of the profile, although the deficiency of the AWS approach yields artificial Se profiles for larger p in Figure 1. The directional effect results from the strong p-dependence of Se, besides the topological difference in the electron distribution among different channels. The former effect is effective for low-energy ions ($E_0 < 100\,\text{keV}$), whereas the latter effect is dominant for high-energy ions ($E_0 > 100\,\text{keV}$). These effects cannot be predicted in the single collision approximation with a target of AWS.

Because a localized electron density centered at each target atom is used, we modify the expression of Se in the Oen-Robinson model. We replace $r_0(E)$ by p, for convenience, and include the energy dependence by replacing the invariant factor k by $s(E)$. Finally, we obtain the expression for Se:

$$Se(p, E) = A(E) \exp(-s(E) \cdot p/a) \qquad (7)$$

Both $A(E)$ and $s(E)$ are determined by least squares analysis after exact calculations. The directional effect is accounted for by different values of $A(E)$ and associated $s(E)$, whereas the phase effect is responsible for the bending profile of Se.

The directional effect, which reflects the topography of the crystal channels, appears more definitely for channeled ions with the highest energy. The factor $A(E)$ exhibits a clear E_0-dependence. Its value also depends on the atomic species and on the axial channels concerned. As a matter of fact, the factor $A(E)$ is scaled by the effective charge introduced by Brandt (1982), as far as core electrons are concerned. The electron distribution sensed by channeled ions depends strongly on the topology of the relevant channels; the value of $A(E)$ for $\langle 110 \rangle$ channeled ions is almost the same as that for random ions, whereas it is much smaller for $\langle 100 \rangle$ channeled ions (by a factor of about two thirds).

Since the p-dependence of Se due to valence electrons is quite different from that due to core electrons, we cannot use a common s value for both kinds of electrons. The factor s due to core electrons increases with E_0, whereas that due to valence electrons is almost invariant (it is less than 0.3). The different values of s reflect the different topologies of the electron distribution sensed by channeled ions for the various directions considered.

Because of the deficiency in calculating ρ_{sol} for channeled ions by using the AWS approach, we can only determine exactly the p-dependence of Se in the case of core electrons. From experience, we know that a decreasing function of Se with respect to p works to account for the range distribution of implanted ions (especially for reproducing the tail of channeling peaks), in a rather wide energy range. We adopt only the database of Se(p, E) due to core electrons from the cluster calculation. Then we use computer simulations to define ρ_{sol} corresponding to valence electrons. That is, instead of calculating the whole ρ_{sol} prior to simulation, the consequential Se(p, E) due to valence electrons is sought via simulations, with the criterion that the proper decreasing function of Se reproduces the depth profiles of implanted ions. As a consequence, ρ_{sol} is spherical within a Wigner-Seitz cell, whereas the declivity of the decreasing function of ρ_{sol} presents a bending at the ion core. The ρ_{sol} inside the ion core is the same as that provided by the AWS approach based on DHFS atoms.

The Monte Carlo simulation code used for computer calculations is based on the ACOCT code (Yamamura and Takeuchi, 1987). The AMLJ potential is chosen for $V(R)$ in the calculation of Sn; it is derived from first principles using LDA. Therefore, the present calculations are valid even for collision partners with very high-Z atoms.

IV. Electronic Stopping Power of Chemical Compounds

The Bragg rule, which assumes the stoichiometric contribution from composite atoms to the total Se in compounds, is valid for ions with energy higher than ≈ 1 MeV/amu (Sattler and Dearnaley, 1967; Sattler and Vook, 1968). It cannot be applied to the case of low-energy ions. We are most concerned with examining the Se of compounds theoretically and seeking an alternative to the Bragg rule for low-energy ions.

1. AXIAL CHANNELING IN ZINC-BLENDE

A good example of such a study is provided by the case of zinc-blende crystals, since the low index axial channels ($\langle 100 \rangle$ and $\langle 110 \rangle$) are made of two kinds of monatomic strings. The cross sections are square and deformed hexagons, respectively. Therein, the same kind of monatomic strings is located alternately and strings are faced at a diagonal site (diagonally paired-atomic strings). An ion entering such a channel with a very small angle with respect to one atomic string is repelled alternately by the paired-atomic strings, so that the ion exhibits a harmonic-like motion. In this case the cluster must have much longer main segments than usual. For

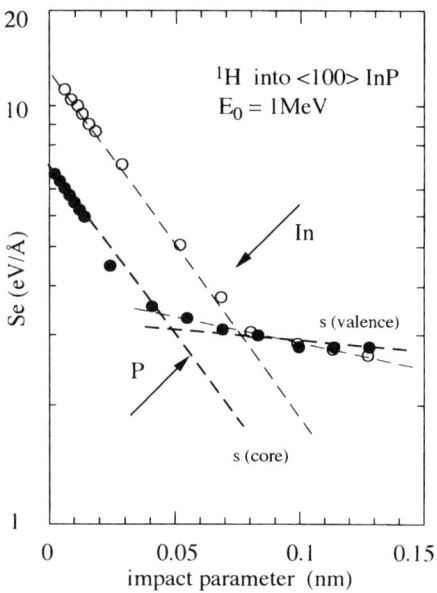

FIG. 2. Values of Se obtained for different atomic species in a diatomic compound. In and P indicate the contribution arising from the monatomic strings made of indium (In, by open circles) and phosphorus (P, by solid circles) atoms, respectively. Broken lines can be expressed by Eq. (7). (Reprinted from Nakagawa (1994). "Inelastic Energy for Channeled Ions in Compound Semiconductor." In *Adv. Mat. '93, III, B Composites, Grain Boundaries and Nanophase Materials*, Elsevier, Amsterdam, pp. 905–908.)

example, the wavelengths associated with 10 keV and 1 MeV boron ions in $\langle 100 \rangle$ silicon are 15.3 nm and 58.84 nm, respectively. By tracing the motion over a quarter of the wavelength, we are able to extract the p-dependence of Se due to each kind of monatomic string. Thus it is possible to consider only one kind of screening length, that is, between the projectile and each atomic species, on describing the p-dependence of Se.

Figure 2 provides an example of the difference in the p-dependence of Se in compounds when ions pass preferentially in the vicinity of different kinds of monatomic strings (Nakagawa, 1994). The broken lines are adapted to Eq. (7). In Figure 2, 2 *modes* are exhibited: the symbols P and In indicate Se sensed by 1 MeV ^{1}H ions moving through the $\langle 100 \rangle$ channel of a InP crystal, on a (100) plane constituted only of P or In atoms, respectively. As a matter of fact, the narrowest $\langle 100 \rangle$ channel allows 2 modes (Fig. 2) and the widest $\langle 110 \rangle$ channel allows 4 modes, due to the deformed channel window from an accurate hexagon. For these six modes, sets of values of $A(E)$ and $s(E)$ were obtained. The strongest p-dependence of Se was found for the case of $\langle 110 \rangle$ channeling, especially at the vicinity of monatomic strings made of heavier atoms, whereas the weakest p-dependence was obtained in the $\langle 100 \rangle$ channel nearby a monatomic string made of lighter atomic species.

2. COMBINATION RULE AS AN ALTERNATIVE TO THE BRAGG RULE FOR THE ELECTRONIC STOPPING POWER OF COMPOUNDS

Evidence of the breakdown of the Bragg rule was provided by Wilson (1987a). He observed a clear difference in the ranges of 200 keV helium ions channeled into $\langle 110 \rangle$ germanium and into $\langle 110 \rangle$ gallium arsenide. The difference was about 200 nm in maximum ranges. If provided the same size of ion cores for gallium, germanium, and arsenic, the mean electron densities of gallium arsenide and germanium become equal, since the lattice constants are the same. Then the difference in the maximum ranges obtained cannot be explained. This result means that the Bragg rule breaks down. Generally speaking, the reason for such a difference can be ascribed to the nonuniform distribution of target electrons in a crystalline solid. Thus the local electron distribution, which reflects the details of chemical bonding between atoms in a solid, should be introduced.

Sets of values of $A(E)$ and $s(E)$ could be obtained with a cluster calculation using the LDA for Se in axial channeling for any binary compound. Ions implanted in a crystal actually experience the intermediate strength of the p-dependent Se during the entire slowing-down process. Therefore, we assume as an a priori rule that the p-dependence of Se due to core electrons results from a certain average (Nakagawa, 1994):

$$S_e(p, E) = \left(\frac{A_1(E) + A_2(E)}{2} \right) \exp \left\{ -\frac{1}{2} \left(\frac{s_1(E)}{a_{i1}} + \frac{s_2(E)}{a_{i2}} \right) \right\} \quad (8)$$

where the suffixes 1 and 2 indicate the results of the strongest and weakest p-dependence of Se, respectively. The screening length a_{ij} inside the exponential term specifies the interaction between the projectile(i) and each atom interacting preferentially at a given moment. Note that this equation is valid only for core electrons, since we have no theoretical basis to define the p-dependence of Se for valence electrons. The p-dependence of Se due to valence electrons is defined by computer simulations. The present study of a zinc-blende crystal can be extended to the case of other compound targets, as long as they can form a crystal containing monatomic strings forming the widest and narrowest channels.

V. Electronic Stopping Power and Range Profiles via Computer Simulations

The main advantage of using computer simulations is to pick up the four-dimensional (space and time) information all along the slowing-down process, for any virtual scheme of collision systems in solids constructed within the computational framework. One can check the influence of any factor defining the phenomena observed by artificial changes on related parameters. From the technological viewpoint regarding ion beams, simulations are expected to assist the material design, that is, how to introduce the desired function onto materials. However, it is necessary to retain the theoretical grounds as much as possible for the promising developments.

As far as the determination of the ion projected ranges R_p is concerned, several analytical evaluations are possible: (1) development in terms of moments as was done in the Winterbon theory; (2) application of the well-known LSS theory to the case of heavy ion implantation into lighter materials at low energy, where Sn dominates the collision events; (3) application of empirical methods to the case of aligned implantations (Rimini, 1994) at higher energy, where Se is dominant during the major part of the ion slowing-down process. These approaches are suitable to account for the reported data; they can hardly provide reliable results in the case of unknown ion-target combinations.

Usually, BCA adopts the established orbital equation by eliminating the variable of time. This leads to tracing the sequential collision events according to the real-space development, and not according to the real-time development. This is the reason BCA is often coupled with LFA. However, the real-time development is inevitable when the LFA breaks, for example, when one must treat synergistic phenomena such as cluster implantation (Insepov, Sosnowski, and Yamada, 1994) or cluster emission in sputtering (Betz *et al.*, 1994, 1995, 1996). In this case, the Newton equation is solved numerically beyond BCA, as a time-difference equation, in a procedure called a molecular dynamics approach, which was first established by Gibson *et al.* (1960). In our simulation, multicollision is taken into account beyond BCA and the p-dependence of Se is defined by nonlinear molecular dynamics.

1. INFLUENCE OF THE ELECTRONIC STOPPING POWER ON RANGE PROFILES

So far we have proposed an intuitive combination rule for the calculation of Se in compounds, leading to Eq. (8), which is available as far as core electrons are concerned. The energy dependence of the parameters $A(E)$ and $s(E)$ was evaluated numerically by solving a difference equation with respect to the ion velocity under the cooperative influence of many target atoms of the crystal. Implicitly, we have also assumed a similar decreasing function for Se in the case of valence electrons.

First, by varying the s value due to valence electrons, computer simulations reveal the strength of the influence of the p-dependence of Se on range profiles. It is important to check whether or not a presumed s value for valence electrons reproduces measured depth profiles. Figure 3 demonstrates the strong influence of the p-depencence of Se due to valence electrons on the depth profiles in the case of ^{1}H (proton) implantation into $\langle 110 \rangle$ gallium arsenide. In the calculations, the s value for core electrons at

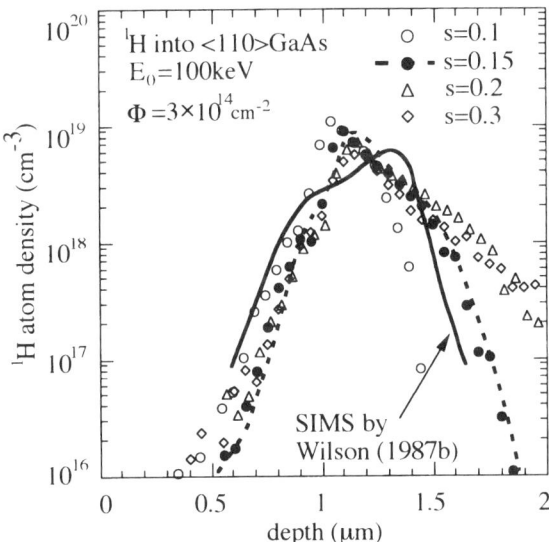

FIG. 3. Influence of the s-value for valence electrons on the range profile of 100 keV ^{1}H ions into $\langle 110 \rangle$ gallium arsenide (GaAs). The solid line refers to a secondary ion mass spectroscopy (SIMS) experiment (Wilson, 1987b) with a fluence of $\Phi = 3 \times 10^{14}\,\mathrm{cm}^{-2}$. (Reprinted from Nakagawa (1995). "Channeling Implantation into Chemical Compounds." *Nucl. Instrum. Meth.* **B96**, 173–178, with permission from the author and Elsevier Science, Amsterdam).

the relevant energy is not a free parameter but is defined from Eq. (8). From the data presented in Figure 3, the most probable s value for valence electrons is 0.15, which is much less than that for core electrons. This value of $s = 0.15$ is consistent with the cluster calculation performed in advance. The consequence of the larger value of s is larger maximum ranges due to the reduction of Se. However, varying the s value of valence electrons does not induce a drastic shift of the peak range; instead it leads to a modification of the damping profiles in the deeper region. The case of $s = 0.15$, denoted by solid circles in Figure 3 (guiding the eyes by a broken line), yields the best range profiles. Furthermore, Figure 4 compares the depth profiles of 100 keV ^{1}H ions implanted into gallium arsenide in different directions. The calculations were also performed with $s = 0.15$ for valence electrons. Figure 4 shows that the directional effect is well reproduced, since calculations and secondary ion mass spectroscopy (SIMS) data (Wilson, 1987b) agree well.

The application of the Bragg rule, which assumes uniform $\rho(r)$, is not compatible with LDA. This conclusion can be checked by comparing the

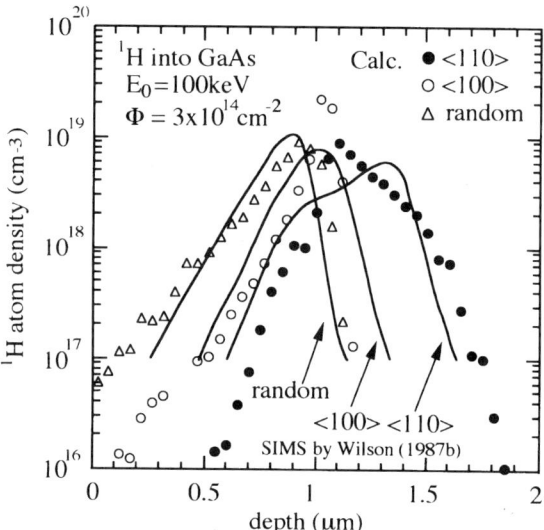

FIG. 4. Range profiles of 100 keV ^{1}H ions implanted into crystalline gallium arsenide (GaAs) in different directions. Calculations were performed with an s value of 0.15 for valence electrons, and an s value for core electrons as defined in Eq. (8). Solid lines hold for the same experiment as that shown in Figure 3. SIMS, secondary ion mass spectroscopy.

results obtained for a monatomic and a diatomic crystalline target with similar lattice constants, average mass, and electron distributions. This is the case of germanium and gallium arsenide. No significant differences are obtained in the case of swift ion implantation (Sattler and Dearnaley, 1967; Sattler and Vook, 1968). In the case of low-energy implantation, however, a relatively strong repulsion due to target atoms prevent ions from experiencing the same energy dissipation, expecially at the vicinity of nuclei of the different atomic species. This is the local distribution of ρ_{sol} that is of importance. The depth profiles of 200 keV helium ions implanted into $\langle 110 \rangle$ germanium and $\langle 110 \rangle$ gallium arsenide presented in Figure 5 are well reproduced by Eq. (8) for core electrons and by using a value of $s = 0.15$ for valence electrons. This proves the usefulness of cluster calculations in the case of ion implantation into compound targets.

The values of projected ranges R_p are predominantly determined by Se at the energy considered. However, Sn is also very significant for the shape of the distribution profiles. Figure 6 illustrates this feature. It presents the depth dependence of the stopping powers for ^1H and ^2H ions implanted into gallium arsenide at $E_0 = 1$ MeV. In this work, Se and Sn are defined as the energy lost per unit length by the incident ions in a thin slab (thickness $\approx R_p/100$) that is parallel to the surface of the sample. The resulting

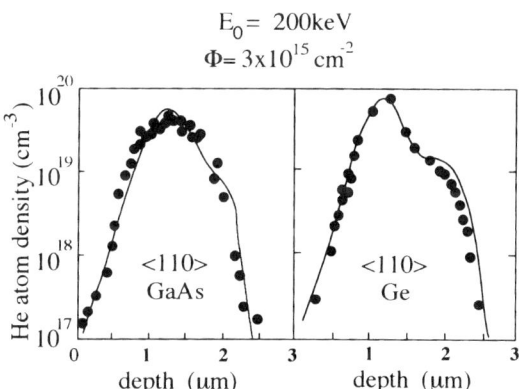

FIG. 5. Range profiles of 200 keV helium (He) ions implanted into $\langle 110 \rangle$ gallium arsenide (GaAs) and $\langle 110 \rangle$ germanium (Ge) at a total fluence of $\Phi = 3 \times 10^{15}$ cm^{-2}. Full circles hold for computer simulations; solid lines hold for secondary ion mass spectroscopy (SIMS) data by Wilson (1987a). (Reprinted from Nakagawa (1995). "Channeling Implantation into Chemical Compounds." *Nucl. Instrum. Meth.* **B96**, 173–178, with permission from the author and Elsevier Science, Amsterdam.)

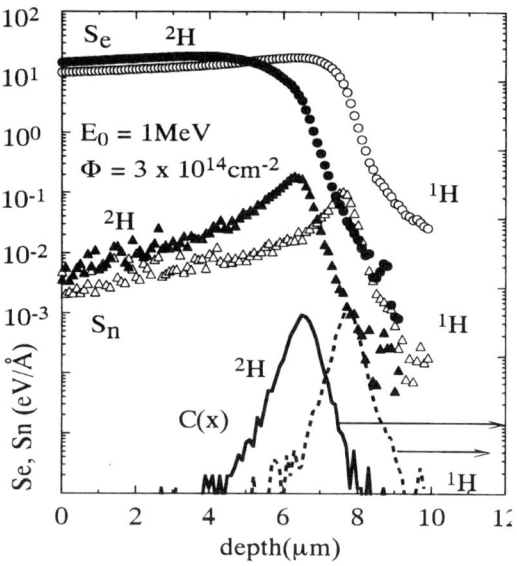

FIG. 6. Stopping power Se (circles) and Sn (triangles) versus the depth x for ^{1}H (open symbols) and ^{2}H (solid symbols) ions implanted into gallium arsenide in a random direction. The curves represent the range distribution $C(x)$ of ^{1}H and ^{2}H ions.

depth profiles, $C(x)$, of the two isotopes are added to Figure 6. The different values of Sn for ^{1}H and ^{2}H ions beneath the surface connect with those of Se. As can be seen in Figure 6, the energy loss is mainly due to Se in the major part of the ion path, whereas the shape of the ion profile fits that of Sn, especially at a greater depth than R_p. This implies the significance of the LDA for Sn for determining the range profiles.

2. SOLID EFFECTS ON THE ELECTRONIC STOPPING POWER FROM COMPUTER SIMULATIONS

The directional effect is shown in Figure 7 by means of two-dimensional Se maps (Nakagawa, 1995). The value of Se of 100 keV ^{1}H ions implanted into (100) gallium arsenide up to a certain depth depends on the incident direction, that is, $\langle 100 \rangle$-aligned or random incidence. A value of $s = 0.15$ for valence electrons was used in the calculations. In the diagrams in Figure 7 the ordinate is the total inelastic energy loss divided by the depth, and the abscissa presents the lateral coordinate of the two-dimensional p defined when the ion enters the solid. From these maps it can be seen that the

3 SOLID EFFECT ON THE ELECTRONIC STOPPING

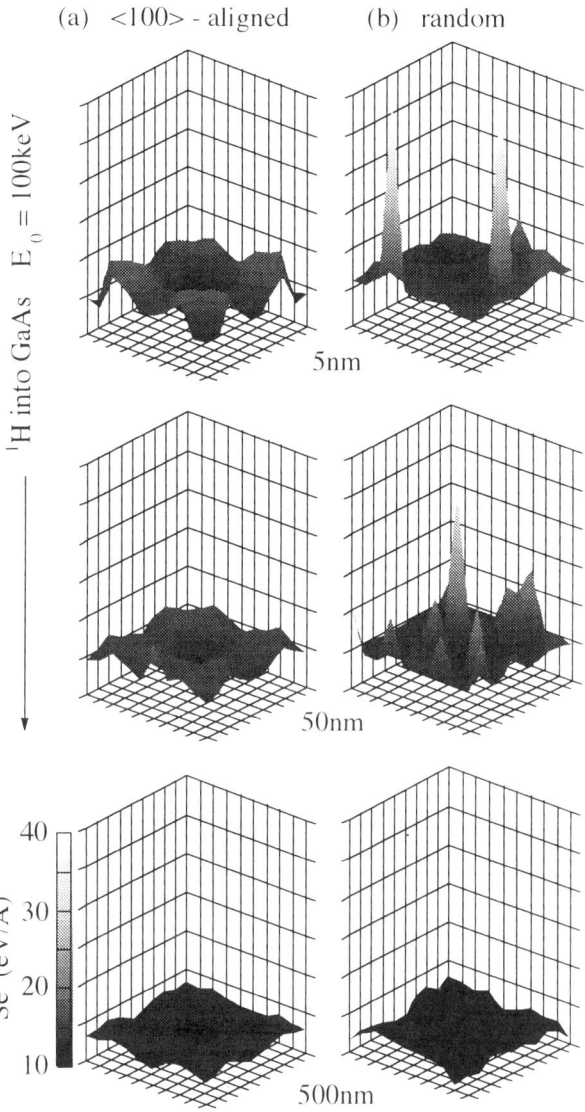

FIG. 7. Contour maps of the total inelastic energy loss divided by the mean depth (at 5, 50, and 500 nm). The mesh in the abscissa indicates the relative location of incident ions with respect to an axial channel. The left diagrams (a) concern the $\langle 100 \rangle$-aligned directions; the right diagrams (b) concern a random orientation. (Reprinted from Nakagawa (1995). "Channeling Implantation into Chemical Compounds." *Nucl. Instrum. Meth.* **B96**, 173–178, with permission from the author and Elsevier Science, Amsterdam).

contour due to channeled ions reflects well the geometry of the axial channel over a rather long period, whereas random ions suffer very huge stopping, as is exhibited by some accidental peaks. This is due to the fact that random ions encounter core electrons more frequently than do channeled ions, and that a much weaker (but significant) p-dependence of Se for valence electrons is observed as compared with core electrons.

An important question must be addressed: How long does a projectile keep the initial memory of aligned implantation? In other words: In what energy range is the p-dependence of Se significant? When E_0 is very low, the memory of aligned implantation is assumed to vanish very soon, because the minimum value of p to hold channeling is rather large. The higher the value of E_0, the wider the range of the distance over which channeled ions keep their initial memory. In an extreme situation, due to the little influence of target atoms on an ion trajectory, the difference between aligned and random implantations in the p-dependence of Se is strongly reduced.

Figure 8 compares the ratio of Se for ^{1}H implanted into gallium arsenide in a random direction to that implanted in a $\langle 100 \rangle$-aligned direction. Each plot arises from the average inelastic energy loss per unit length in a given slice of the crystal. Arrows indicate the position of R_p of channeled ions measured by SIMS (Wilson, 1987b). The crossing point of each curve with a horizontal broken line (ratio = 1) indicates the depth at which the two types of ions have the same mean energy. It must be noted that the difference between $\langle 100 \rangle$-aligned and random implantations remains over long distances and is more important as E_0 increases, since the crossing occurs at a greater depth in the case of high-energy implantation.

VI. Concluding Remarks

This article has examined the influence of the "solid effects" on the electronic stopping power Se for ions implanted into a crystalline target. These solid effects — namely, the phase effect, which is related to the nature of the irradiated medium, and the directional effect, which expresses its crystallographic character — cause the material dependence of the radiation effects. This chapter has been devoted mainly to the study of the impact parameter, p, dependence of Se, which particularly concerns the different contributions arising from valence and core electrons. The solid effects mostly have been accounted for by a factor s in Eq. (7) that expresses the dependence of Se on p. As a matter of fact, the factor s is not common to

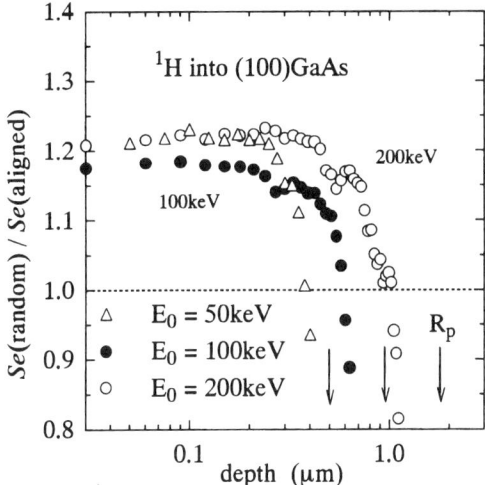

FIG. 8. Calculated ratios of Se for ^{1}H randomly implanted into (100) gallium arsenide (GaAs) to that for the $\langle 100 \rangle$-aligned case, as a function of the depth of the crystal. Arrows indicate the depth of R_p for channeled ions.

valence and core electrons; thus the function can be a hybrid function with respect to p. The value s due to core electrons depends strongly on the ionic energy in the case in which the ionic energy $E > 100 \,\text{keV}$, whereas that due to valence electrons is almost invariant with E.

First, cluster calculations were performed to study the p-dependence of Se in solids. Nonlinear molecular dynamics, which takes into account Se as a frictional force due to target electrons, was applied in a cluster representing a bunch of axial channels. Two kinds of solid effects were found, which cannot be predicted by a binary collision model. The first is the different behavior of Se for ions incident into different directions due to core electrons, which is dominant for high-energy ions ($E > 100 \,\text{keV}$). The second is the topological characteristics of each channel mostly due to valence electrons, which is particularly revealed with low-energy ions ($E < 100 \,\text{keV}$). The borderline between the influence of valence and core electrons was almost coincident to the bending point of the curve describing Se with respect to p. A few problems concerning cluster calculations have not been discussed in detail in this chapter. The most serious ones seem to be: (i) formulation of a more realistic electron distribution in solids including compounds; and (ii) consideration of higher-order corrections to Se in terms of Z_1^3 or Z_1^4. A more concrete p-dependence of Se has been investigated by means of computer simulations, due to the necessity to accurately account for the contribution of valence electrons.

How strong is the influence of solid effects on Se? This question has been examined in the case of a diatomic crystalline solid. The study of axial channeling into low index directions of zinc-blende crystals has allowed us to introduce a rule describing the p-dependent Se, available even for low-energy ions. This method gives results different from those obtained by using the Bragg rule, which averages respective Se of atoms of each species according to the stoichiometry of the compound. In fact, it has been proven that our simple rule is available for core electrons but cannot be applied to the case of valence electrons. To this purpose we have performed a computer simulation by using a similar but weaker p-dependence of Se for valence electrons than that used for core electrons.

Second, we have used Monte Carlo simulations to calculate the range profiles of ions implanted into binary III–V semiconductors. A hybrid function of Se versus p, with a bending point around the ion core of the target atom concerned, reproduces the directional effect obtained in the case of 100 keV ^{1}H implantation into gallium arsenide in different orientations. The best values for core electrons reproducing the experimental data are much smaller than those of core electrons, and they are consistent with that predicted by cluster calculations. Another crucial result obtained from the calculations presented in this work is that the difference in the size of ion cores was found to be essential for the determination of the range profiles. The difference in the range distribution of the same ion into germanium and gallium arsenide was reproduced by taking into account the different sizes of ion cores of relevant atoms and the bending profile of Se with respect to p. From these results, we draw three conclusions. (1) The strength of the p-dependence of Se due to valence electrons, which is quite different from that of core electrons, determines the details of the range profiles of implanted ions. (2) The description presented in this paper is a good alternative to the Bragg rule, particularly for ions with incident energy smaller than 1 MeV/amu. (3) Cluster calculations prior to simulations provide good predictability for unknown colliding systems.

Finally, one of the most important needs in the field of ion beam technology is a device design that makes use of high-energy ion beam implantation into compound crystals (i.e., in the MeV range). To this purpose, the use of computer simulations, especially for unknown compound crystals, has proved to be essential. At high energy, the directional effect of Se becomes significant and strongly influences the lateral distribution of implanted ions. In such a case, a study of the energy straggling in compounds is inevitable (Nakagawa et al., 1995). An experimental study of the energy straggling in bulk materials should help to refine the knowledge of the material dependence of the radiation effect. The device design could be better developed by taking into account the correlation between the directional effect and the energy straggling.

ACKNOWLEDGMENTS

Firstly, deepest thanks go to Professor J. P. Biersack, Hahn Meitner Institut, Berlin, for giving me the opportunity to begin this research on the electronic stopping power under the auspices of the ESPRIT program. Sincere thanks are due to Professor L. Thomé, CSNSN-IN2P3-CNRS, Orsay, for his discerning discussions, and to Dr. W. Eckstein, Max Plank Institut in Garching, for his valuable discussions. I am also grateful to Professor E. Rimini and his colleagues at the University of Catania, for providing their stimulating experimental results prior to publication, and to Professors Y. Yamamura and H. Saito, Okayama University of Science, for their great help in providing simulations and inclusive discussions.

REFERENCES

Arnau, A., Bauer, P., Kastner, F., Salin, A., Ponce, V. H., Fainstein, P. D., and Echenique, P. M. (1994). "Phase Effect in the Energy Loss of Hydrogen Projectiles in Zinc Targets," *Phys. Rev.* **49**, 6470–6480.
Azuma, T., Komaki, K., Yamagata, M., Yamazaki, Y., Sekiguchi, M., Hattori, T., and Hasegawa, T. (1996). "Trajectory Dependent Charge-State Distribution of Planar Channeled Heavy Ions." *Nucl. Instrum. Meth.* **B115**, 310–318.
Besenbacher, F., Andersen, J. U., and Bonderup, E. (1980). "Straggling in Energy Loss of Energetic Hydrogen and Helium Ions," *Nucl. Instrum. Meth.* **168**, 1–15.
Betz, G., Kirchner, R., Husinsky, W., Füdenauer, F., and Urbassek, H. M. (1994). "Molecular Dynamics Study of Sputtering of Cu(111) under Ar Ion Bombardment," *Radiation Effects Def. Solids* **130/131**, 251–266.
Betz, G., and Husinsky, W. (1995). "Molecular Dynamics Studies of Cluster Emission in Sputtering," *Nucl. Instrum. Meth.* **B102**, 281–292; see also Kornich, G. V., Betz, G., and King, B. V. (1996). *Nucl. Instrum. Meth.* **B115**, 467–467.
Biersack, J. P. (1982). "New Projected Range Algorithm as Derived from Transport Equations," *Z. Phys. A* **305**, 95–101.
Biersack, J. P., and Haggmark, L. G. (1980). "A Monte Carlo Computer Program for the Transport of Energetic Ions in Amorphous Targets," *Nucl. Instrum. Meth.* **174**, 257–269.
Bohr, N. (1948). "The Penetration of Atomic Particles Through Matter," *Kgl. Danske Videnskab. Selskab. Mat-Fys. Medd.* **18**(8).
Brandt, W. (1982). "Effective Charge of Ions and the Stopping Power of Dense Media," *Nucl. Instrum. Meth.* **194**, 13–19.

Bulgakov, Yu. V., Nikolaev, V. S., and Shulga, V. I. (1974). "The Experimental Determination of the Impact Parameter Dependence of Inelastic Energy Loss of Channeled Ions," *Phys. Lett.* **46A**, 477–478.

Carlson, T. A., Lu, C. C., Tucker, T. C., Nestor, C. W., Jr., and Malik, F. B. (1970). "Eigenvalues, Radial Expectation Values, and Potentials for Free Atoms from $Z = 2$ to 126 as Calculated from Relativistic Hartree-Fock-Slater Atomic Wave Functions." In *ORNL-4614 UC-34-Physics.*

Chu, W.-K. (1978). "Basic Physical Concepts." In *Backscattering Spectrometry* (W-K. Chu, J. W. Meyer, and M.-A. Nicolet, eds.), Academic Press, New York.

Datz, S., Moak, C. D., Noggle, T. S., Appleton, B. R., and Lutz, H. O. (1969). "Potential Energy and Differential-Stopping-Power Functions from Energy Loss Spectra of Fast Ions Channeled in Gold Single Crystals," *Phys. Rev.* **179**, 315–326.

Echenique, P. M., Nieminen, R. M., and Ritchie, R. H. (1981). "Density Functional Calculation of Stopping Power of an Electron Gas for Slow Ions," *Solid State Commun.* **37**, 779–781.

Eckstein, W. (1991). *Computer Simulation of Ion-Solid Interactions*, Springer-Verlag, Berlin.

Fano, U., and Lichten, W. (1965). "Interpretation of Ar^+-Ar Collisions at 50 keV," *Phys. Rev. Lett.* **14**, 627–629.

Firsov, O. B. (1957). "Interaction Energy of Atoms for Small Nuclear Separations," *Soviet Physics-JETP* **5**, 1192–1196.

Firsov, O. B. (1958). "Calculation of the Interaction Potential of Atoms." *Soviet Physics-JETP* **6**, 534–537.

Gemmel, D. S. (1974). "Channeling and Related Effects in the Motion of Charged Particles Through Crystals," *Rev. Mod. Phys.* **46**(1), 129–227.

Gertner, I., Meron, M., and Rosner, B. (1978). "Electronic Energy Loss of Ions in Solids in the Energy Range $10\text{-}10^4$ keV/Nucleon," *Phys. Rev. A* **18**, 2022–2029.

Gibson, J. B., Goland, A. N., Milgram, M., and Vineyard, G. H. (1960). "Dynamics of Radiation Damage," *Phys. Rev.* **120**, 1229–1253.

Gilbert, T. L., and Wahl, A. C. (1967). "Single-Configuration Wavefunctions and Potential Curves for the Ground States of He_2, Ne_2, and Ar_2," *J. Chem. Phys.* **47**, 3425–3438.

Gordon, R. G., and Kim, Y. S. (1972). "Theory for the Forces Between Closed-Shell Atoms and Molecules," *J. Chem. Phys.* **56**, 3122–3133.

Inokuti, M. (1991). "How is Radiation Energy Absorption Different Between the Condensed Phase and the Gas Phase?," *Radiation Effects Def. Solids* **117**, 143–162.

Insepov, Z., Sosnowski, M., and Yamada, I. (1994). "Molecular-Dynamics Simulation of Metal Surface Sputtering by Energetic Rare-Gas Cluster Impact." In *Adv. Mat. '93, IV, Laser and Ion Beam Modification of Materials* (I. Yamada *et al.*, eds.), Elsevier, Amsterdam, pp. 111–118.

L'Hoir, A., Andriamonje, S., Anne, R., de Castro Faria, N. V., Chevallier, M., Cohen, C., Dural, J., Gaillard, M. J., Genre, R., Hage-Ali, M., Kirsch, R., Farizon-Mazuy, B., Mory, J., Moulin, J., Poizat, J. C., Quéré, Y., Remillieux, J., Schmaus, D., and Toulemonde, M. (1990). "Impact Parameter Dependence of Energy Loss and Target-Electron-Induced Ionization for 27 MeV/u Xe^{35+} Incident Ions Transmitted in [110] Si Channels," *Nucl. Instrum. Meth.* **B48**, 145–155; see also Vickridge, I., L'Hoir, A., Gyulai, J., Cohen, C., and Abel, F. (1990). "$^{27}Al(p,\gamma)^{28}$Si Narrow Resonance in Channeling: A Measurement of Inelastic Energy Transfers at Small Impact Parameters," *Europhys. Lett.* **13**, 635–640.

La Ferla, A., Galvagno, G., Raineri, V., Setola, R., Rimini, E., Carnera, A., and Gasparotto, A. (1992). "Axial Channeling of Boron Ions into Silicon," *Nucl. Instrum. Meth.* **B66**, 339–344.

Latta, B. M., and Scanlon, P. J. (1974). "Orbital-Integral Corrections to the Lindhard Atomic Stopping Power for Classically Scattered Heavy Atoms," *Phys. Rev. A* **10**, 1638–1645.

Lindhard, J. (1976). "The Barkas effect — or Z^3, Z_1^4-Corrections to Stopping of Swift Charged

Particles," *Nucl. Instrum. Meth.* **132**, 1–5.
Lindhard, J., Scharff, M., and Schiøtt, H. E. (1963). "Range Concepts and Heavy Ion Ranges," *Kgl. Danske Videnskab. Selskab. Mat-Fys. Medd.* **33**(14).
Lindhard, J., and Winther, A. (1964). "Stopping Power of Electron Gas and the Equipartition Rule," *Kgl. Danske Videnskab. Selskab. Mat-Fys. Medd.* **34**(4).
Mann, A. and Brandt, W. (1981). "Material Dependence of Low-Velocity Stopping Powers," *Phys. Rev. B* **24**, 4999–5003.
Moruzi, V. L., Janak, J. F., and Williams, A. R. (1978). *Calculated Electronic Properties of Metals*, Pergamon Press, New York.
Nakagawa, S. T. (1991). "A Realistic Interatomic Potential in Solids," *Radiation Effects Def. Solids* **116**, 21–28.
Nakagawa, S. T. (1993a). "Impact Parameter Dependence of the Electronic Stopping Power of Channeled Ions," *Phys. Stat. Sol. (b)* **178**, 87–98.
Nakagawa, S. T. (1993b). Solid-Effect on the Electronic Stopping and Application to Range Estimation," *Nucl. Instrum. Meth.* **B80/81**, 7–11.
Nakagawa, S. T. (1994). "Inelastic Energy Loss for Channeled Ions in Compound Semiconductor." In *Adv. Mat. '93, III, B Composites, Grain Boundaries and Nanophase Materials*, Elsevier, Amsterdam, pp. 905–908.
Nakagawa, S. T. (1995). "Channeling Implantation into Chemical Compounds," *Nucl. Instrum. Meth.* **B96**, 173–178.
Nakagawa, S. T., and Biersack, J. P. (1991). Unpublished material.
Nakagawa, S. T., Thomé, L. Saito, H., and Clerc, C. (1996). "Impact Parameter Dependence of the Energy Straggling of ^{1}H and ^{2}H into GaAs," *Nucl. Instrum. Meth.* **B115**, 345–347
Nakagawa, S. T., and Yamamura, Y. (1988a). "Depth Profiling and Stoichiometric Changes due to High-Fluence Ion Bombardments," *Nucl. Instrum. Meth.* **B33**, 780–783.
Nakagawa, S. T., and Yamamura, Y. (1988b). "Interatomic Potential in Solids and Its Application to Range Calculations," *Radiation Effects* **105**, 239–256.
Northcliffe, L. C., and Schilling, R. F. (1970). "Range and Stopping-Power Tables for Heavy Ions." *Nucl. Data Tables A* **7**, 233–463.
Oen, O. S., and Robinson, M. T. (1976). "Computer Studies of the Reflection of Light Ions from Solids." *Nucl. Instrum. Meth.* **132**, 647–653.
Rimini, R. (1994). Trends in Ion Implantation Technology of Semiconductors. In *Adv. Mat '93, IV, Laser and Ion Beam Modification of Materials* (I. Yamada et al., eds.), Elsevier, Amsterdam, pp. 41–46.
Robinson, M. T. (1994). "The Binary Collision Approximation: Background and Introduction," *Radiation Effects Def. Solids* **130/131**, 3–20.
Robinson, M. T., and Torrens, I. M. (1974). "Computer Simulation of Atomic-Displacement Cascades in Solids in the Binary-Collision Approximation," *Phys. Rev. B* **9**, 5008–5024.
Sabelli, N. H., Benedek, R., and Gilbert, T. L. (1979). "Ground-State Potential Curves for Al_2 and Al_2^{6+} in the Repulsive Region," *Phys. Rev. A* **20**, 677–688.
Sattler, A. R., and Dearnaley, G. (1967). "Channeling in Diamond-Type and Zinc-Blende Lattices: Comparative Effects in Channeling of Protons and Deuterons in Ge, GaAs, and Si," *Phys. Rev.* **161**, 244–252.
Sattler, A. R., and Vook, F. L. (1968). "Channeling in Zinc-Blende Lattices: Energy-Loss Studies for Hydrogen and Helium Ions in InAs, GaSb, AlSb, and InSb," *Phys. Rev.* **175**, 526–532.
Schiwietz, G. (1990). "Coupled-Channel Calculation of Stopping Powers for Intermediate-Energy Light Ions Penetrating Atomic H and He Targets," *Phys. Rev. A* **42**, 296–306.
Sigmund, P. (1969). "Theory of Sputtering. I. Sputtering Yield of Amorphous and Polycrystalline Targets," *Phys. Rev.* **184**, 383–416.

Thomé, L. and Garrido, F. (1995). "The Use of Markers and Nuclear Microanalysis to Monitor Atomic Transport Induced by Ion Bombardment in Solids," *Modern Phys. Lett. B* **9**, 163–186.

Wilson, K. L., and Baskes, M. I. (1978). "Deuterium Trapping in Irradiated 316 Stainless Steel," *J. Nucl. Material* **76/77**, 291–297.

Wilson, R. G. (1987a). "Ranges and Depth Distributions of 200-keV He Ions Channeled in Si, Ge, and GaAs Crystals," *J. Appl. Phys.* **61**, 2489–2491.

Wilson, R. G. (1978b). "Depth Distributions and Range and Shape Parameters for ^{1}H and ^{2}H Implanted into Si and GaAs in Random and Channeling Orientations," *J. Appl. Phys.* **61**, 2826–2835.

Wilson, W. D., and Bisson, C. L. (1971). "Inert Gases in Solids: Interatomic Potentials and their Influence on Rare-Gas Mobility," *Phys. Rev. B* **3**, 3984–3992.

Yamamura, Y., and Takeuchi, W. (1987). "Monocrystal Sputtering by the Computer Simulation Code ACOCT," *Nucl. Instrum. Meth.* **B29**, 461–470.

Ziegler, J. F. (1988). *Ion Implantation: Science and Technology*, 2nd ed. (J. F. Ziegler, ed.). Academic Press, Boston.

Ziegler, J. F., Biersack, J. P., and Littmark, U. (1985). *Stopping and Ranges of Ions in Solids*, Vol. 1. Pergamon Press, New York.

CHAPTER 4

Ion Beams in Amorphous Semiconductor Research

G. Müller

DAIMLER BENZ AG
FORSCHUNG UND TECHNIK
MÜNCHEN, GERMANY

S. Kalbitzer

MAX PLANCK INSTITUT FÜR KERNPHYSIK
HEIDELBERG, GERMANY

G. N. Greaves

DARESBURY LABORATORY
WARRINGTON, ENGLAND

I. INTRODUCTION 85
II. ION BEAM PRODUCTION OF AMORPHOUS SILICON 87
   1. *Impact of Disorder on Doping, Electronic Transport, and Optical Properties* . 87
   2. *Irreversible Ordering Phenomena in Undoped Amorphous Material* 95
   3. *Structural Relaxation and Thermal Crystallization of Doped Material* 99
   4. *Effects of Hydrogenation and Fluorination* 105
III. ION BEAM DOPING OF PLASMA-DEPOSITED AMORPHOUS SILICON 106
   1. *Gas Phase and Ion Implantation Doping* 106
   2. *Doping Mechanism in Hydrogenated Material* 110
   3. *Generation and Annealing of Implantation Damage* 115
IV. STRUCTURAL AND CONFIGURATIONAL CHANGES IN AMORPHOUS SILICON 121
   1. *Changes in Pure Material* 121
   2. *Changes in Hydrogenated Material* 122
   References 123

I. Introduction

During the past decade the technology of hydrogenated amorphous silicon (a-Si:H) and its alloys with carbon and germanium (a-Si$_{1-x}$C$_x$:H, a-Si$_{1-x}$Ge$_x$:H) has matured. Like crystalline silicon (c-Si), a-Si:H can be doped, and it exhibits a pronounced field effect and an efficient photoconductivity. The same is also true to a lesser extent for its alloys. Due to these

Structural transformations between c-Si, a-Si and a-Si:H

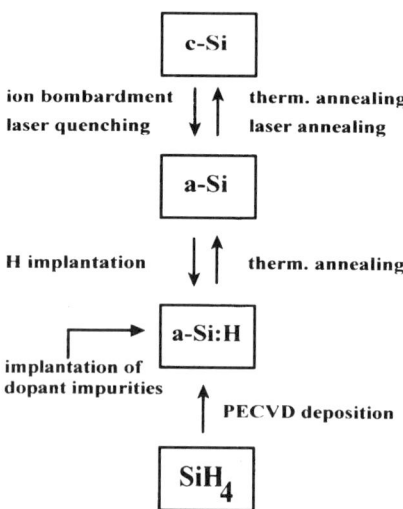

FIG. 1. Preparation of amorphous silicon (a-Si) in pure and hydrogenated (a-Si:H) form by different experimental techniques. Note the equivalence of ion implantation to plasma deposition (Müller, 1993; Böhringer et al., 1983, 1985) PECVD, plasma-enhanced chemical vapor deposition; c-Si, crystalline silicon; SiH_4, silane; therm., thermal.

properties, these materials now provide the basis for the fabrication of thin-film solar cells and for a variety of large-area optoelectronic devices (Kanicki, 1991). Due to the presence of several disorder elements in the amorphous network, such as distorted Si-Si bonds, dangling bonds, and Si-H_n groups ($1 < n < 3$), a-Si:H also exhibits surprising and qualitatively new physical properties that are unknown from its crystalline counterpart.

This chapter reviews some of those experiments in which ion beam techniques have been used to address central issues of the physics and technology of amorphous semiconductor materials. As illustrated in Figure 1, the work described subsequently follows two main lines: In the first approach, monocrystalline silicon-on-sapphire films (SOS; $d = 0.1$ to 0.5 µm) have been ion-beam disordered and hydrogen (H) has been introduced successively to obtain thin films of a-Si:H on optically transparent and electrically insulating substrates. This ion beam synthesis approach allows us to separate disorder- and hydrogen-related effects on the electrical and optical properties. Also, the impact of disorder on the bonding of dopant impurities can be studied. This first group of experiments will be described in Part II. In Part III, we turn to experiments in which plasma-deposited a-Si:H films serve as starting material into which substitutional and interstitial dopant impurities are implanted. By applying a variety of analytical methods to such a-Si:H films, a picture of the doping mechanism, clues to their metastability, and last but not least, a detailed understanding of radiation damage phenomena have emerged. The concluding part, Part IV, gives a synopsis of this work dealing with structural and configurational changes in pure and hydrogenated random Si networks.

II. Ion Beam Production of Amorphous Silicon

1. IMPACT OF DISORDER ON DOPING, ELECTRONIC TRANSPORT, AND OPTICAL PROPERTIES

a. Damage Accumulation in Crystalline Silicon Targets

A unique feature of ion implantation is that crystalline semiconductor materials can be transformed to the amorphous state in a quasi-continuous manner. In principle, therefore, it is possible to investigate changes in the electronic transport and optical properties as increasing amounts of disorder are introduced into the c-Si lattice. Such experiments have become possible using monocrystalline Si thin films on insulating and optically transparent sapphire substrates. In Figure 2 the results of some amorphization studies are summarized in which light ion beams (O^+, Si^+) have been used to introduce defects into c-Si targets. In these particular experiments, electron spin resonance (ESR) measurements and Rutherford backscattering spectrometry and channeling techniques (RBS/C) were used to characterize the increase of radiation damage with ion fluence (Brower and Beezhold, 1972; Prisslinger *et al.*, 1975; Müller and Kalbitzer, 1980; Müller, 1993)).

The data in Figure 2 clearly bring out that, in the case of light-ion bombardment of c-Si, amorphous material develops in three stages: in the low-dose regime (i) the irradiation first produces a variety of vacancy-type and vacancy-impurity–type defects, while leaving the long-range order of the c-Si lattice intact. After reaching a state in which the crystalline material is critically predamaged, further irradiation into the intermediate-dose regime (ii) causes the long-range order to relax and the point defects of the c-Si lattice to transform into the dangling bond defects characteristic of a-Si. In the high-dose regime (iii), finally, the material is completely amorphous and its properties are stable under prolonged irradiation, that is, the "damage saturation" regime is reached.

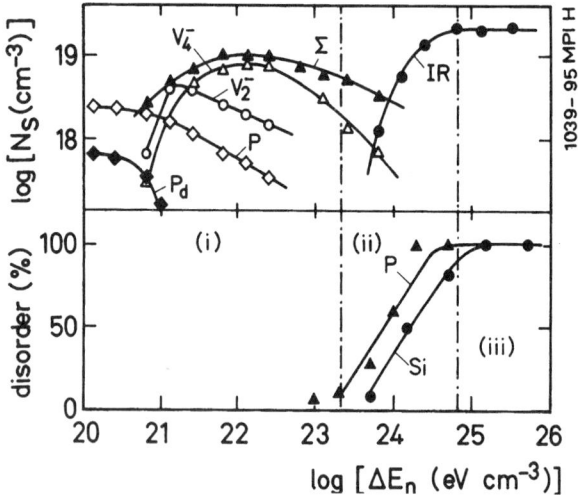

FIG. 2. Electron spin density N_s (top) and dechanneled ion beam fraction equivalent to the degree of disorder (bottom) of ion-bombarded n-type crystalline silicon (c-Si) specimens as a function of the energy deposited into atomic displacements ΔE_n. V_2^-, negatively charged divacancy; V_4^-, negatively charged four vacancy; Σ, Σ-centers; IR, isotropic resonance of $g =$ 2.0055; P, nondegenerate phosphorus; P_d, degenerate phosphorus. See the text for stages (i), (ii), and (iii). (Reprinted from Müller and Kalbitzer (1980). "The Crystalline-to-Amorphous Transition in Ion-Bombarded Silicon." *Philos. Mag.* **B41**, 307–325, with permission from the author and Taylor and Francis, London.)

The most interesting point brought out by the data in Figure 2 is that it is possible to produce both monocrystalline and amorphous material with high and roughly equal densities of point defects. As a consequence, the physical properties of both kinds of semiconductor materials should be largely controlled by the presence of these point defects. A comparison of their properties, however, should reveal most clearly those properties of a-Si that are due to the loss of the long-range order still present in the defective c-Si lattice (regime (i)).

b. Impact of Disorder on Optical Properties

From an optical point of view the transition regime (ii) is the most interesting because there, along with the loss of the long-range order, significant optical changes also take place (Bhatia, Krätschmer, and Kalbitzer (1988). Figure 3 summarizes optical absorption data for the c- and the a-phases of Si. In its crystalline phase Si exhibits an indirect bandgap of

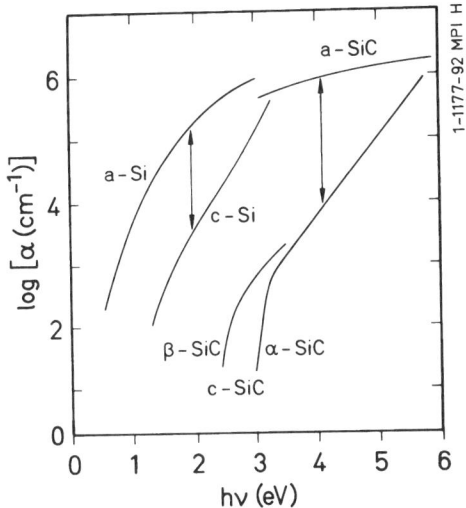

FIG. 3. Optical absorption of the crystalline phases of silicon (c-Si) and crystalline silicon carbide (c-SiC) and the amorphous phases of Si (a-Si) and SiC (a-SiC) as a function of photon energy. α-SiC and β-SiC denote the hexagonal and cubic polytypes of c-SiC, respectively. The arrows denote reversible transitions between these phases induced either by ion bombardment (c → a) or by thermal annealing (a → c). (Reprinted from Müller (1993). "The Contribution of Ion Beam Techniques to the Physics and Technology of Amorphous Semiconductors," *Nuclear Instruments and Methods in Physics Research* **B80/81**, 957–965, with permission from the author and Elsevier Science, Amsterdam.)

1.1 eV. Due to the indirect gap, the optical absorption immediately above the bandgap is small, increasing gradually as the photon energy is increased toward the direct bandgap energy of 3.3 eV. At this photon energy the optical absorption constant is large enough to render films with thicknesses on the order of a few tens of nm almost completely opaque. The comparison in Figure 3 shows that, within the photon energy interval between the indirect bandgap and the onset of direct transitions, the optical absorption of ion-bombarded a-Si is considerably enhanced. In addition, a sizable optical absorption is also observed below the bandgap energy of c-Si. Whereas the latter effect is due to a tailing of the valence and conduction band states into the bandgap region, the former is due to a relaxation of the crystalline k-selection rule for optical transitions in the a-state. Also shown in Figure 3 are similar data for silicon carbide (SiC). As is well known, c-SiC is a wide-bandgap semiconductor that can exist in a large number of polytypes. Depending on polytype, these materials exhibit indirect bandgaps ranging from 2.2 to 3.3 eV. Upon ion beam amorphization, essentially the same kind of behavior can be observed as in a-Si: a huge optical absorption enhancement in the range of sub-bandgap photon energies and, to a lesser extent, within the range of indirect band-to-band transitions in c-SiC ($h\nu < 6$ eV).

As is well known, k-conservation in crystals is caused by the long-range symmetries of the corresponding lattice structures. Considering the data of Figure 3, it is therefore relevant to ask to which extent these symmetries are broken when semiconductor lattices become disordered by irradiation with energetic ion beams. This problem has been investigated in considerable detail in the case of a-Si (Richter and Breitling, 1957; Fortner and Lannin, 1988). Radial distribution functions indicate that in the a-phase the tetrahedral short-range order is more or less preserved up to the second-nearest-neighbor shell. Over larger distances, however, the Si atoms appear to be randomly arranged within the a-Si network. Long-range symmetry is therefore broken on a length scale comparable with the unit cell of the c-Si lattice. In this situation, the uncertainty in the crystal momentum becomes comparable with the crystal momentum itself and k-conservation can no

FIG. 4. Crystalline-amorphous (c-a) contrast pattern written into a fine-grained polycrystalline silicon carbide (SiC) film on sapphire. The transparent and opaque regions represent c-SiC and a-SiC, respectively. The minimum feature size is 2 µm. The continuous line in the lower part represents the size of a 20 keV neon ion microbeam used for the pattern writing. (Reprinted from Müller (1993). "The Contribution of Ion Beam Techniques to the Physics and Technology of Amorphous Semiconductors," *Nuclear Instruments and Methods in Physics Research* **B80/81**, 957–965, with permission from the author and Elsevier Science, Amsterdam.)

longer dominate optical transitions in the a-phase. A strongly enhanced optical absorption therefore takes place above the indirect bandgap of c-Si.

Some years ago, it was suggested that the irradiation-induced relaxation of the k-selection rule might have important applications (Kalbitzer, 1987). Due to the higher optical absorption of the a-phase, crystalline-amorphous (c-a) contrast patterns can be produced with feature sizes in the submicron range in cases in which irradiation is performed with a program-controlled ion microbeam. Figure 4 gives an example of such a pattern that has been written into a fine-grained polycrystalline SiC film on sapphire (SiCOS) (Ruttensperger et al., 1991). The optical contrast in these films is sufficient to selectively expose and develop standard photoresists (Krötz et al., 1994). With the c-a contrast in SiC occurring in the ultraviolet range, this technique might prove useful for the fabrication of photomasks for submicron lithography (Derst et al., 1989). Attractive features of this technique are its conceptual simplicity and the reversibility of the c-a contrast formation. The latter feature allows for pattern repair by local laser annealing and rewriting with an ion microbeam. Additional applications of this technique could be in high-density optical storage or in the fabrication of integrated optical elements.

c. Impact of Disorder on Electronic Transport and Doping

The optical effects produced by the relaxation of the k-selection rule, in principle, can be observed in a continuous random network (CRN), that is, in a disordered network in which all its valence electrons are taken up in covalent bonds. Model building experience and Raman scattering experiments indicate that ideal random Si networks can be built by introducing two disorder elements into the c-Si lattice: relatively small distortions of the tetrahedral bond angle ($\Delta\theta \approx 8$ to 13 degrees), and relatively large deviations in the dihedral angle ($\Delta\phi \approx 180$ degrees), that is, in the relative orientation of adjoining tetrahedra (Overhof and Thomas, 1989). Due to the combined action of both disorder elements, the six-fold rings of Si atoms present in the c-Si lattice become distorted and a small fraction of the Si atoms become incorporated into fivefold or sevenfold rings. These latter features prevent the a-Si network from relaxing spontaneously to the c-state. The height of this kinetic barrier can be inferred from the activation energy for solid-phase epitaxial recrystallization of about 2.7 eV (Csepregi et al., 1977). The introduction of these disorder elements leads to an exponential tailing of the conduction and valence band states into the bandgap. The electronic states in these tails are localized at random positions in the CRN. Within the CRN these states act as shallow traps limiting the band mobility of charge carriers.

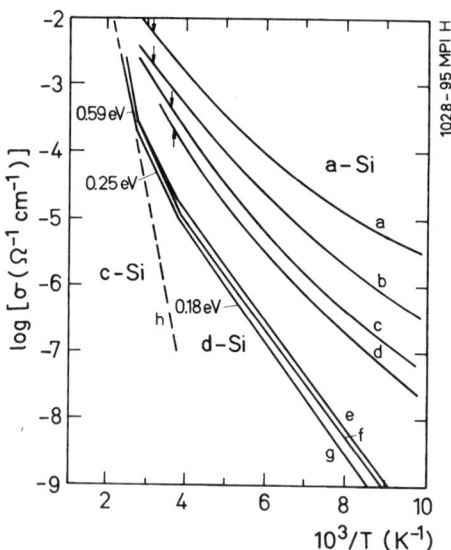

FIG. 5. Temperature-dependence of electrical conductivity of Neon-irradiated silicon thin films on sapphire. (a to d) Amorphized silicon (a-Si); (e to g) damaged silicon (d-Si); and (h) crystalline intrinsic silicon (c-Si). Ion fluences for samples (a to g): 100, 50, 20, 10, 5, 2, 1×10^{14} Ne/cm^2. Ne, neon. (Reprinted from Müller and Kalbitzer (1980). "The Crystalline-to-Amorphous Transition in Ion-Bombarded Silicon." *Philos. Mag.* **B41**, 307–325, with permission from the author and Taylor and Francis, London.)

The ESR data in Figure 2 show that a CRN is an incomplete description of a real a-Si network. Real films contain dangling bond defects associated with threefold coordinated, neutral Si atoms (Si_3^0). At these sites the fourth valence electron has failed to form a chemical bond. This electron gives rise to a localized defect level in the center of the a-Si mobility gap. As revealed from Figure 2, these defects evolve as the final result of a complex defect accumulation process. Figure 5 shows how the accumulation of such defects influences the electrical conductivity of the irradiated material. We see that, upon starting the irradiation, the electrical conductivity of c-Si rapidly approaches its intrinsic value ($\sigma \approx 10^{-5} \Omega^{-1} cm^{-1}$ at $T = 300$ K). This increase occurs because of the generation of compensating vacancy and vacancy-impurity-type defects within an otherwise perfectly ordered c-Si network. When the irradiation is continued into the intermediate dose regime (ii), the conductivity increases again, until the amorphous saturation level of about $2 \times 10^{-3} \Omega^{-1} cm^{-1}$ is approached. This increase occurs due to the onset of thermally activated tunneling transitions between irradiation-induced dangling bond defects. An important characteristic of this

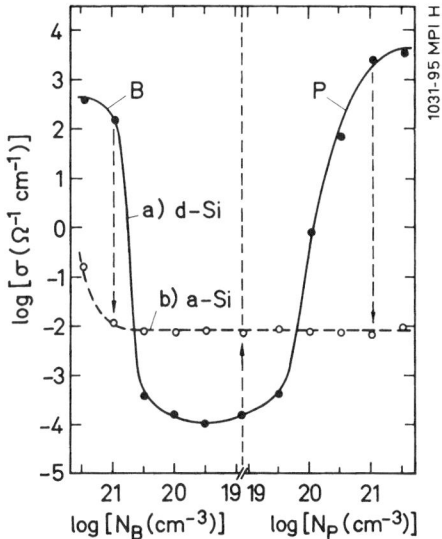

FIG. 6. Room-temperature conductivity as a function of boron (B) and phosphorus (P) dopant concentration for ion-beam-damaged crystalline silicon (d-Si) and amorphized silicon (a-Si). (Reprinted from Müller and Kalbitzer (1980). "The Crystalline-to-Amorphous Transition in Ion-Bombarded Silicon," *Philos. Mag.* **B41**, 307–325, with permission from the author and Taylor and Francis, London.)

kind of transport is an upward curved line in an Arrhenius plot, $\ln \sigma \propto 1/T$, which is caused by a continuous decrease of the conductivity activation energy as the temperature is lowered. This kind of transport was first interpreted by Mott (1969) in terms of a variable-range hopping mechanism. Accurate measurements on ion bombardment a-Si have revealed that the temperature-dependence of the variable-range hopping conductivity $\sigma(T)$ does not strictly follow a $\ln(\sigma) \propto T^{1/4}$ law, as predicted by Mott, due to the presence of a nonconstant density of localized defect levels within the a-Si mobility gap (Pfeilsticker, Kalbitzer, and Müller, 1978; Ruttensperger, Müller, and Krötz, 1993). In addition, Coulomb correlation effects have been shown to influence the low-temperature hopping transport (Vögele, Kalbitzer, and Böhringer, 1985).

Disregarding the fact that the electronic transport through dangling bond defects is characterized by a special form of $\sigma(T)$, the overall effect of point defects within an amorphous and a predamaged crystalline matrix should be the same: both dangling bond and vacancy-type defects should act as efficient recombination centers for photogenerated excess carriers (Stolk *et al.*, 1994). In addition, they should reduce the doping efficiency in the defective Si networks by trapping doping-induced excess carriers from the conduction and valence band states. Figure 6 summarizes electrical conductivity data obtained on a series of SOS specimens doped to high boron (B) and phosphorous (P) concentrations prior to ion bombardment with neon (Ne) ions at room temperature (Müller and Kalbitzer, 1980). In order to compare the behavior of a defective single crystal with that of a completely amorphous layer, the Ne irradiation was interrupted after applying a dose of about 10^{14} Ne ions/cm^2 to take electrical measurements and then continued to completely amorphize the predamaged samples with an areal dose of about 5×10^{15} Ne/cm^2.

At this first ion dose, the irradiated layers contain a high concentration of vacancy-type defects, however, the long-range symmetries of the c-Si network are left intact. Curve (a) in Figure 6 shows that under these conditions the conductivity in moderately doped material is relatively low, that is, on the order of the intrinsic conductivity of c-Si, as long as the dopant concentration does not exceed a certain threshold value. Above this threshold value, which is on the order of magnitude of the ESR spin density (Fig. 2), a large and rapid increase in the electrical conductivity occurs. In the limit of high dopant concentrations ($N_i \approx 3 \times 10^{21}$ cm^{-3}), a quasi-metallic conductivity is reached, as expected, for degenerate band conduction. This suggests that as long as the long-range symmetries of the c-Si lattice are intact, the majority of the dopant atoms also tend to remain on substitutional network sites. In this situation, the large number of defects in the predamaged material can be filled or depleted with electrons and high levels of n- or p-type conductivity can be achieved. As can be seen further from Figure 6, the variation of the conductivity with dopant concentration is unsymmetric with respect to p- and n-type doping. From the fact that about 10 times higher dopant concentrations are needed to obtain degenerate band conduction in defective p-type c-Si, the density of localized defect states in the predamaged c-Si obviously is much higher in the lower half of the bandgap.

This situation changes drastically when predamaged Si is amorphized by further irradiation. In this event, the degenerate band conduction in the heavily doped samples disappears and roughly a constant level of variable-range hopping conduction is observed irrespective of the dopant concentration incorporated (curve (b) of Fig. 6). Obviously, in this latter case, filling or emptying of the localized defect states has not been possible and a conversion to heavily doped n- or p-type material does not occur. According to the spin density and the RBS/C measurements reported in Figure 2, because the defect density in the predamaged crystalline and the amorphized silicon is on the same order of magnitude, the conclusion must be drawn that the vast majority (>90%) of dopant impurities no longer occupy electronically active lattice sites after the long-range symmetries of the silicon network have been destroyed.

This latter interpretation is in line with earlier predictions, which stated that "amorphous semiconductors cannot be doped." The corresponding arguments were presented in a classic paper by Mott (1967) in which he argued that substitutional doping in amorphous semiconductors is impossible because of the inherent flexibility of random network structures. In the special case of a P impurity in the random Si network, this means that the impurity is likely to form three covalent bonds to nearest-neighbor Si atoms in order to fill its valence shell with a maximum number of 8 electrons (8-N rule, N = number of valence electrons). In this bonding configuration, a P impurity would remain neutral and therefore electrically inactive. In the following we denote such configurations by P_3^0 with the lower index standing for the coordination number and the upper one for the charge state of the P impurities. Due to the rigidity of long-range ordered c-Si lattices, a P impurity must behave quite differently: There it is forced into a substitutional network site in which it can donate one of its five valence electrons to the conduction band, forming four covalent bonds to its nearest-neighbor Si atoms With the notation introduced above, such an electrically active configuration is denoted by P_4^+.

2. Irreversible Ordering Phenomena in Undoped Amorphous Material

As lattice damage is generated as the unavoidable by-product of every ion implantation process, many studies have been devoted to the process of damage annealing in ion-bombardment-disordered Si. In this context, the thermal properties of a-Si have been studied in considerable detail. In particular, its melting temperature (Thompson et al., 1984), its irreversible structural relaxation (Donovan et al., 1985, 1989; Roorda et al., 1989, 1991), and its epitaxial recrystallization onto an underlying c-Si substrate (Csepregi et al., 1977) have attracted major attention. Some of the relevant information gained is contained in Figure 7, which plots the excess free enthalpy of a-Si and liquid (l-) Si relative to that of c-Si as a function of the lattice temperature T (Donovan et al., 1985). Figure 7 reveals three important points: (1) a-Si possesses an excess free enthalpy relative to c-Si on the order of $\Delta G \approx 0.1$ eV/atom, (2) a-Si has a melting temperature of about 250 K lower than the melting temperature of c-Si, and (3) the excess free enthalpy and the melting point of a-Si depend on the state of structural relaxation reached during low-temperature annealing (Roorda and Sinke, 1992).

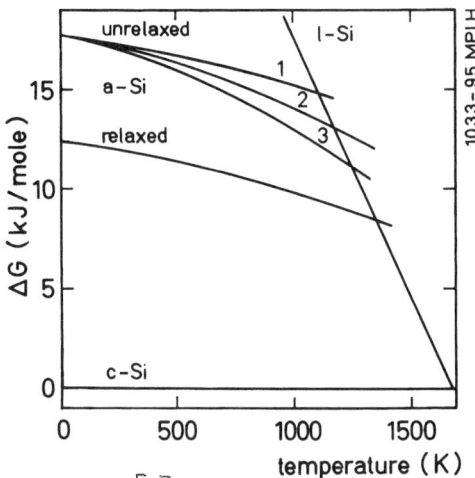

FIG. 7. Temperature-dependence of the excess free enthalpy of amorphous silicon (a-Si) and liquid silicon (l-Si) relative to crystalline silicon (c-Si). The melting point of a-Si, as defined by the intersection of the a-Si and l-Si curves, is distinctly lower than that of c-Si. Its exact value depends on the degree of relaxation induced during annealing, as indicated by the sequence of the corresponding curves. See the text for 1–3. (Reprinted from Müller and Krötz (1993). "Structural Equilibium in Pure and Hydrogenated Amorphous Silicon." *Mat. Res. Soc. Symp. Proc.* **297**, 237–248, with permission from the author and Materials Research Society, Pittsburgh, PA.)

The data in Figure 7 clearly reveal that, in the whole range of temperatures up to its melting point, a-Si is thermodynamically unstable with respect to c-Si. The release of this excess enthalpy can be monitored when ion bombardment a-Si specimens are heated in a calorimeter with a c-Si specimen in the reference cell (Donovan et al., 1985; Roorda et al., 1989, 1991). Such calorimetric experiments show that as-amorphized a-Si specimens irreversibly release about one third of their excess enthalpy before crystallization sets in at temperatures above 600°C. Specimens of a-Si that have been pre-annealed before measurement, only release heat after the preannealing temperature has been exceeded. On the whole, these observations have revealed that a-Si settles into successively lower energy states by activation over increasingly higher kinetic barriers. As the crystallization threshold is approached, the activation energy for structural relaxation approaches the activation energy for epitaxial crystallization of about 2.7 eV (Csepregi et al., 1977). This phenomenon of precrystallization heat release has become generally known as "structural relaxation" of a-Si.

The microscopic processes that accompany the irreversible structural relaxation of a-Si networks have been characterized more closely by using

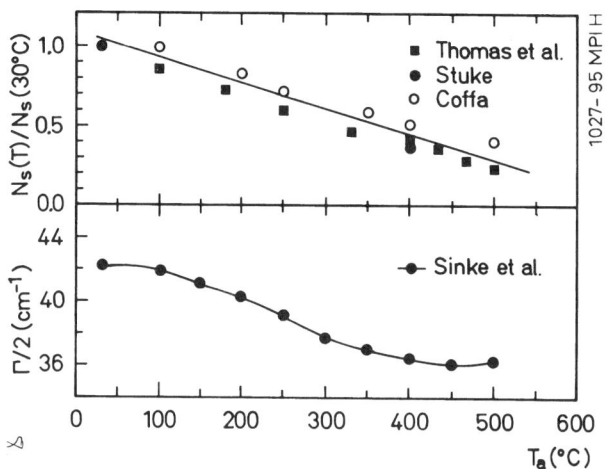

FIG. 8. Irreversible structural changes in amorphous silicon (a-Si) as a function of the annealing temperature T_a: top, electron spin density; bottom, half-width of the Raman transverse optical (TO) band. (Reprinted from Müller, Krötz, Kalbitzer, and Greaves (1994). "Reversible and Irreversible Structural Changes in Amorphous Silicon." *Philos. Mag.* **B69**, 177–196, with permission from the author and Taylor & Francis, London.)

a number of different techniques. Figure 8 shows the annealing-induced changes in ion-bombardment a-Si as seen by ESR (Stuke, 1977; Thomas et al., 1978; Coffa, 1992) and Raman scattering measurements (Sinke et al., 1988). On the one hand, the changes in electron spin density probe properties that are due to the presence of dangling bond defects in the a-Si network. These results suggest a decrease of roughly a factor of three in the dangling bond density as the a-Si is fully relaxed by annealing. On the other hand, the Raman scattering data probe a property related to the band-tail state distribution in a-Si. They show that the half-width $\Gamma/2$ of the Raman transverse optical (TO) peak is decreased by the annealing. This quantity has been related to the average bond-angle distortion $\Delta\Theta_b$ in the a-Si network (Beeman, Tsu, and Thorpe, 1985), yielding

$$\Gamma/2 = 3.2\Delta\Theta_b + 7.4 \qquad (1)$$

where the units are in cm^{-1} and degree for the half-width and angle, respectively. With this formula the previous data suggest a narrowing of the average bond-angle distortion from about 12 degrees in the as-amorphized state to approximately 9 degrees in the fully relaxed state. More recently, rapid thermal annealing experiments have been carried out in an attempt at determining the ultimate order in a-Si films (de Wit et al., 1992). Performing annealing experiments for different lengths of time and temperatures up to 850°C, these authors concluded that $\Gamma/2$ in a-Si cannot be reduced to values lower than about 35 cm^{-1} without inducing crystallization. According to Eq. (1), this corresponds to a minimum value of the bond-angle distortion in a-Si of about 8.5 degrees. Moreover, detailed kinetic experiments on such material revealed that the bond-angle relaxation is mediated by the diffusive motion of dangling bond defects within the a-Si network. Annealing of dangling bond defects takes place when diffusing defects annihilate via mutual recombination (Stolk et al., 1994).

The changes in electron spin density and Raman half-width are also reflected in changes in the localized density-of-state (LDOS) distribution. In order to obtain such information we have recently performed photothermal deflection spectroscopy (PDS) measurements on a series of a-Si films prepared by ion bombardment of thin c-Si layers (Ruttensperger, Müller, and Krötz, 1993). In order to allow optical absorption measurements to be performed over a major photon energy range, hetero-epitaxially grown c-Si films on optically transparent sapphire substrates were used. In fitting the

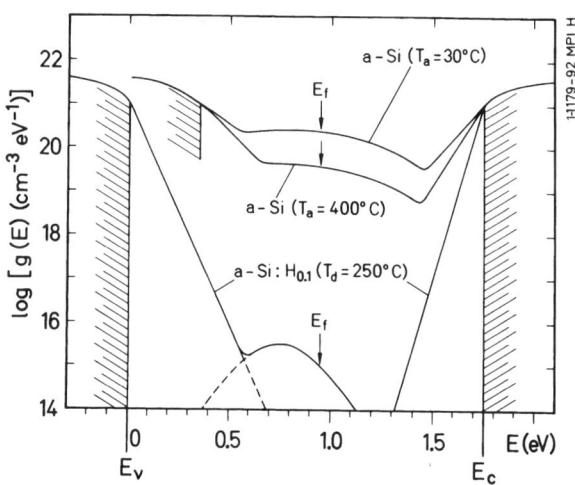

FIG. 9. Localized density-of-state distribution in amorphized silicon (a-Si) (upper curves) and in hydrogenated amorphous silicon (a-Si:H) (lower curve). States above E_c and below E_v are delocalized. Localized tail states arise from fluctuations in the Si-Si bond angles, midgap states from dangling bond defects. Arrows denote the position of E_f in undoped material. Note that E_g (a-Si) $\cong$ 1.40 eV is markedly smaller than is E_g (a-Si:H$_{0.1}$) $\cong$ 1.75 eV. T_a, annealing temperature. T_d, deposition temperature. (Reprinted from Müller (1993). "The Contribution of Ion Beam Techniques to the Physics and Technology of Amorphous Semiconductors," Nuclear Instruments and Methods in Physics Research B80/81, 957–965, with permission from the author and Elsevier Science, Amsterdam.)

optical absorption data, the usual assumption of exponential band tails and a Gaussian dangling bond peak near midgap was made. In addition, it was required that the best-fitting LDOS distributions should also reproduce the ESR data of Figure 8 with $N_S(30°C) = 5 \times 10^{19}$ cm^{-3}. The LDOS distributions so obtained are displayed in the two upper curves of Figure 9, showing that LDOS distributions with sharper band tails and fewer dangling bond defects emerge as the a-Si is irreversibly relaxed toward more well-ordered, lower-energy states. Similar measurements have also been carried out by a Dutch group. This group did not explicitly take into account changes in the widths of the conduction and valence band tails. Using Tauc plots as a means of optical characterization, these authors concluded that structural relaxation tends to widen the optical bandgap from about 1.05 eV in the as-amorphized state up to about 1.45 eV in the fully relaxed state (Berntsen *et al.*, 1993; Stolk *et al.*, 1993).

3. Structural Relaxation and Thermal Crystallization of Doped Material

The data of Figure 6 have already shown that electronic transport in as-amorphized Si is of the variable-range-hopping–type and are fairly insensitive to doping. This fact implies that more than 90% of the dopant impurities take on electrically inactive bonding configurations within the a-Si network. The data of Figure 9, however, also show that low-temperature annealing is effective in reducing the density of compensating defect states. An interesting experiment therefore is to anneal a-Si samples and to look for the presence of residual doping effects. Figure 10 summarizes the results of such experiments, showing that interesting changes occur when the a-Si is relaxed by thermal annealing (Müller and Kalbitzer, 1978; Geissler-Pfeilsticker, 1980; Coffa *et al.*, 1993; Müller *et al.*, 1996). Whereas a low level of variable-range hopping conduction ($\sigma_{RT} \cong 10^{-4} \Omega^{-1}$cm^{-1}) is observed after thermal relaxation of undoped and lightly doped specimens ($N_i \leqslant 10^{20}$ cm^{-3}), a significant variation of the hopping conductivity with impurity concentration becomes evident when higher concentrations of substitutional B and P impurities are incorporated prior to amorphization. With high concentrations of these impurities, the variable-range hopping conduction can be increased up to levels on the order of $10^{-2}\Omega^{-1}$cm^{-1}. Substitutional As and Ga impurities, however, are found to be surprisingly inactive, producing no conductivity enhancements at all.

At first sight, the data of Figure 10 suggest that, although relatively inefficient, a-Si can be doped by substitutional B and P impurities. Further insight into these dopant-induced conductance changes was obtained by performing sub-bandgap optical absorption measurements using the

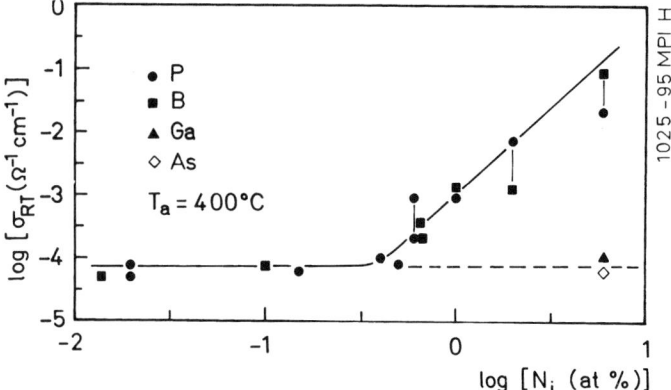

FIG. 10. Electrical conductivity at room temperature as a function of dopant impurity concentration in thermally relaxed amorphous silicon (a-Si). The observed transport behavior is of the variable-range hopping type. P, phosphorus; B, boron, Ga, gallium; As, arsenic; T_a, annealing temperature. (Reprinted from Müller, Hellmich, Krötz, Kalbitzer, Greaves, Derst, Dent, and Dobson (1996). "Dopant-Defect Interactions in Hydrogen-Free Amorphous Silicon," Philos. Mag. B73, with permission from the author and Taylor & Francis, London.)

photo-thermal deflection spectroscopy (PDS) technique (Müller et al., 1996). Some of the optical absorption data obtained on doped and thermally relaxed a-Si are shown in Figure 11. A comparison with the electrical conductivity data of Figure 10 reveals that high levels of conductivity enhancement are always associated with high levels of sub-bandgap optical absorption. An interesting observation in this context is that the conductivity and optical absorption enhancements seem to depend on the atomic size of the implanted dopant impurities with B leading to the largest, P to moderate, and Ga and As to very marginal enhancements only. Considering the correlation between enhanced defect densities and enhanced levels of variable-range hopping conduction, it is suggested that the observed conductivity changes are due to enhanced defect level densities in the vicinity of the Fermi energy rather than due to a shift of the Fermi energy through a given localized density-of-state distribution that had been reduced by thermal relaxation.

More microscopic information about the bonding of substitutional dopant impurities in random Si networks was obtained by x-ray absorption fine structure (XAFS) measurements (Greaves et al., 1992, 1993; Dent et al., 1993, 1994). Because of the inaccessibility to measurement of convenient K-edge energies for the most commonly used B and P impurities, these latter investigations had to be confined to the heavier and less frequently used As and Ga impurities. In order to avoid effects of dopant–dopant interaction, dopant concentrations close to the limit of XAFS detectability were used

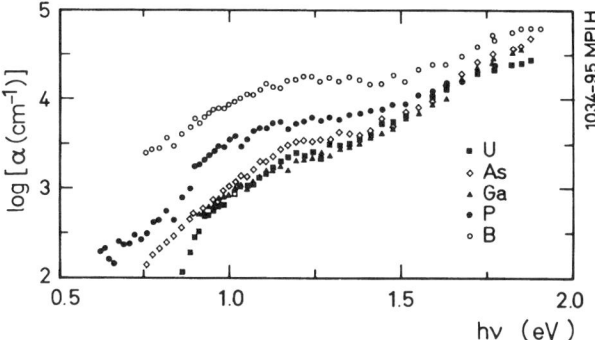

FIG. 11. Optical absorption spectra of thermally relaxed amorphous silicon (a-Si) films ($T_a = 400°C$) containing the donor species arsenic (As) and phosphorus (P) and the acceptor species gallium (Ga) and boron (B) at concentrations of $N_i = 3 \times 10^{21}$ cm^{-3}. The curve denoted by (U) refers to an undoped reference specimen. (Reprinted from Müller and Kalbitzer (1980). "The Crystalline-to-Amorphous Transition in Ion-Bombarded Silicon," *Philos. Mag.* **B41**, 307–325, with permission from the author and Taylor & Francis, London.)

(5×10^{19} impurities/cm^3). Such measurements provide several pieces of information: (i) the average number of nearest-neighbor silicon atoms, (ii) the average nearest-neighbor distances, and (iii) the statistical spread in these distances. As examples, in Figure 12(a) we show the Fourier transform of the k^3-weighted XAFS function of As sites in as-amorphized Si. On the one hand, these data clearly do not indicate any significant order beyond the first shell of silicon neighbors. On the other hand, Figure 12(b) shows the same function for arsenic impurities in thermally recrystallized Si ($T_a \approx 700°C$). As expected, the data in Figure 12(b) clearly reveal a high degree of order beyond the first shell of silicon neighbors.

Figure 13 displays information about the average coordination number N_{av} extracted from such As and Ga impurity XAFS spectra. Whereas in the as-amorphized state the athermal bonding behavior is impurity-specific, both impurities exhibit a very strong tendency toward occupying threefold coordinated alloying sites in the thermally relaxed state (As_3^0, Ga_3^0). Another piece of information is that the bonding disorder around the impurity sites is reduced as the material is relaxed by annealing at successively higher temperatures (Müller et al., 1984). Annealing beyond 600°C, finally, causes the a-Si films to recrystallize epitaxially onto the underlying Si substrates (Prisslinger et al., 1975; Csepregi et al., 1977). Figure 13 shows that upon exceeding this threshold temperature, N_{av} increases, however, without yet reaching the expected value of four at these temperatures.

In context with this last problem, it is instructive to consider the data summarized in Table I. Table I compares the coordination numbers and

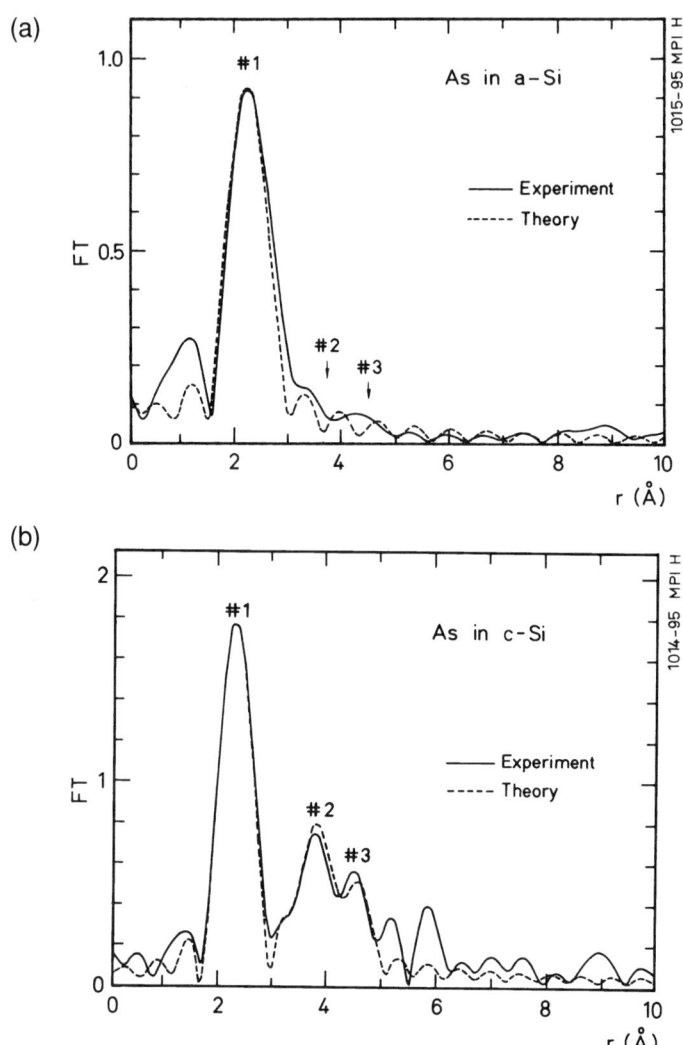

FIG. 12. (a) Fourier transform of the k^3-weighted XAFS function of arsenic (As) sites in as-amorphized a-Si. (b) Fourier transform of the k^3-weighted XAFS function of an As impurity in thermally recrystallized silicon (Si). The numbers 1, 2, and 3 indicate the positions of neighboring shells in the Si matrix; XAFS, x-ray absorption fine structure. (Reprinted from Müller, Hellmich, Krötz, Kalbitzer, Greaves, Derst, Dent, and Dobson (1996). "Dopant-Defect Interactions in Hydrogen-Free Amorphous Silicon," *Philos. Mag.* **B73**, with permission from the author and Taylor & Francis, London.)

4 ION BEAMS IN AMORPHOUS SEMICONDUCTOR RESEARCH

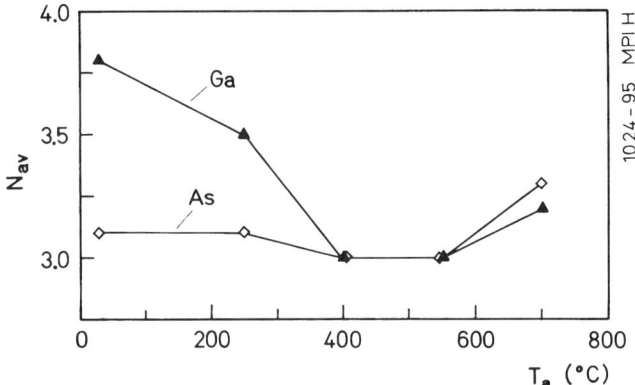

FIG. 13. Average coordination number (N_{av}) of arsenic (As) and gallium (Ga) impurities as a function of the annealing temperature (T_a) in amorphous and thermally recrystallized silicon (c-Si). Recrystallization occurs at $T_a \sim 600°C$. (Reprinted from Müller, Hellmich,. Krötz, Kalbitzer, Greaves, Derst, Dent, and Dobson (1996). "Dopant-Defect Interactions in Hydrogen-Free Amorphous Silicon." *Philos. Mag.* **B73**, with permission from the author and Taylor & Francis, London.)

shell radii of the first-, second-, and third-nearest-neighbor shells around As impurities in two different kinds of As-doped c-Si. The first case deals with a c-Si specimen doped by thermal in-diffusion of As impurities and the second with a specimen doped by means of As implantation at room temperature and thermally recrystallized afterward at 700°C. The comparison of both data sets reveals that, up to an annealing temperature of 700°C,

TABLE I

AVERAGE COORDINATION NUMBERS OF FIRST-, SECOND-, AND THIRD-NEAREST SILICON (Si) NEIGHBORS OF ARSENIC (As) IMPURITIES IN DIFFERENT KINDS OF CRYSTALLINE SILICON (c-Si) SPECIMENS

	0.04 at% As Thermally Diffused into c-Si[a]			0.1 at% As Ion Implanted into a-Si, ($T_a = 700°C$)[b]		
		Shell Radius			Shell Radius	
	Coordination Number	(A) ±0.01	$2\sigma^2(A^2)$ ±0.001	Coordination Number	(A) ±0.01	$2\sigma^2(A^2)$ ±0.001
Shell 1	4	2.40	0.003	3.3	2.37	0.008
Shell 2	12	3.82	0.021	3.6	3.82	0.024
Shell 3	12	4.47	0.018	3.6	4.47	0.019

[a] Local environment of As impurities thermally diffused into a c-Si specimen.
[b] Local environment of As ions implanted into amorphous silicon (a Si) thermally recrystallized at 700°C.

a considerable fraction of the As impurities has not yet entered well-ordered substitutional dopant sites (As_4^+). In such thermally recrystallized material, N_{av} in the first shell is smaller than four and this same shell also appears to be slightly more disordered. In the second- and third-nearest-neighbor shells, however, significantly fewer Si atoms are detected than in the thermally diffused sample.

In a recent publication Müller et al. (1996) have shown that this unexpected behavior can be explained by assuming the impurity XAFS function to be a mixture of two different kinds of impurity sites: (1) threefold coordinated dopant sites (As_3^0, Ga_3^0) surrounded by locally amorphous environments and (2) fourfold coordinated dopant sites (As_4^+, Ga_4^-) surrounded by perfectly crystalline environments. From the relative contribution of both kinds of sites, a local level of crystallinity can be defined. An analysis of the Si XAFS, however yields the global level of crystallinity developed at a certain annealing temperature T_a (Dent et al., 1993, 1994; Kalbitzer et al., 1995). Figure 14 summarizes these Si and impurity XAFS data, bringing out very clearly the effect of a locally retarded crystallization in the immediate neighborhood of substitutional As and Ga dopant impurities.

Taken together, these experiments have revealed three main facts. (1) In thermally relaxed a-Si, the vast majority of substitutional dopant impurities are taken up into electrically inactive, threefold coordinated, alloying configurations as was expected on the basis of the 8-N rule (Mott, 1967). (2)

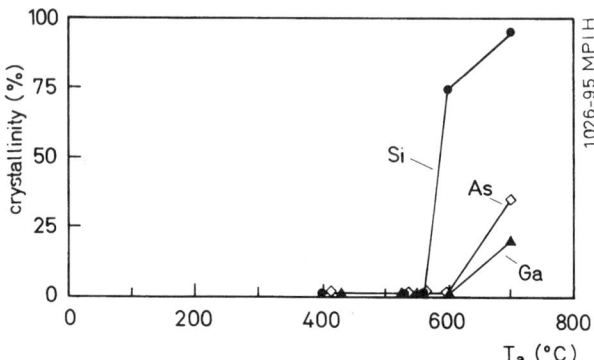

FIG. 14. Degree of local crystallinity around arsenic (As) and gallium (Ga) impurities as well as silicon (Si) network sites as a function of annealing temperature (T_a). The network ordering around the impurity sites is seen to be retarded. (Reprinted from Müller, Hellmich, Krötz, Kalbitzer, Greaves, Derst, Dent, and Dobson (1996). "Dopant-Defect Interactions in Hydrogen-Free Amorphous Silicon." Philos. Mag. **B73**, with permission from the author and Taylor & Francis, London.)

Threefold coordinated impurity sites tend to slow down the thermal relaxation of random Si networks. (3) The crystallization of a-Si networks is retarded or even inhibited in the immediate neighborhood of some of the dopant sites. Note that this last explanation is at variance with former postulations of compensating impurity centers, for example, of interstitial dopant components such as bismuth (Bi^+) and amphoteric Si lattice defects, or both.

4. EFFECTS OF HYDROGENATION AND FLUORINATION

Concerning electronic applications, defect-rich material with variable-range hopping transport is practically useless. Defect passivation techniques, therefore, are most important to the successful application of a-semiconductors in devices. The dangling bond defects present in amorphized Si have been successfully passivated by Böhringer et al. (1983, 1985) by means of hydrogen implantation and subsequent thermal annealing. Isotropic resonance measurements confirmed that the implanted H is taken up in the form of isolated Si-H bonds. Upon forming such bonds, localized dangling bond

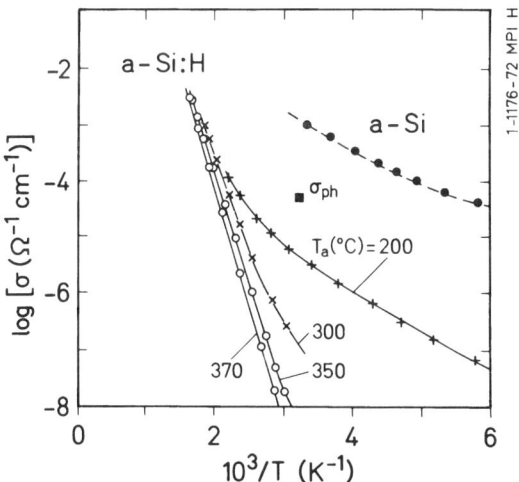

FIG. 15. Temperature-dependence of the conductivity of amorphous silicon (a-Si) films in the as-amorphized state (dashed line) and after hydrogen (H) implantation and different states of annealing (full lines). The single point labeled σ_{ph} refers to the photoconductivity of the sample annealed at 370°C. T_a, annealing temperature; ○, 350 and 370°C; ×, 300°C; +, 200°C. (Reprinted from Müller, (1993). "The Contribution of Ion Beam Techniques to the Physics and Technology of Amorphous Semiconductors," *Nuclear Instruments and Methods in Physics Research* **B80/81**, 957–965, with permission from the author and Elsevier Science, Amsterdam.)

levels are removed from the bandgap and Si-H-derived states appear in the valence band density of states (Ley, 1984). At the same time, some widening of the mobility gap is observed, as shown in the lower curve of Figure 9. Figure 15 shows that in this way variable-range hopping transport in a-Si can be reduced to the extent that singly activated band transport takes over. Similar effects were also observed after fluorine (F) implantation (Böhringer, Liu, and Kalbitzer, 1983; Böhringer, Mannsperger, and Kalbitzer, 1985). F implantation produced a defect passivation thermally more stable than did the conventional introduction of H.

The same authors also showed that the successful passivation of dangling bond defects has a number of interesting consequences: (i) the passivated material is photoconductive, (ii) it can be substitutionally doped to exhibit p- or n-type conduction, with the conductivity varying over more than 8 orders of magnitude, and (iii) upon prolonged illumination, the H-passivated films suffer from metastable changes in their electrical transport properties. These latter changes were reversible upon low-temperature annealing in the dark. All these properties had been observed previously in plasma-deposited a-Si:H films (Spear and LeComber, 1976; Staebler and Wronski, 1977) and had often been attributed to the special growth environment present during plasma-enhanced chemical vapor deposition. The successful step-by-step production of a-Si:H by means of an entirely different technique, however, convincingly demonstrated that the low defect densities, the doping effect, and the metastability of plasma-deposited films are intrinsic properties of a-Si:H and, very likely, are only related to the presence of bonded H in the a-Si:H network.

III. Ion Beam Doping of Plasma-Deposited Amorphous Silicon

1. GAS PHASE AND ION IMPLANTATION DOPING

The first successful demonstration of doping in a-Si:H is due to Spear and LeComber (1976). These authors used a plasma-enhanced chemical vapor deposition technique to deposit a-Si:H films from a silane (SiH_4) glow discharge. These authors demonstrated that low-defect-density a-Si:H can be sensitively doped by the addition of small amounts of phosphine (PH_3) or diborane (B_2H_6) to the silane (SiH_4) discharge. This was one of the major breakthroughs in this field because it turned a-Si:H from a subject of basic investigation into a material of major applied interest. An overview of the many device applications of a-Si:H that have since been demonstrated can be found in the book by Kanicki (1991).

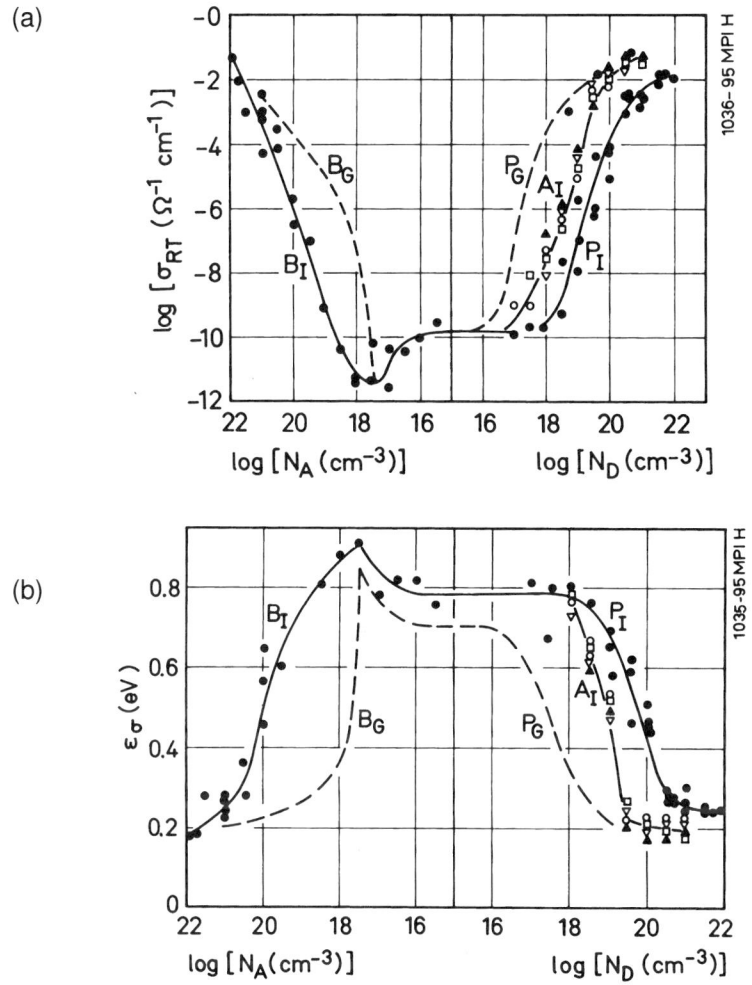

FIG. 16. (a) Room temperature conductivity σ_{RT} as a function of dopant impurity concentration. Ion implantation doping: P_I phosphorus, A_I alkali (sequence from top at $N_i = 10^{19}$ cm^{-3}: potassium, rubidium, cesium, sodium; B_I boron; film thickness: $d \sim 0.1$ μm). Gas-phase doping: P_G and B_G in hydrogenated amorphous silicon (a-Si:H) films of $d \sim 1$ μm. (b) Conductivity activation energy ε_σ as a function of dopant impurity concentration derived from the conductivity data of (a). □, boron (B) or phosphorus (P); ○. sodium (Na); ▽, potassium (K); ●, rubidium (Rb); ▲, cesium (Cs). (Reprinted from Kalbitzer, Müller, LeComber, Spear (1980). The Effects of Ion Implantation on the Electrical Conductivity of Amorphous Silicon, *Philos. Mag.* **B41**, 439–456, with the permission from the author and Taylor and Francis, London).

The demonstration of gas-phase doping was soon followed by a similar demonstration of doping using the ion implantation technique (Müller et al., 1977; Spear et al., 1979; LeComber et al., 1980; Kalbitzer et al., 1980). Figure 16(a) summarizes some of these early data that had been obtained by doping relatively thin films of a-Si:H ($d \sim 0.1$ μm) with the help of a low-energy ion implanter. These latter data show that ion implantation with substitutional B and P impurities allows the room temperature conductivity σ_{RT} of a-Si:H samples to be controlled over some 10 orders of magnitude. In addition, it shows that n-type doping is also possible with interstitial alkaline impurities. Figure 16(b) demonstrates that these conductivity changes are brought about by a variation of the conductivity activation energy ε_σ. From the magnitude of this variation, it was inferred that substitutional doping allows the Fermi energy to be shifted throughout a 1.2-eV wide interval of the localized density-of-states distribution within the 1.7 eV mobility gap. The dashed lines in Figures 16(a) and 16(b), finally, show how σ_{RT} and ε_σ vary in relatively thick films of gas-phase-doped material ($d \sim 1$ μm). On the

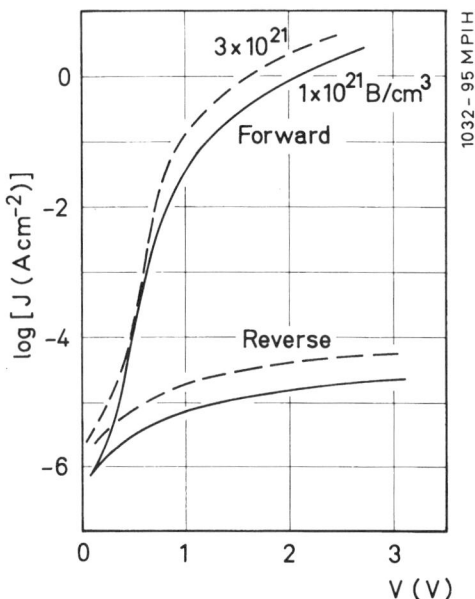

FIG. 17. Electrical characteristics of a p-n junction formed by boron (B) ion implantation into an n-type hydrogenated amorphous silicon (a-Si:H) film on stainless steel substrate. (Reprinted from Müller (1993). "The Contribution of Ion Beam Techniques to the Physics and Technology of Amorphous Semiconductors," Nuclear Instruments and Methods in Physics Research B80/81, 957–965, with permission from the author and Elsevier Science, Amsterdam.)

whole, these early experiments indicated that ion implantation doping allows for the same range of control over the electronic transport properties as does gas-phase doping, although at a considerably smaller doping efficiency. Later investigations into the implantation doping process showed that there are two main reasons for this apparent lack of efficiency: (1) enhanced densities of compensating defect levels in the surface and interfacial regions of a-Si:H films (Müller et al., 1983: Mannsperger, et al., Kalbitzer, and Müller, 1986; Luft and Tsuo, 1993) and (2) an incomplete annealing of the implantation-induced bond-angle distortions. The second effect results in enhanced densities of localized band-tail states in implantation-doped material. We will return to this subject in Section 3 of Part III.

In the meantime, our understanding of the ion implantation process in a-Si:H has progressed well. To date, the efficiency of interstitial doping approaches the efficiency of gas-phase doping in thick a-Si:H films with PH_3 (Desalvo et al., 1993). Significant steps in this development have been the proper account of thickness-dependant effects (Müller et al., 1983;

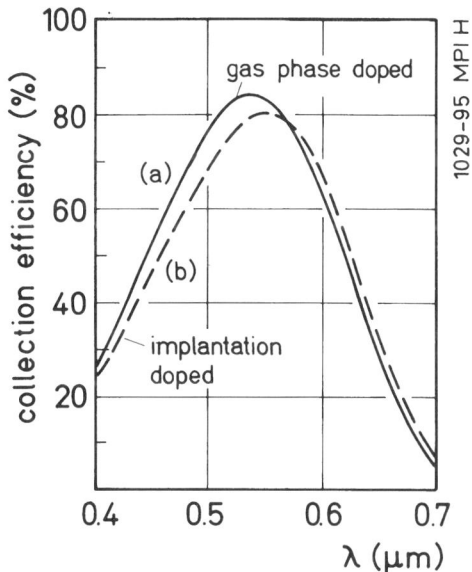

FIG. 18. Spectral response of solar cells with gas-phase (a) and ion-implanted (b) entrance windows. The thickness of the implanted n^+ hydrogenated amorphous silicon (a-Si:H) entrance window amounts to about 15 nm, the air mass 1 (AM1) efficiency to about 6%. (Reprinted from Müller (1993). "The Contribution of Ion Beam Techniques to the Physics and Technology of Amorphous Semiconductors," Nuclear Instruments and Methods in Physics Research B80/81, 957–965, with permission from the author and Elsevier Science, Amsterdam.)

Mannsperger, Kalbitzer, and Müller, 1986), the search for optimum annealing temperatures, and the application of posthydrogenation techniques (Fonseca, Galloni, and Nylandsted-Larsen, 1993. Galloni *et al.*, 1990). So far, several types of devices have been demonstrated that include ion-implantation-doped a-Si:H layers. These include high-current p-n junctions (LeComber *et al.*, 1980; Kalbitzer *et al.*, 1980), thin-film solar cells (Spear *et al.*, 1981), and field-effect transistors. Some of the results obtained on ion-implantation-produced a-Si:H devices are shown in Figures 17 and 18.

2. Doping Mechanism in Hydrogenated Material

The discussion in Section 3 of Part II has shown that the majority of substitutional dopant impurities tends to become alloyed into random Si networks forming threefold coordinated, electrically inactive impurity sites. Due to the limitations involved in the evaluation of XAFS spectra and the high densities of compensating defect centers in chemically pure a-Si, a small fracton (< 1-10%) of electrically active impurity sites could not be ruled out. However, the data displayed in Figure 9 have shown that the introduction of bonded H into random Si networks allows the density of localized defect states to be reduced by several orders of magnitude. The preceding paragraph, finally, has shown that under these conditions substitutional doping in random Si networks can indeed be observed. As a consequence, the 8-N rule in its original form (Mott, 1967) cannot be universally true. Therefore, the open questions that remain are the following: Which mechanism actually supports the residual doping effect in a-Si:H? To which extent does the residual doping effect rely on the presence of bonded hydrogen within hydrogenated random Si networks?

The first hint to a more adequate description of doping came from the observation that in gas-phase-doped material, doping is always associated with the formation of dangling bond defects. The other relevant observation was that the doping efficiency η, that is, the fraction of electrically active dopant sites, decreases monotonically as more dopant impurities are introduced. On the basis of these results, in 1982, Street proposed that the 8-N rule be generalized to include ionized states of the dopant impurities as well (Street, 1982). In this form, the 8-N rule stated that in addition to forming electrically inactive P_3^0 sites, P impurities can also enter the a-Si:H network in the form of electrically active P_4^+ sites. This occurs, however, only at the expense of formation of compensating dangling bond defects (Si_3^-), which accommodate the donated electrons in energy levels well below the P_4^+ donor energy levels. In this way, the formation of neutral P_4^0 levels is prevented, which is forbidden by the 8-N rule. In order to account for the variations of the defect density and the doping efficiency η with the total dopant density, Street further proposed that the individual configurations are related to each other via a thermal equilibrium process established by the reversible valence alternation reaction:

$$P_3^0 + Si_4^0 \leftrightarrow P_4^+ + Si_3^- \qquad (2)$$

The law of mass action then yields $[Si_3^-] \propto [P]^{0.5}$ and $\eta \propto [P]^{-0.5}$ in agreement with experiment. In turn, the equilibrium was assumed to be established at the surface of the growing film and to be quenched at a temperature close to the temperature of the substrate onto which the film was grown.

Soon thereafter (Müller et al., 1984), the same functional dependencies for $[Si_3^-]$ and η were found in plasma-deposited a-Si:H films doped by ion implantation (Figures 19 and 20). These results indicated that, in addition to surface equilibration, bulk equilibration was also possible. Another important result of the implantation experiments is that interstitial (alkali)

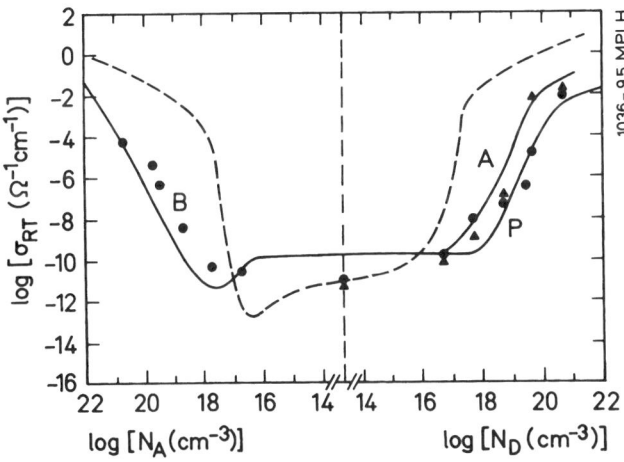

FIG. 19. Ion implantation doping effect in plasma-deposited hydrogenated amorphous silicon (a-Si:H). Dashed line: calculated conductivity as a function of donor and acceptor concentration with an electron–hole mobility of 10/1 cm^2/Vs; full line–data points: electrical conductivity as a function of implanted dopant density (A, alkali; P, phosphorus; B, boron). A comparison of the calculated and measured curves reveals an inverse square-root relation of the doping efficiency η with implanted impurity concentration. ●, boron (B) or phosphorus (P); ▲, alkali (A). (Reprinted from Müller (1993). "The Contribution of Ion Beam Techniques to the Physics and Technology of Amorphous Semiconductors," *Nuclear Instruments and Methods in Physics Research* **B80/81**, 957–965, with permission from the author and Elsevier Science, Amsterdam.)

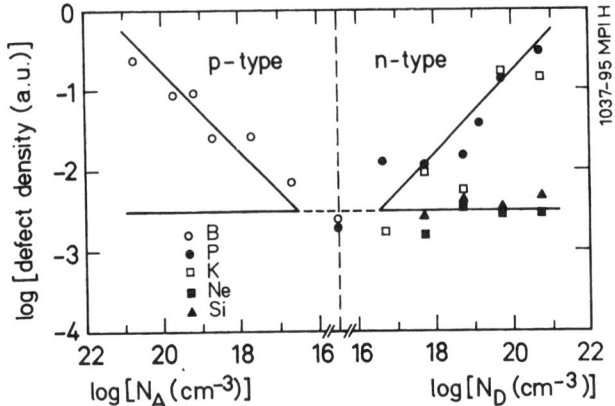

FIG. 20. Dangling bond density as a function of implanted dopant density. Note the null effect on rare gas or self-ion implantation. B, boron; P, phosphorus; K, potassium; Ne, neon; Si, silicon. (Reprinted from Müller (1993). "The Contribution of Ion Beam Techniques to the Physics and Technology of Amorphous Semiconductors," *Nuclear Instruments and Methods in Physics Research* **B80/81**, 957–965, with permission from the author and Elsevier Science, Amsterdam.)

doping produces the same increase in the defect density as does substitutional doping. This result suggests that autocompensation doping according to the reaction in (2) occurs as a result of two independent reactions (Müller, Kalbitzer, and Mannsperger, 1986):

$$\text{doping:} \quad P_3^0 \leftrightarrow P_4^+ + e^- \quad (3a)$$

$$\text{defect production} \quad e^- + Si_4^0 \leftrightarrow Si_3^- \quad (3b)$$

Both reactions are linked when the donated electron becomes trapped in a conduction band-tail state. In this sequence the later reaction describes the addition of a valence electron to a normally coordinated Si atom and its re-coordination according to the generalized 8-N rule. Another interpretation of reaction (3b) (Stutzmann, 1987) is the decay of a weak Si-Si bond into two threefold coordinated, negatively charged dangling bond defects (Si_3^-), triggered by the band-tail trapping of two electrons:

$$2e^- + 2Si_4^0 \leftrightarrow 2Si_3^- \quad (3c)$$

Alkali doping, finally, is described by replacing the reaction in Eq. (3a)

$$A^0 \leftrightarrow A^+ + e^- \tag{3d}$$

where A is lithium, sodium, potassium, rubidium and cesium.

In connection with the reaction in (3d), an interesting experiment was carried out by Schölch et al. (1987). They implanted lithium (Li) ions into the high-field regions of a-Si:H p-i-n junctions. After annealing these junctions for increasing lengths of time, nuclear reaction profiling techniques were employed to determine the resulting Li depth profiles. These profiles revealed both the expected diffusional broadening and a drift in the electrical junction fields. From the form of the observed profiles it was inferred that the entire Li population is subject to ion drift. This proved that the implanted Li is not divided into two different populations, one being permanently active and the other electrically inactive. Rather, it was inferred that each Li atom can assume either charge state for a certain fraction of time. Depending on the time spent in the positive charge state, an effective charge $q^* < q$ resulted (q: elementary charge). Although q^* always turned out to be smaller than q, q^* was found to be much higher in p^+-i than in n^+-i junctions. This latter effect is fully consistent with the predictions of the autocompensation model of doping and due to the counterdoping action of the boron acceptors in p^+-i junctions.

Considering the reactions in (3a) and (3d), we see that these merely act as suppliers of electrons to the reaction in (3b). Just as other potential sources of electrons are electron accumulation regions in devices and photo-generated electrons, reactions (3b) and (3c) also account for the metastable increase in the defect density in field-effect transistors (Powell, van Berkel, and Deane, 1991; Krötz and Müller, 1991; Liu and Spear, 1991; Müller and Krötz, 1993) biased into accumulation, electron injection through the tip of a scanning tunneling electron microscope (Hartmann et al., 1991), and in pin solar cells subjected to prolonged illumination (Staebler and Wronski, 1977). Ion implantation experiments, therefore, also have provided the basic clues for the formulation of a generalized autocompensation model (Müller, Kalbitzer, and Mannsperger, 1986; Krötz et al., 1991) which, in addition to doping, also provides an understanding of the metastable structural changes in a-Si:H devices.

Over the past few years a major issue has been to understand how the previously described coordination changes can be accommodated within the a-Si:H network. The extensive work of the Xerox® group suggests that all these reactions implicitly rely on the diffusion of H bonded within the a-Si:H matrix (Street et al., 1987; Jackson and Kakalios, 1988; Kakalios and Street,

1988; Kakalios and Jackson, 1988). Parallel work on H in c-Si seems to substantiate this latter point of view (Johnson, 1991). A direct consequence is that the speed of equilibration is limited by the H diffusion rate in the a-Si:H matrix. As H diffusion requires a thermal activation energy of about 1.0 to 1.5 eV, rapid equilibration is only possible at temperatures close to the deposition temperature of the a-Si:H films, which is about 250 to 300°C for electronic-grade material. At regular device operating temperatures, of about room temperature, the a-Si:H network seems to be frozen and large changes in carrier density can be accommodated without corresponding changes in the amorphous network structure.

Similar to the a-Si case, XAFS measurements were also carried out on substitutionally doped a-Si:H films. The most significant outcome of these experiments was that the local environments of substitutional dopant impurities in a-Si and in a-Si:H are fairly similar (Greaves, 1992, 1993). In both materials, two different bonding configurations could be detected: relatively well-ordered substitutional (As_4^+, Ga_4^-) dopant and less well-ordered alloying sites (As_3^0, Ga_3^0). Some particularly interesting results concern the effect of electronic compensation. To this end, two different sets of XAFS measurements were performed on ion implanted Ga acceptors. In the first case, Ga impurities were implanted into undoped a-Si:H films and the implantation damage was removed by low-temperature annealing. In the second case the same implantation and annealing sequence was applied to an a-Si:H film that was heavily P-doped during deposition ($[PH_3]/[SiH_4] \sim 10^4$ vppm). Subsequently, the Ga coordination was measured in both kinds of a-Si:H films. The data summarized in Table II show that the average Ga coordination in the p-type films is close to three, whereas it is almost four in the compensated a-Si:H films. Quite obviously, electrons donated by the P donors recombine with holes supplied by the Ga acceptors

TABLE II

COORDINATION OF GALLIUM (Ga) ACCEPTORS IN SINGLE-DOPED AND IN COMPENSATED HYDROGENATED AMORPHOUS SILICON (a-Si:H) FILMS. (Prior to Ga implantation, the a-Si:H:P films were doped with 1% PH_3/SiH_4 during deposition.)

Impurity	Material	Implanted Concentration (at%)	Coordination Number	Interatomic Distance (A)	Debye-Waller Factor $2\sigma^2(A^2)$
Ga	a-Si:H	0.1	3.3 ± 0.2	2.39 ± 0.01	0.013 ± 0.001
Ga	a-Si:H:P	0.1	3.7 ± 0.2	2.39 ± 0.01	0.013 ± 0.001
Ga	a-Si:H:P	0.5	3.9 ± 0.2	2.41 ± 0.01	0.012 ± 0.001
Ga	a-Si:H:P	2.0	3.4 ± 0.2	2.41 ± 0.01	0.14 ± 0.001

and thus cause additional Ga impurities to assume active acceptor sites. The high average coordination number in compensated a-Si:H shows that the activation can be fairly complete, reaching activation levels close to 100%. Consistent with this interpetation the average Ga coordination level again decreases to 3.4 as the material becomes overcompensated by high-dose Ga implants.

3. GENERATION AND ANNEALING OF IMPLANTATION DAMAGE

The first investigations of implantation damage in a-Si:H are due to Müller and LeComber (1980). They noted that the process of damage accumulation in a-Si:H exhibits many similarities to the accumulation process in c-Si. With regard to damage annealing, the most important observation was that a hardening of the radiation damage occurs when certain dose levels are exceeded. Exceeding these thresholds, a complete recovery of the pre-irradiation state was no longer possible when the annealing temperature of the a-Si:H films was limited to their original deposition temperature. In this early investigation, the reason for the damage hardening could not yet be identified.

More recently, the problem of radiation damage in a-Si:H was reinvestigated using electrical conductivity and PDS measurements as a means of investigation (Stitzl, Krötz, and Müller, 1991a, 1991b). The idea behind these experiments was that in a hydrogenated random Si network, ion bombardment should have two main effects. (1) It should break existing Si-Si bonds and thus increase the native or doping-induced dangling bond density. (2) It should increase the number of distorted Si-Si bonds and thus widen the Urbach tails. Annealing, however, should tend to restore the pre-irradiation state. As an example, in Figure 21 we show sub-bandgap optical absorption spectra obtained on a lightly doped n-type a-Si:H film. After an initial measurement, the film first was subjected to high-dose Ar^+ irradiation and annealed afterward. Figure 21 shows very clearly the irradiation-induced increase in the dangling bond density and the widening of the Urbach tail. Concerning the annealing effect, only partial relaxation to the pre-irradiation state can be observed.

Similar measurements were carried out on a large number of a-Si:H samples taken from a single specimen of lightly doped n-type material (10 vppm PH_3/SiH_4). Prior to measurement, the individual samples were irradiated with different ion species, ion doses, and ion energies to vary the relative contributions of nuclear and electronic stopping total energy loss. The data so obtained are summarized in Figure 22 in which the conductivity of the irradiated a-Si:H films and the corresponding dangling bond densities

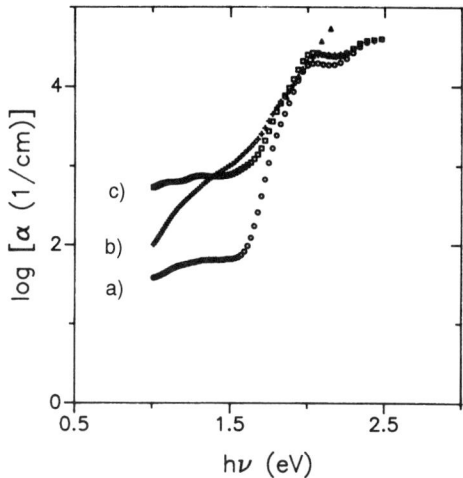

FIG. 21. Sub-bandgap optical absorption of n+ hydrogenated amorphous silicon (a-Si:H) films after different treatments: (a) as-deposited state, (b) after room-temperature ion bombardment with 10^{15} Ar+/cm²; and (c) after annealing at 250°C for 30 min. ○, as-deposited; □, after annealing at 250°C for 30 min; +, after room-temperature ion bombardment. (Reprinted from Stitzl, Krötz, and Müller (1991a). "Accumulation and Annealing of Implantation Damage in a-Si:H." *Appl. Phys. A* **53**, 235–240, Figure 4, with permission of the author and Springer-Verlag. Copyright Springer-Verlag.)

are plotted as functions of the energy transferred into atomic displacements. These plots reveal a fairly universal trend for all kinds of radiation, with measurable damage levels setting in at approximately 10^{-3} eV/target atom, and damage saturation occurring at energy transfers on the order of 10 eV/target atom. At saturation, all kinds of radiation induce dangling bond densities on the order of 10^{19} cm^{-3}. On the whole, these data suggest that, similar to the case of c-Si, implantation damage is predominantly accumulated via atomic displacement collisions.

Figure 23 brings out a second important aspect of the implantation damage problem, demonstrating that, in addition to introducing dangling bond defects, heavy ion irradiation also introduces comparable densities of increasingly strained Si-Si bonds. This conclusion is also supported by more recent Raman measurements. It was found that Si+ implantations into a-Si:H increase the Raman half-width $\Gamma/2$ from about 34 to 44 cm^{-1} as the radiation dose is increased into the saturation range (Bernsten *et al.*, 1993; Stolk *et al.*, 1993). Using Eq. (1), it is inferred that along with the production of dangling bond defects, heavy ion irradiation also increases the average bond-angle distortion in the a-Si:H network from 8 to about 12 degrees.

A third exciting aspect of the radiation damage problem is brought out

4 ION BEAMS IN AMORPHOUS SEMICONDUCTOR RESEARCH

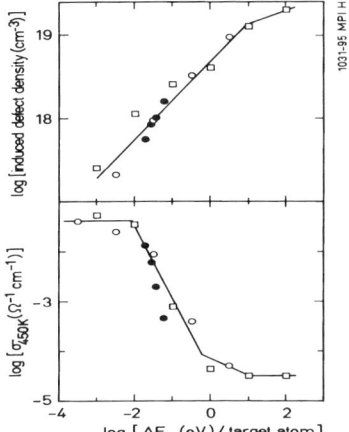

FIG. 22. Top, Photothermal deflection spectroscopy–derived dangling bond densities as a function of the energy dissipated into atomic displacement collisions. ○, hydrogen ion dose variation; ●, hydrogen ion energy variation; □, argon ion dose variation; ■, various boron ion and phosphorus ion irradiations. Bottom, Electrical conductivity as measured at 450 K. (Reprinted from Stitzl, Krötz, and Müller (1991a). "Accumulation and Annealing of Implantation Damage in a-Si:H." *Appl. Phys. A* **53**. 235–240, Figure 1, with permission of the author and Springer-Verlag. Copyright Springer-Verlag.)

by the conductivity data of Figure 22. These data show that σ_{450K} reduces from almost $10^{-1}\Omega^{-1}\text{cm}^{-1}$ to about $3 \times 10^{-5}\Omega^{-1}\text{cm}^{-1}$, as increasing densities of compensating dangling bond defects are introduced in the course of the irradiations.[1] A closer analysis of the data of Figure 22 reveals that a much lower conductivity on the order of $10^{-9}\Omega^{-1}\text{cm}^{-1}$ should be reached in the event that defect compensation alone is operative (Stitzl, Krötz, and Müller, 1991b). A possible way to resolve this problem is to assume that the irradiations activate previously inactive donor impurities. A mechanism for such an activation is provided by the autocompensation reaction in (3a). In cases in which the irradiations make the bonded H within the a-Si:H matrix sufficiently mobile, the density of electrically active donor impurities should then adjust to the reduced number of electrons in the conduction band. Quantitative evaluations along this line have yielded the data plotted in Figure 24 (Stitzl, Krötz, and Müller, 1991b). We can see that

---

[1] At temperatures in excess of about 400 K the conductivity of the irradiated a-Si:H samples was found to be singly activated. The conductivity data presented in Figure 22(a) therefore reflect changes in the Fermi energy position with increasing doses of radiation.

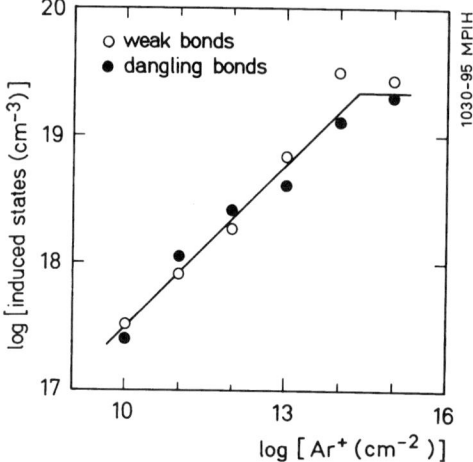

FIG. 23. Weak and dangling bond densities as derived from photothermal deflection spectroscopy of argon-irradiated samples of hydrogenated amorphous silicon (a-Si:H). (Reprinted from Stitzl, Krötz, and Müller (1991a). "Accumulation and Annealing of Implantation Damage in a-Si:H." *Appl. Phys. A* **53**, 235–240, Figure 2, with permission of the author and Springer-Verlag. Copyright Springer-Verlag.)

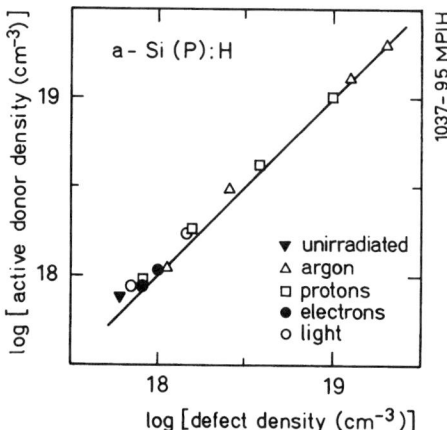

FIG. 24. Calculated density of active donors as a function of the total, that is, the equilibrium and irradiation-induced dangling bond density in a-Si(P):H. (Reprinted from Stitzl, Krötz, and Müller (1991a). "Autocompensation Doping in Light-Soaked and in Radiation-Damaged a-Si:H," *Appl. Phys. A* **52**, 335–338, Figure 4, with permission of the author and Springer-Verlag. Copyright Springer-Verlag.)

for each dangling bond defect induced, a previously inactive donor impurity is activated, until eventually 100% electrical activity has been reached. In considering these data it should be noted that donor activation is a fairly commonly observed phenomenon when bias-annealing experiments are performed on doped a-Si:H films (Street et al., 1987, Liu and Spear, 1991; Krötz and Müller, 1991). Another example of dopant activation has been revealed in the course of the Li doping experiments of the Heidelberg group (Schölch et al., 1987).

Concerning applications of the ion implantation doping technique, an important question is how far the effects of radiation damage can be reversed by annealing. To this end the same set of irradiated n-type samples was heated to a fixed temperature T_a and subsequently subjected to increasingly longer annealing times t_a. In order to assess the quality of the annealing process, the conductivity of the a-Si:H samples at $T = 450$ K was measured. The so obtained values of $\sigma(T_a, t_a)$ were fitted to stretched exponential decays, which previously had been observed in the decay of metastable structural states in a-Si:H (Street et al., 1987; Jackson and Kakalios, 1988; Kakalios and Street, 1988; Kakalios and Jackson, 1988):

$$\sigma(T_a, t_a) = \sigma_s + (\sigma_0 - \sigma_s) \cdot \exp(-(t_a/\tau)^\beta)$$

Fits to this equation yielded four important parameters that allowed different aspects of the implantation damage to be quantified:

- σ_0 is the initial conductivity after irradiation, which measures the amount of implantation damage accumulated under the particular irradiation conditions.
- σ_s is the conductivity after very long annealing times, which measures the reversibility of the annealing process.
- τ is the time within which the induced defect concentrations relax toward their annealed values; as usual, τ is assumed to be thermally activated: $\tau = \tau_0 \cdot \exp(E_a/kT)$.
- β is a dispersion parameter which arises from the time-dependence of the hydrogen diffusion coefficient.

Some of the data obtained in this way are summarized in Figure 25. We see that σ_s no longer returns to the pre-irradiation value of σ_{450K} when a threshold energy transfer of about $3 \cdot 10^{-2}$ eV/target atom is exceeded in the irradiations. A further increase in the energy transfer into atomic displacements is seen to significantly increase the extent of irreversibility. The data for the annealing activation energy Q, which characterizes the energy barrier for the reversible damage fraction, clearly bring out the fact that the increase

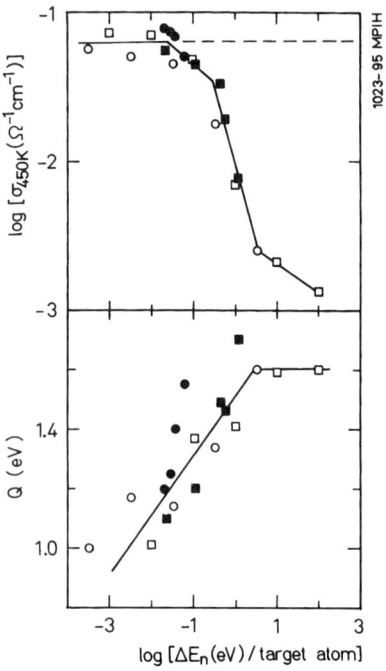

FIG. 25. Annealing parameters as obtained from fits to stretched exponential decay curves as a function of the energy deposited into atomic displacement collisions. Top, Saturation conductivity as measured at 450 K after very long annealing times. Bottom, Annealing activation energy for the reversible damage fraction. Q(eV), activation energy; ■, various B^+ and P^+ irradiations; □, Ar^+ dose variation; ●, H^+ energy variation; ○, H^+ dose variation. (Reprinted from Stitzl, Krötz, and Müller (1991a). "Accumulation and Annealing of Implantation Damage in a-Si:H." *Appl. Phys. A* **53**, 235–240, Figure 3, with permission of the author and Springer-Verlag. Copyright Springer-Verag.)

in the irreversible damage fraction is also accompanied by a significant hardening of the reversible damage fraction.

From these results it is suggested that in a-Si:H, two different annealing mechanisms with distinctly different activation energies are operative: from the fact that the observed $\sigma(T_a, t_a)$ curves can be fitted to stretched exponential decays, a hydrogen-diffusion-mediated annealing process is inferred (Street *et al.*, 1987). The likely annealing mechanism is a localized event in which a diffusing H atom becomes trapped at an isolated dangling bond defect within the a-Si:H network. Depending on the amount of energy initially transferred in the irradiation, H diffusion lengths in the range from a single lattice constant up to about 65 nm are inferred. The data presented previously have shown that in addition to dangling bond defects, the irradiation also tends to introduce additional bond-angle strain. In turn, this strain leads to an increase in the Urbach tail width. Concerning the annealing of this bond-angle strain, recent Raman scattering work (Bernsten *et al.*, 1993; Stolk *et al.*, 1993) has shown that this kind of radiation damage can only be partially annealed at temperatures below the deposition temperature of the a-Si:H films ($T_d \leq 250°C$). A complete relaxation of the bond-angle strain, however, is possible by increasing the annealing temperature toward the crystallization threshold ($T_a \approx 550°C$). Such a gradual annealing of the bond-angle strain is characteristic of the structural relaxation process that already has been described in Section 2 of Part II. As stated there, the structural relaxation process is characterized by a wide spectrum of annealing activation energies, ranging from 0.5 eV to about 2.7 eV. The activation energy spectrum of these intrinsic long-range processes must be compared with the H-related ones, which according to Figure 23 are characterized by activation energies in the range of from 0.9 to 1.8 eV. As a consequence, a full relaxation of the bond-angle strain cannot be obtained without suffering diffusive H losses. A possible way of dealing with this situation is by employing two-step annealing procedures that consist of annealing at temperatures in excess of the deposition temperature and by performing rehydrogenation processes afterward. In this way, an Italian group was able to obtain higher alkali doping efficiencies, which approach the efficiency of substitutional gas-phase doping with PH_3 (Desalvo, *et al.*, 1993).

IV. Structural and Configurational Changes in Amorphous Silicon

1. CHANGES IN PURE MATERIAL

The work reported on previously can be summarized as follows: In a-Si networks, the long-range order is broken on a length scale comparable with a single lattice constant of the c-Si lattice (0.45 nm). This kind of symmetry breaking gives rise to a number of phenomena that are qualitatively new and unknown from the crystalline counterpart of Si. Prominent examples among others are the relaxation of the k-selection rule in optical transitions, the occurrence of variabe-range hopping transport at the Fermi energy, and a change in the doping mechanism. All these effects would be observable in a disordered but otherwise completely rigid Si network.

In this context it is important to note that both chemically pure and hydrogenated random Si networks are metastable structures that are thermodynamically unstable with respect to c-Si. Our discussions have shown that such networks are able to support two different kinds of structural change: structural relaxation and configurational equilibration (Müller and Krötz, 1993; Müller et al., 1994). The first process can be observed in both materials, indicating that it is supported by intrinsic degrees of freedom of the random Si networks. The second effect, however, can only be observed in hydrogenated random Si networks, indicating that it is H related.

Turning to the irreversible structural relaxation process first, bond-angle strain is relieved in such processes by dangling bond diffusion, and dangling bond defects are eventually annealed out by mutual recombination (Stolk et al., 1994). These structural changes are relatively long-range in nature and are characterized by a wide spectrum of activation energies. These effects are seen most clearly in a-Si where activation energies can range from below 0.5 eV (Battaglia, 1993) up to the activation energy for solid-phase epitaxial recrystallization of about 2.7 eV (Roorda et al., 1989, 1991). Similar signs of structural relaxation can also be observed in the annealing of a-Si:H previously subjected to heavy ion bombardment (Bernsten et al., 1993; Stolk et al., 1993).

2. CHANGES IN HYDROGENATED MATERIAL

Hydrogenation of random Si networks removes localized defect states from the bandgap and thus enables low-mobility band transport to take over. In such defect-passivated material, a residual doping effect is observed that can successfully be described within the generalized model of autocompensation doping (Street, 1982; Müller, Kalbitzer, and Mannsperger, 1986; Krötz, et al., 1991). A characteristic feature of this process is that it involves a strong coupling between the electronic and configurational degrees of freedom of the random Si network. The efficiency of this latter coupling critically relies on the availability of diffusing H atoms within the a-Si:H network (Street et al., 1987). At normal device operating temperatures the bonded H atoms are frozen in at their trapping sites within the a-Si:H networks, and in this state, essentially normal semiconductor behavior can be observed. Elevated temperatures on the order of 100 to 200°C or ionizing radiation, however, can cause the bonded H in the a-Si:H networks to become sufficiently mobile and the coupling of electronic and configurational degrees of freedom to become effective. The considerations of Section 3 of Part III have shown that, in addition to the relatively long-range structural relaxation processes, such local changes also play an important role in the generation and annealing of radiation damage in a-Si:H networks (Stitz, Krötz, and Müller, 1991a, 1991b; Bernsten et al., 1993; Stolk et al., 1993; Wagner, Gleskova, and Nakata, 1995).

REFERENCES

Battaglia, A. (1993). Private communication.
Beeman, D., Tsu, R., and Thorpe, M. G. (1985). "Structural Information from the Raman Spectrum of Amorphous Silicon," Phys. Rev. B **32**, 874–878.
Bhatia, K. L., Krätschmer, W., and Kalbitzer, S. (1988). "Optical Absorptivity of Ion-Beam Irradiated Silicon," Appl. Phys. A **45**, 69–72.
Bernsten, A. J. M., Stolk, P. A., van der Weg, W. F., and Saris, F. W. (1993). "Annealing of Ion-Implanted Hydrogenated Amorphous Silicon: Stable and Removable Damage," Mat. Res. Soc. Symp. Proc. **297**, 303–308.
Böhringer, K., Liu, X. H., and Kalbitzer, S. (1983). "Electronic Properties of a-Si:H and a-Si:F Films Produced by Ion Implantation," J. Phys. C **16**, L1187–L1192.
Böhringer, K., Mannsperger, H., and Kalbitzer, S.l (1985). "Electronic Transport in a-Si:H Produced by Ion Implantation," Nucl. Instrum. Meth. **B7/8**, 299–301.
Brower, K. L., and Beezhold, W. (1972). "Electron Paramagnetic Resonance of the Lattice Damage in Oxygen-Implanted Silicon," J. Appl. Phys. **43**, 3499–3506.
Coffa, S. (1992). Private communication.
Coffa, S., Priolo, F., Poate, J. M., and Glarum, S. H. (1993). "Conductivity Changes and Impurity-Defect Interaction in Ion Implanted Amorphous Silicon," Nucl. Instrum. Meth. Phys. Res. **B80/81**, 603–606.
Csepregi, L., Kennedy, E. F., Gallagher, T. G., and Mayer, J. W. (1977). "Reordering of Amorphous Layers of Si Implanted with ^{31}P, ^{75}As and ^{11}B Ions," J. Appl. Phys. **48**, 4234–4239.
Cullis, A. G., and Chew, N. G. (1984). "Melting Temperature and Explosive Crystallisation of Amorphous Silicon during Pulsed-Laser Irradiation," Phys. Rev. Lett. **52**, 2360–2363.
de Wit, L., Roorda, S., Sinke, W. C., Saris, F. W., Bernsten, A. J. M., and van der Weg, W. F. (1992). "Structural Relaxation of Amorphous Silicon Induced by High-Temperature Annealing," Mat. Res. Soc. Symp. Proc. **205**, 3–8.
Dent, A. J., Dobson, B. R., Greaves, G. N., Kalbitzer, S., Derst, G., and Müller, G. (1993). "The Changing Environments of Dopants in Amorphous Silicon at Various Stages of Annealing," Mat. Res. Soc. Symp. Proc. **307**, 21–26.
Dent, A. J., Greaves, G. N., Derst, G., Kalbitzer, S., and Müller, G. (1994). "Tracing the Changing Environment of Lattice Sites in the Recrystallisation of Amorphized Silicon," Physica B, **208–209**, 503.
Derst, G., Wilbertz, Ch., Bhatia, K. L., Krätschmer, W., and Kalbitzer, S. (1989). "The Optical Properties of SiC for Crystalline/Amorphous Contrast Formation," Appl. Phys. Lett. **54**, 1722–1725.
Desalvo, A., Zignani, F., Galloni, R., Rizzoli, R., and Ruth, M. (1993). "Doping of Amorphous Silicon by Potassium Ion Implantation," Philos. Mag. **B67**, 131–142.
Donovan, E. P., Spaepen, F., Poate, J. M., and Jacobson, D. C. (1989). "Homogeneous and Interfacial Heat Releases in Amorphous Silicon," Appl. Phys. Lett. **55**, 1516–1518.
Donovan, E. P., Spaepen, F., Turnbull, D., Poate, J. M., and Jacobson, D. C. (1985). "Calorimetric Studies of Crystallisation and Relaxation of Amorphous Si and Ge," J. Appl. Phys. **57**, 1795–1804.

Fonseca, F. J., Galloni, R., and Nylandsted-Larsen, A. (1993). "Electrical Activation of Potassium and Phosphorus Ions Implanted in Hydrogenated Amorphous Silicon," *Philos. Mag.* **B67**, 107-115.
Fortner, J., and Lannin, J. S. (1988). "Radial Distribution Functions of Amorphous Silicon," *Phys. Rev. B* **39**, 5527-5530.
Galloni, R., Tsuo, Y. S., Baker, D. W., and Zignani, F. (1990). "Doping and Hydrogenation by Ion Implantation of Glow-Discharge-Deposited Amorphous Silicon Films," *Appl. Phys. Lett.* **56**, 241-253.
Geissler-Pfeilsticker, R. (1980). PhD thesis, University of Heidelberg, Heidelberg, Germany.
Greaves, G. N., Dent, A. J., Dobson, B. R., Kalbitzer, S., and Müller, G. (1993). "Environments of Ion-Implanted Dopants in Amorphous Silicon at Various Stages of Annealing," *Nucl. Instrum. Meth. Phys. Res.* **B80/81**, 966-972.
Greaves, G. N., Dent, A. J., Dobson, B. R., Kalbitzer, S., Pizzini, S., and Müller, G. (1992). "Environments of Ion-Implanted As and Ga Impurities in Amorphous Silicon," *Phys. Rev.* **B45**, 6517-6533.
Hartmann, E., Behm, R. J., Krötz, G., Müller, G., and Koch, F. (1991). "Writing Electronically Active Nanometer-Scale Structures with a Scanning Tunneling Microscope," *Appl. Phys. Lett.* **59**, 2136-2138.
Jackson, W. B., and Kakalios, J. (1988). "Kinetics of Carrier-Induced Metastable Defect Formation in Hydrogenated Amorphous Silicon." In *Amorphous Silicon and Related Materials* (H. Fritzsche, ed.) World Scientific, Singapore, pp. 247-295.
Johnson, N. M. (1991). "Hydrogen in Crystalline and Amorphous Silicon," *J. Non-Cryst. Solids* **137&138**, 11-16.
Kakalios, J., and Jackson, W. B. (1988). "The Hydrogen Glass Model." In *Amorphous Silicon and Related Materials* (H. Fritzsche, ed.) World Scientific, Singapore, pp. 207-245.
Kakalios, J., and Street, R. A. (1988). "Thermal Equilibrium Effects in Doped Hydrogenated Amorphous Silicon." In *Amorphous Silicon and Related Materials* (H. Fritzsche, ed.) World Scientific, Singapore, pp. 165-205.
Kalbitzer, S. (1987). "Ionographic Patterns with Amorphous/Crystalline Contrast," *Appl. Phys.* **A44**, 153-155.
Kalbitzer, S., Greaves, G. N., Dent, A. J., Derst, G., and Müller, G. (1995). "The Environment of Lattice Sites during Thermal Epitaxial Regrowth of Ion-Beam Amorphized Silicon," *Nucl. Instrum, Meth.* **B97**, 312-315.
Kalbitzer, S., Müller, G., LeComber, P. G., and Spear, W. E. (1980). "The Effects of Ion Implantation on the Electrical Properties of Amorphous Silicon," *Philos. Mag.* **B41**, 439-456.
Kanicki, J. (1991). *Amorphous and Microcrystalline Semiconductor Devices*. Artech House, Boston, London.
Krötz, G., and Müller, G. (1991). "Structural Equilibration in Intrinsic, Single-Doped and Compensated TFTs — Experiments and Calculations," *J. Non-Cryst. Solids* **137&138**, 163-166.
Krötz, G., Müller, G., Derst, G., Wilbertz, Ch., and Kalbitzer, S. (1994). "Thin-Film SiC as an Optical and Optoelectronic Material," *Diamond and Related Materials* **3**, 917-921.
Krötz, G., Wind, J., Stitzl, H., Müller, G., Kalbitzer, S., and Mannsperger, H. (1991). "Experimental Tests of the Autocompensation Model of Doping," *Philos. Mag.* **B63**, 101-121.
LeComber, P. G., Spear, W. E., Müller, G., and Kalbitzer, S. (1980). "Electrical and Photoconductive Properties of Ion Implanted Amorphous Silicon," *J. Non-Cryst. Solids* **35&36**, 327-332.

Ley, L. (1984). "Photoemission and Optical Properties." In *The Physics of Hydrogenated Amorphous Silicon; Topics in Applied Physics*. Springer-Verlag, Berlin, Heidelberg, New York, Tokyo, pp. 61–161.

Liu, E. Z., and Spear, W. E. (1991). "An Investigation of the Phosphorus Doping Mechanism in a-Si by Sweep-Out Experiments," *J. Non-Cryst. Solids* **137&138**, 167–170.

Luft, W., and Tsuo, Y. S. (1993). *Hydrogenated Amorphous Silicon Alloy Deposition Processes*. Marcel Dekker, New York, p. 26.

Mannsperger, H., Kalbitzer, S., and Müller, G. (1986). "Doping Efficiencies of Gas-Phase and Ion Implantation Doped a-Si:H," *Appl. Phys.* **A41**, 253–258.

Mott, N. F. (1967). "Electrons in Disordered Structures," *Adv. Phys.* **16**, 49–144.

Mott, N. F. (1969). "Conduction in Non-Crystalline Materials III: Localised States in the Pseudogap and Near Extremities of Conduction and Valence Bands," *Philos. Mag.* **19**, 835–852.

Müller, G., Kalbitzer, S., Spear, W. E., and LeComber, P. G. (1977). "Doping of Amorphous Silicon by Ion Implantation." In *Proc. 7th Internatl. Conference on Amorphous and Liquid Semiconductors* (W. E. Spear, ed.) CICL, University of Edinburgh, pp. 442–446.

Müller, G., and Kalbitzer, S. (1978). "Doping of Amorphous Silicon in the Hopping Transport Regime," *Philos. Mag.*, **B38**, 241–254.

Müller, G., and Kalbitzer, S. (1980). "The Crystalline-to-Amorphous Transition in Ion-Bombarded Silicon, *Philos. Mag.* **B41**, 307–325.

Müller, G., and LeComber, P. G. (1980). "Implantation Damage in Amorphous Silicon," *Philos. Mag.* **B43**, 419–431.

Müller, G., Winterling, G., Kalbitzer, S., and Reinelt, M. (1983). "Hydrogen Incorporation, Doping and Thickness-Dependent Conductivity in Glow-Discharge-Deposited a-Si:H Films," *J. Non-Cryst. Solids* **59&60**, 469–472.

Müller, G., Mannsperger, H., Böhringer, K., and Kalbitzer, S. (1984). "Defect Absorption and Substitution of Doping Impurities on Ion Implantation Doped Amorphous Silicon." In *Proc. E-MRS Spring Conference (Strassburg) Poly-, Micro-Crystalline and Amorphous Semiconductors, Les Editions de Physique* (P. Pinard and S. Kalbitzer eds.), Paris, France, pp. 497–502.

Müller, G., Kalbitzer, S., and Mannsperger, H. (1986). "A Chemical Bond Approach to Doping, Compensation and Photo-Induced Degradation in Amorphous Silicon," *Appl. Phys.* **A39**, 243–250.

Müller, G., and Krötz, G. (1993). "Structural Equilibration in Pure and Hydrogenated Amorphous Silicon," *Mat. Res. Soc. Symp. Proc.* **297**, 237–248.

Müller, G. (1993). "The Contribution of Ion Beam Techniques to the Physics and Technology of Amorphous Semiconductors," *Nucl. Instrum. Meth. Phys. Res.* **B80/81**, 957–965.

Müller, G., Krötz, G., Kalbitzer, S., and Greaves, G. N. (1994). "Reversible and Irreversible Structural Changes in Amorphous Silicon," *Philos. Mag.* **B69**, 177–196.

Müller, G., Hellmich, W., Krötz, G., Kalbitzer, S., Greaves, G. N., Derst, G., Dent, A. J., and Dobson, B. R. (1996). "Dopant-Defect Interactions in Hydrogen-Free Amorphous Silicon," *Philos. Mag.* **B73**, 245–259.

Overhof, H., and Thomas, P. (1989). *Hydrogenated Amorphous Semiconductors*, Springer Tracts in Modern Physics, Vol. 114, Springer-Verlag, Berlin, Heidelberg, New York, Tokyo.

Pfeilsticker, R., Kalbitzer, S., and Müller, G. (1978). "Observation of Variable-Range Hopping at 'Natural' Phonon Frequencies," *Z. Phys.* **B31**, 233–235.

Powell, M. J., van Berkel, C., and Deane, S. C. (1991). "Instability Mechanisms in Amorphous Silicon Thin-Film Transistors and the Role of the Defect Pool," *J. Non-Cryst. Solids* **137&138**, 1215–1220.

Prisslinger, R., Kalbitzer, S., Kräutle, H., Grob, J. J., and Siffert, P. (1975). "Recovery of Silicon Layers Damaged by Low-Energy Ion Bombardment." In *Ion Implantation in Semiconductors* (S. Namba, ed.), Plenum Press, New York, pp. 547–552.
Richter, H., and Breitling, H. (1957). "Struktur des Amorphen Germaniums und Siliziums," *Z. Natf.* **A13**, 988–996.
Roorda, S., Doorn, S., Sinke, W C., Scholte, P. M. L. O., and v. Loenen, E. (1989). "Calorimetric Evidence for Structural Relaxation in Amorphous Silicon," *Phys. Rev. Lett.* **62**, 1880–1883.
Roorda, S., Sinke, W. C., Poate, J. M., Jacobson, D. C., Dierker, S., Dennis, B. S., Eaglesham, D. J., Spaepen, F., and Fuoss, P. (1991). "Structural Relaxation and Defect Annihilaton in Pure Amorphous Silicon," *Phys. Rev. B* **44**, 3702–3725.
Roorda, S., and Sinke, W. C. (1992). "Defects, Entropy and Melting of Amorphous Silicon," *Mat. Res. Soc. Symp. Proc.* **205**, 9–14.
Ruttensperger, B., Krötz, G., Müller, G., Derst, G., and Kalbitzer, S. (1991). "Crystalline-Amorphous Contrast Formation in Thermally Crystallized SiC," *J. Non-Cryst. Solids* **137&138**, 635–638.
Ruttensperger, B., Müller, G., and Krötz, G. (1993). "Density-of-State Distribution and Variable-Range Hopping Transport in Amorphous Silicon Prepared by Ion Bombardment," *Philos. Mag.* **B68**, 203–214.
Schölch, H. P., Weiser, M., Kalbitzer, S., and Müller, G. (1987). "Effective Charge of Li in a-Si:H by Transport Measurements," *Solid State Commun.* **64**, 1419–1422.
Sinke, W. C., Warabisako, T., Miyao, M., Tokuyama, T., Roorda, S., and Saris, F. W. (1988). "Transient Structural Relaxation in Amorphous Silicon," *J. Non-Cryst. Solids* **99**, 308–323.
Spear, W. E., and LeComber, P. G. (1976). "Electronic Properties of Substitutionally Doped Amorphous Si and Ge," *Philos. Mag.* **33**, 935–949.
Spear, W. E., LeComber, P. G., Kalbitzer, S., and Müller, G. (1979). "Doping of Amorphous Silicon by Alkali-Ion Implantation," *Philos. Mag.* **B39**, 159–165.
Spear, W. E., Gibson, R. A., Yang, D., LeComber, P. G., Müller, G., and Kalbitzer, S. (1981). "Some New Developments in the Field of Amorphous Silicon Solar Cells," *J. Phys. (Paris)* **42**, C4-1143.
Staebler, D. L., and Wronski, C. R. (1977). "Reversible Conductivity Changes in Discharge-Produced Amorphous Si," *Appl. Phys. Lett.* **31**, 292–294.
Stitzl, H., Krötz, G., and Müller, G. (1991a). "Accumulation and Annealing of Implantation Damage in a-Si:H," *Appl. Phys.* **A52**, 235–240.
Stitzl, H., Krötz, G., and Müller, G. (1991b). "Autocompensation Doping in Light-Soaked and in Radiation-Damaged a-Si:H," *Appl. Phys.* **A53**, 335–338.
Stolk, P. A., Berntsen, A. J. M., Saris, F. W. and van der Weg, W. F. (1993). "Separating the Contributions of Hydrogen and Structural Relaxation to Damage Annealing in a-Si:H," *Mat. Res. Soc. Symp Proc.* **297**, 127–132.
Stolk, P. A., Saris, F. W., Bernsten, A. J. M., van der Weg, W. F., Barklie, R., Sealy, L. T., Krötz, G., and Müller, G. (1994). "Contribution of Defects to Electronic, Structural and Thermodynamic Properties of Amorphous Silicon," *J. Appl. Phys.* **75**, 7266–7286.
Street, R. A. (1982). "Doping and the Fermi Energy in Amorphous Silicon," *Phys. Rev. Lett.* **49**, 1187–1190.
Street, R. A., Kakalios, J., Tsai, C. C., and Hayes, T. M. (1987). "Thermal Equilibrium Processes in Amorphous Silicon," *Phys. Rev. B* **35**, 1316–1333.
Stutzmann, M. (1987). "Weak Bond–Dangling Bond Conversion in Amorphous Silicon," *Philos. Mag.* **B56**, 63–70.
Stuke, J. (1977). "ESR in Amorphous Germanium and Silicon." In *Proc. 7th Internat. Conference on Amorphous and Liquid Semiconductors* (W. E. Spear ed.) CICL, University of Edinburgh, pp. 406–418.

Thomas, P. A., Brodsky, M. H., Kaplan, D., and Lepine, D. (1978). "Electron Spin Resonance of Ultrahigh Vacuum Evaporated Amorphous Silicon: in-situ and ex-situ Studies," *Phys. Rev. B* **18**, 3059–3073.
Thompson, M. O., Galvin, G. J., Mayer, J. W., Peercy, P. S., Poate, J. M., Jacobson, D. C., Cullis, A. G., and Chew, N. G. (1984). "Melting Temperature and Explosive Crystallisation of Amorphous Silicon during Pulsed-Laser Irradiation," *Phys. Rev. Lett.* **52**, 2360–2363.
Wagner, S., Gleskova, H., and Nakata, Jun-ichi (1995). "Equilibration and Stability in Undoped Amorphous Silicon." In *Proc. 16th Internat. Conference on Amorphous Semiconductors*, in press.
Vögele, V., Kalbitzer, S., and Böhringer, K. (1985). "Observation of Correlation Effects in the Hopping Transport in Amorphous Silicon," *Philos. Mag.* **B52**, 153–168.

CHAPTER 5

Sheet and Spreading Resistance Analysis of Ion Implanted and Annealed Semiconductors

Jumana Boussey-Said

LABORATOIRE DE PHYSIQUE DES COMPOSANTS À SEMICONDUCTEURS
ENSERG, GRENOBLE, FRANCE

I. INTRODUCTION	129
II. SHEET RESISTANCE MEASUREMENT	130
1. Collinear Four-Point Probe Method	131
2. Sheet Resistance of Arbitrarily Shaped Samples	133
III. SPREADING RESISTANCE PROBES PROFILING	135
1. Principle of Spreading Resistance Profiling Measurements	136
2. Extraction of Resistivity Profiles from Spreading Resistance Raw Data	138
IV. APPLICATIONS	143
1. High-Dose and High-Energy Arsenic-Implanted Silicon Layers	143
2. Low-Energy Boron and Boron Fluoride Implantation in a Preamorphized Silicon Substrate	152
V. SUMMARY	161
References	162

I. Introduction

Ion implantation is now a well-established technique for highly controlled introduction of dopants into semiconductors. Its increasing use in very large scale integrated (VLSI) device fabrication attests to its fundamental advantages over conventional thermal diffusion techniques: a high degree of reproducibility and precise control of dopant purity, dose, and spatial distribution can be achieved by ion implantation doping techniques (Ziegler, 1992). However, the bombardment of the silicon lattice with high-energy accelerated ions (typically up to 200 keV) leads to the creation of a high density of defects in the implanted substrate. Several experiments have been conducted during the last decades in order to determine the nature and the spatial distribution of such defects under various experimental conditions of ion implantation (Narayan and Holland, 1984; Prussin, David, and Tauber,

1982). It is now well known that when the point defect density induced by implantation is higher than a critical value, ion implantation may form a continuous amorphous layer in the surface of the implanted target (Dennis and Hale, 1977; Battaglia and Campisano, 1993).

From a technological point of view, subsequent thermal annealing must be done to remove ion-implantation-induced defects, to recrystallize the amorphous layer, and to achieve electrical activation of the implanted impurities (Gibbons, 1972). The efficiency of such a thermal treatment must be investigated in the manufacturing research and development area. Therefore, structural and electrical parameters related to the density of the induced damage are extracted and compared before and after annealing. Cross-section transmission electron microscopy (XTEM) and Rutherford backscattering spectrometry (RBS) techniques are commonly used to investigate the thermal recovery of the implanted layer by evaluating its degree of crystallinity, and Secondary Ion Mass Spectroscopy (SIMS) provides dopant distribution. Nevertheless, these techniques are highly expensive to install and require specially trained technicians for operation and maintenance. However, the more important property of a final VLSI device is still its electrical behavior (e.g., leakage current, threshold voltage, and drain current) tightly correlated to the free carrier concentration. In order to model or extract device parameters from electrical characteristics, a knowledge of dopant electrical activation and of dopant mobility at different depths from the surface of the sample is very important. Such a profiling is generally obtained by SIMS, stripping Hall effect techniques, and spreading resistance measurements, all of which, unfortunately, are destructive.

This chapter gives an overview of the use of sheet resistance and spreading resistance probe profiling in the field of ion-implantation-induced defects characterization. In Part II, we review the principle and the necessary equipment for sheet resistance and spreading resistance profiling measurements. Emphasis is on the spreading resistance raw data conversion in order to obtain resistivity profiles. In Part III, we present some interesting results that have been obtained by these electrical analysis techniques (e.g., low-temperature arsenic activation, electrical activation ratio of boron in a preamorphized silicon substrate, and residual defects and their effects on the redistribution of impurities.

II. Sheet Resistance Measurement

The simplest estimate of the sheet resistance of a semiconductor R_s (ohms per square) is by direct measurement of the resistance between two opposite sides of an arbitrary square sample of this material. According to Ohm's law, the resistance of a square of side l is then given by

$$R = \frac{\rho l}{wl} = \frac{\rho}{w} = R_s \qquad (1)$$

where w is the thickness of the sample under consideration, and ρ is its resistivity.

Regarding the difficulty of obtaining such a sample for each ion-implanted semiconductor during process development, other equivalent techniques have been optimized and are now currently applied. In the following sections, we survey the basic theory of these methods and review the more important aspects of their use in the field of the characterization of ion-implanted layers.

1. COLLINEAR FOUR-POINT PROBE METHOD

The more common technique is the four-point probe method illustrated in Figure 1, where a current I delivered by a constant current source is forced through the outer two probes and the voltage decrease V is measured between the two inner probes either with a potentiometer, which draws no current at all, or with a high-impedance voltmeter, which draws very little current. As a consequence, the contact resistance R_c at each metal probe–semiconductor contact is negligible in either case because the voltage decrease across the contact surface is negligible due to the very small current that flows through it (Wieder, 1979).

For equally spaced collinear four-point probes (Fig. 1), the potential V at a distance r from an electrode carrying a current I in a specimen whose thickness, $w \gg s$, and whose boundaries lie at infinity is given by the relationship:

$$V = \frac{\rho \cdot I}{2\pi \cdot r} \qquad (2)$$

For probes resting on a semi-infinite medium, with current entering probe 1 and leaving probe 4, the voltage V becomes

$$V = \frac{\rho \cdot I}{2\pi} \left(\frac{1}{r_1} - \frac{1}{r_2} \right) \qquad (3)$$

where r_1, r_2 are the distances from probes 1 and 4, respectively.

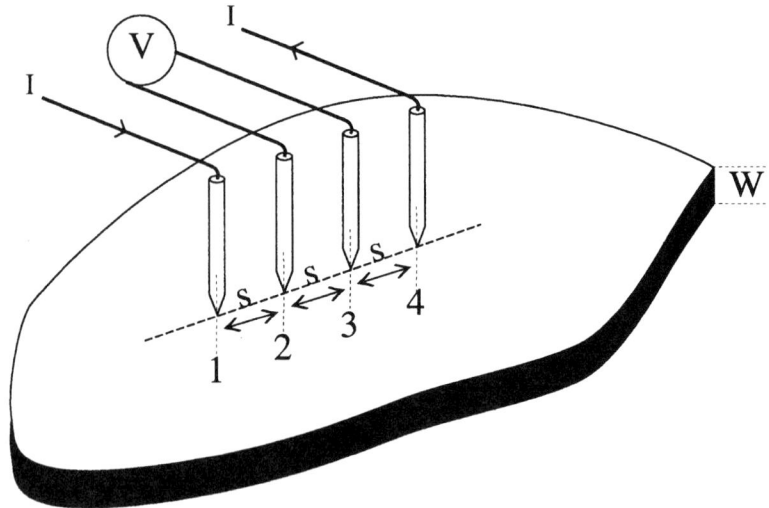

FIG. 1. Configuration of collinear, four-point probe, sheet resistance measurements. I, current; V, potential; $w \gg s$, *thickness*, 1–4, probes.

When applied at probe 2, Eq. (3) leads to the expression of the voltage at this probe:

$$V_2 = \frac{\rho \cdot I}{2\pi}\left(\frac{1}{s} - \frac{1}{2s}\right) \quad (4)$$

and the voltage at probe 3 is

$$V_3 = \frac{\rho \cdot I}{2\pi}\left(\frac{1}{2s} - \frac{1}{s}\right) \quad (5)$$

The potential difference between the inner probes is then given by

$$V = V_2 - V_3 = \frac{\rho \cdot I}{2\pi s} \quad (6)$$

and the semiconductor resistivity is deduced from

$$\rho = 2\pi s \cdot \left(\frac{V}{I}\right) \quad (7)$$

usually expressed in ohms per centimeter with V measured in volts, I in amperes, and s in centimeters. The most common probe spacing of the commercially available apparatus ranges from 0.5 to 1 mm.

In practice, semiconductor wafers are not semi-infinite in extent in either lateral or vertical dimensions so that Eq. (7) must be corrected for finite geometries. For an arbitrarily shaped sample, the resistivity is given by

$$\rho = 2\pi s \cdot \left(\frac{V}{I}\right) \cdot F \qquad (8)$$

where F is a correction factor that takes into account edge and thickness effects and the nature of the surface boundary of the bottom wafer.

For instance, for very thin samples ($w < s/2$) located on a nonconducting substrate, it has been calculated that the correction factor is given by (Albers and Berkowitz, 1985)

$$F = \frac{w/s}{2\ln(2)} \qquad (9)$$

and that the sheet resistance of such samples is

$$R_s = \frac{\rho}{w} = \frac{\pi}{\ln(2)} \frac{V}{I} = 4.532 \cdot \frac{V}{I} \qquad (10)$$

under the additional constraint that the sample must have a lateral width greater than or equal to $40/s$.

2. Sheet Resistance of Arbitrarily Shaped Samples

The theoretical foundation of measurements on irregularly shaped samples is based on conformal mapping, which was developed by Van Der Pauw (1958a, 1958b). He showed how the specific resistivity of a flat sample of arbitrary shape can be measured without knowing the pattern of the current if the following conditions are met: (i) the contacts are at the periphery of the sample, (ii) the contacts are ohmic and "sufficiently small" to be considered point contacts, (iii) the sample is homogeneous and singly connected, and (vi) the sample is uniformly thick.

The resistivity of a flat sample of arbitrary shape with contacts 1, 2, 3, and 4 along the circumference, as shown in Figure 2, can be determined by applying a current I between contact 1 and contact 2 along its thickness

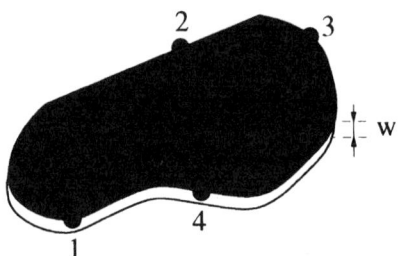

FIG. 2. Arbitrarily shaped sample with four contacts, as described by Van Der Pauw. W, thickness.

dimension and measuring the potential difference V between contacts 3 and 4; this leads to the resistance $R_{12,34}$. In a similar manner, $R_{23,41}$ is obtained and the resistivity is given by

$$\rho = \frac{\pi w}{\ln(2)} \frac{(R_{12,34} + R_{23,41})}{2} \cdot f \qquad (11)$$

where f is a function only of the ratio $R_{12,34}/R_{23,41}$ (i.e., on the commutated current and measured voltage). Figure 3 shows, in graphical form, the dependence of f on the ratio $R_{12,34}/R_{23,41}$. For a symmetrical sample such

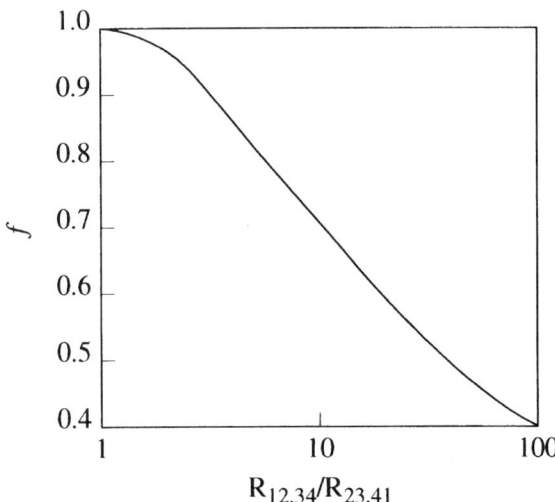

FIG. 3. The Van Der Pauw correction factor as a function (f) of the ratio $R_{12,34}/2_{23,41}$.

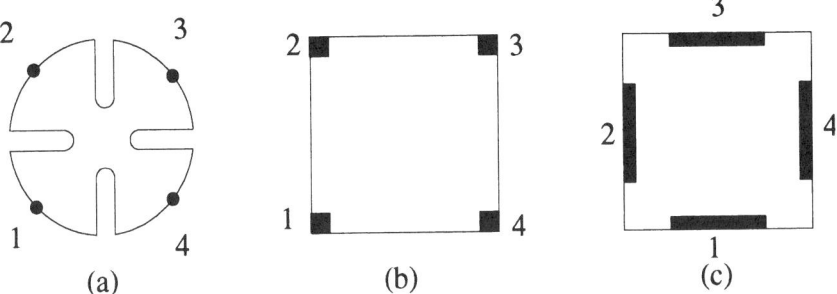

FIG. 4. Various geometries used for Van Der Pauw sheet resistance measurements: (a) cloverleaf, (b) square with contacts at the corners, and (c) square with contacts at the edges 1–4, probes.

as a square, $R_{12,34}/R_{23,41} = 1$, which corresponds to $f = 1$, such that the sheet resistance reduces to

$$R_s = \frac{\pi}{\ln(2)} R_{12,34} = 4.532 R_{12,34} \qquad (12)$$

as in the four-point probe expression in Eq. (10).

One of the advantages of the Van Der Pauw structure is the small sample size compared with the area required for four-point probe measurements. Nevertheless, care must be taken concerning the size and location of the contacts. Real contacts have finite dimensions and may not be exactly at the periphery of the sample (Chwang, Smith, and Crowell, 1974). It has been shown by Van Der Pauw that the error so introduced can be eliminated by the use of a "cloverleaf" configuration, such as that in Figure 4(a). Such a structure occupies too much of the useful area of the sample and makes sample preparation more complicated. For this reason, simple square structures with contacts that can be applied directly to the corners (Fig. 4(b)) or edges (Fig. 4(c)) of thin ion-implanted film are commonly considered desirable.

III. Spreading Resistance Probes Profiling

Whatever the chosen configuration, the conventional four-point probe measurements give only an average resistivity, that when combined with the mobility values determined by Hall effect measurements, leads to an estimation of the free carrier concentration in the characterized sample. This is sufficiently accurate for uniformly doped substrates but is unsuitable for nonuniformly doped samples in which resistivity profiles need to be determined.

The spreading resistance measurement technique is used more extensively to generate a resistivity and a dopant profile. The equipment is in wide commercial use and is manufactured by several companies. Although still expensive, it is rather simple to use, allows a very quick analysis, and provides an abundance of information concerning the resistivity profile and the homogeneity of a doped layer. In Part IV, we report on some results obtained using this technique, showing how it allows us to probe the extent of the created damage, and to monitor the annihilation process of the implantation-induced defects as a function of the annealing conditions (temperature and time).

1. Principle of Spreading Resistance Profiling Measurements

The measurements are made by placing two carefully aligned probes of hardened, highly conductive (tungsten–osmium) alloy on the surface to be characterized. The electrical resistance between the probes is then measured using a small dc (10 mV) bias. The probe–semiconductor contact is a metal–semiconductor contact that generally has nonlinear current-voltage characteristics. However, for applied voltages less than $kT/q \approx 25$ mV, the contact current-voltage characteristics are almost linear.

The displacement of the probes on a beveled surface that is very slightly angled enables spreading resistance values detected on the beveled surface to be converted into the corresponding depth resistance (Ehrstein, 1979) via a simple such relationship:

$$\Delta h = \Delta x \cdot \sin \alpha \qquad (13)$$

where Δx is the probe displacement step on the beveled surface, α is the bevel angle, and Δh is the corresponding vertical step (Fig. 5).

Also, preparation of uniform low-angle bevels requires a great deal of skill and experience. Currently, beveled surfaces are obtained by lapping—using diamond paste—small pieces of semiconductors mounted on a bevel block, as illustrated in Figure 6. The angles so obtained are accurately measured using an alpha-step profilometer. Bevel angles of 10′ to 5 degrees are typical and provide very good depth resolution, which can be on the order of a few tens of angstroms. In addition, providing the sample with an oxide coating

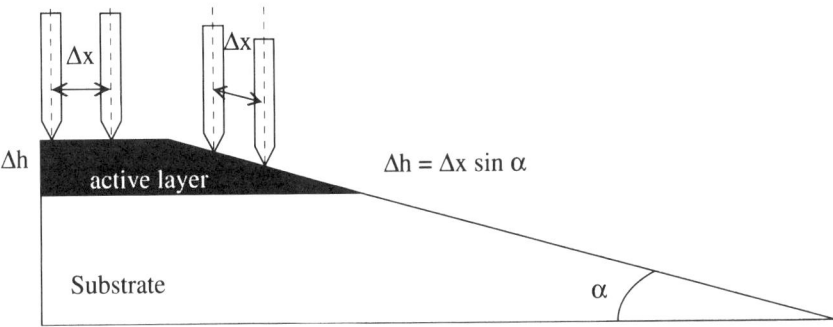

FIG. 5. Spreading resistance depth profile is obtained by probe displacement along the beveled surface with a step Δx, where α is the low bevel angle. Vertical resolution of up to 50 nm can be reached. Δh, corresponding vertical step.

has two advantages: the coating shifts the bevel curvature near the oxide surface (Fig. 7), and it provides a sharp corner at the bevel, which clearly defines the beginning of the profile precisely because the spreading resistance of the oxide is very high. In the case of ion-implanted samples, a thin oxide film is usually already available on the surface of the samples and can be used for the bevel boundary determination.

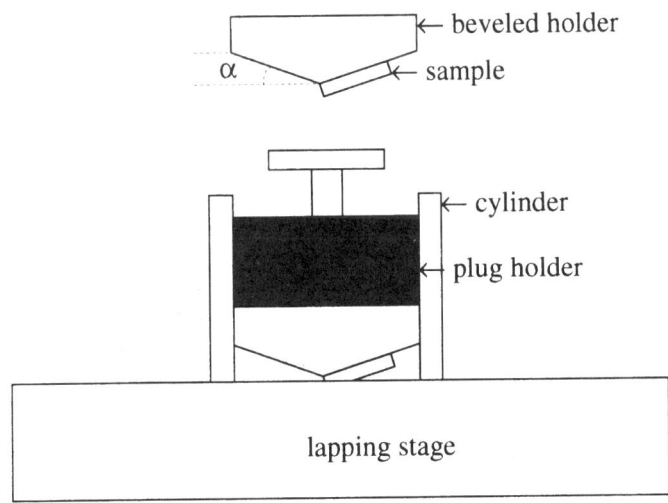

FIG. 6. Low-angle beveled surface preparation for spreading resistance depth profiling.

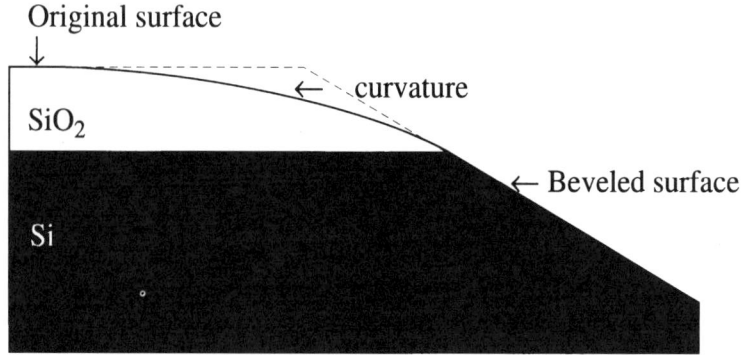

FIG. 7. Bevel curvature defect in the capping silicon dioxide (SiO$_2$) layer.

2. EXTRACTION OF RESISTIVITY PROFILES FROM SPREADING RESISTANCE RAW DATA

In the past, a great deal of attention was focused on the problem of analyzing raw spreading resistance data to give accurate resistivity profiles. In fact, the measured electrical resistance R_m between the probes results from three contributions (Fig. 8):

$$R_m = R_c + SR + R_s \qquad (14)$$

where R_c is the contact resistance at each metal probe–semiconductor contact, SR is the spreading resistance under each probe, and R_s is the bulk

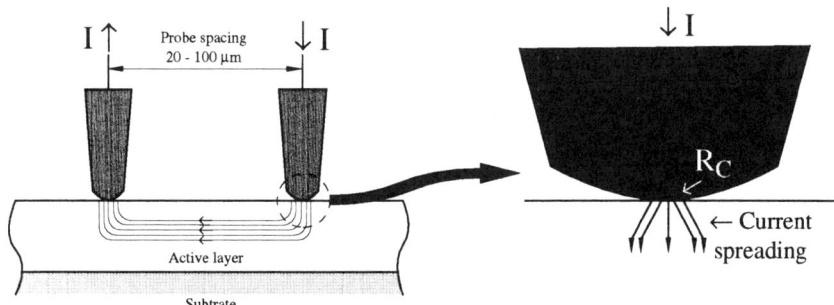

FIG. 8. Two-point probe, as used in spreading resistance profiling measurements. Metal probe–semiconductor contact is equivalent to contact resistance R_c and spreading resistance.

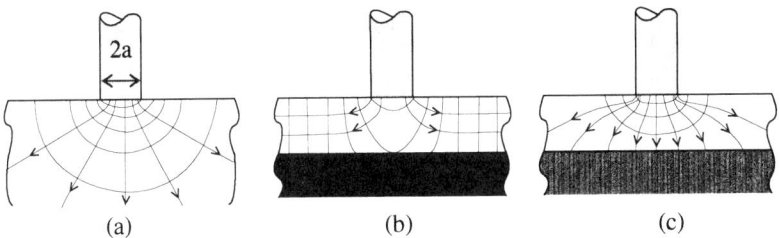

FIG. 9. Lines indicate the distribution of current obtained by injection of current through a cylindrical contact into (a) a semi-infinite homogeneous material, (b) an active layer on an underlying insulating substrate, and (c) an active layer on a conducting underlying substrate.

semiconductor resistance that, for reasonable spacing probe (from 20 to 100 μm) is negligible compared with the component of circuit resistance due to the constriction of the current and its subsequent spreading.

The major difficulty of spreading resistance analysis is to keep the contact resistance value R_c constant and reproducible. Therefore, the system must be periodically calibrated against known resistivity standards (provided by the National Bureau of Standards), and the probes must be periodically reconditioned. Calibration curves giving SR dependence upon semiconductor resistivity, and thus for p- and n-type materials, relative to a given probe spacing and contact radius must be accurately elaborated (Carver et al., 1987).

In the case of a semi-infinite homogeneous medium (Fig. 9(a)), the spreading resistance due to current constriction (and subsequent spreading) injected by a cylindrical form contact is given by

$$SR = \frac{\rho}{4a} \quad (15)$$

where ρ is the semiconductor resistivity and a is the radius of the circle.

In practice, a is not known exactly and is replaced by a scale parameter called the electrical contact radius, which depends on the probe impact on the characterized surface layer and is usually extracted from calibration curves as a function of the semiconductor resistivity and type $a(\rho)$. Besides, a correction factor must be introduced in Eq. (15) to account for the fact that spreading resistance measurements are not made on a semi-infinite homogeneous material, leading to — for the pair of probes — an expression

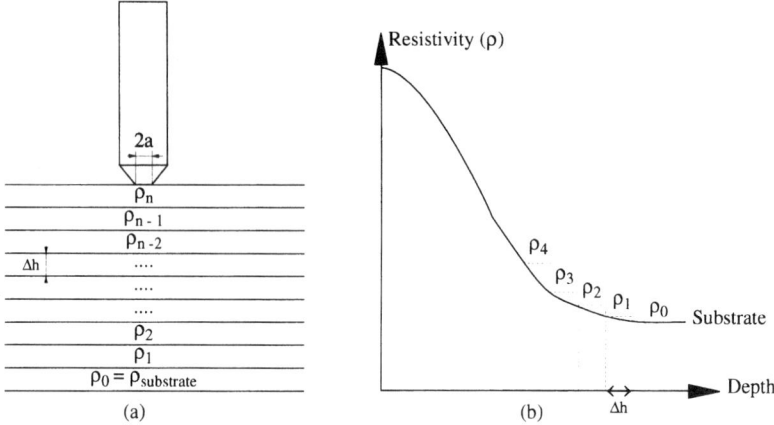

FIG. 10. Multilayer analysis for resistivity profile extraction from spreading resistance profiling measurements. (a) Substrate divided into Δh = thick sublayers. (b) Semiconductor resistivity in constant along each sublayer.

of the form (Mazure, 1986)

$$SR = \frac{\rho}{2a(\rho)} \cdot \text{C.F.} \tag{16}$$

Here, C.F. accounts for any and all aspects of the structure, such as resistivity gradients or layers boundaries (Fig. 9(b) and Fig. 9(c)), which are able to modify the response of the probes. Therefore, when a spreading resistance profile is obtained, each value is correlated with the local resistivity of the characterized semiconductors at this depth, and also with the mean resistivity of the residual layer. In order to extract a local resistivity value, a multilayer model has been used by several authors (Schumann and Gardner, 1969; Berkowitz, and Lux, 1981). The substrate is subdivided into n Δh-thick sublayers (Fig. 10(a)), and the semiconductor resistivity is supposed to be constant along each sublayer (Fig. 10(b)). The Laplace equation for the nth sublayer can be written as

$$\nabla\left(-\frac{1}{\rho_n}\nabla V_n\right) = 0 \tag{17}$$

By applying boundary conditions at each layer, we obtain

$$SR_n = \frac{\rho_n}{2a} \cdot \text{C.F.} \tag{18}$$

5   SHEET AND SPREADING RESISTANCE ANALYSIS   141

where SR_n is the nth measured spreading resistance value and C.F. is the correction factor given by a complex analytical expression:

$$\text{C.F.} = \frac{4}{\pi} \int_0^\infty (1 + 2\theta(t)) \cdot \left(\frac{J_1(t)}{t^2} - \frac{J_0\left(\frac{s}{a} \cdot t\right)}{2t} \right) \sin(t) \cdot dt \qquad (19)$$

where J_0 and J_1 are Bessel functions, s is the probe spacing, a is the electrical contact radius, t is the integration variable, and $\theta(t)$ is the integration factor taking into account the effect of the other sublayers, which can be calculated by a recurrence procedure defined by Choo et al., 1977. Other authors have resolved the Laplace equation under the assumption of multilayers with an exponential variation of the resistivity in the sublayers.

A different approach, usually referred to as local slope analysis, was proposed by Dickey and Ehrstein (1979) to evaluate this correction factor. An approximate expression for the correction factor is derived by considering two limiting cases that can be solved readily. Thus, for a thin layer implanted in a substrate of the opposite conductivity type (or on an insulating substrate),

$$\text{C.F.} = k_1 m \qquad (20)$$

where m is the logarithmic slope of the measured curve (spreading resistance) versus the depth (Fig. 11(a))

$$m = \frac{\Delta \log(SR)}{\Delta h} \qquad (21)$$

Likewise, for a thin layer on a conducting substrate (Fig. 11(b)),

$$\text{C.F.} = -\frac{k_2}{m} \qquad (22)$$

For intermediate cases, Dickey and Ehrstein proposed an interpolation function of the form

$$\text{C.F.} = \frac{k_1 m}{2} + \sqrt{k_1 k_2 + \left(\frac{k_1 m}{2}\right)^2} \qquad (23)$$

where $k_1 = (2a/\pi) \log(s/a)$ and $k_2 - 4/\pi \cdot a$, depending only on probe spacing

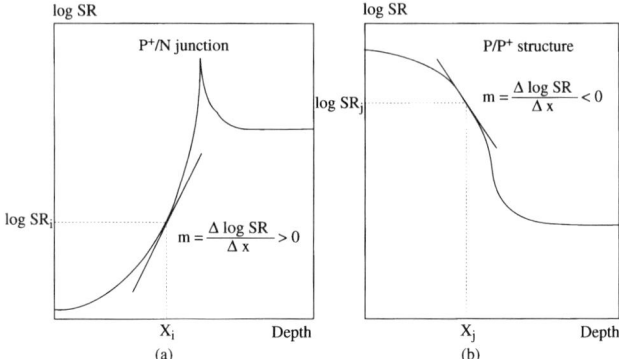

FIG. 11. Variation of spreading resistance (SR) logarithmic local slope for two limiting cases: (a) P^+/N junction and (b) P/P^+ structure characterized by positive and negative m values, respectively. m, logarithmic slope of the measured curve.

and electrical contact radius. Despite the apparent simplicity of this model, comparison between resistivity profiles calculated by Dickey and Ehrstein's procedure and those determined by the Schumann and Gardner approach are quite favorable.

Whatever algorithm is used the resistivity profiles so constructed provide information about local resistivity depth variations only. Because there is no way to measure the mobility and resistivity profiles independently, the doping concentration profiles are inferred from the resistivity profiles by assuming ideal variation of the mobility versus doping concentration.

Another critical feature of the spreading resistance are the artifacts that can arise when determining the junction location. Indeed, in the case in which the impurity is implanted or diffused in a substrate of the opposite type of conductivity, it is well known that carrier spilling occurs, and the junction location given by the spreading resistance profiling must be corrected in order to obtain metallurgical junction. André (1991) has proposed a simple method based on experimental data, whereas other authors have proposed a calculation of the Poisson equation to take this phenomenon into account.

In conclusion, despite the need for complex calculations, and for very skilled operators to obtain reliable profiles, spreading resistance measurements are currently used for the analysis of activation and redistribution of implanted dopants in VLSI devices. The fundamental strengths of the spreading resistance measurements lie in their ability to profile almost any combination of layers with very high resolution and in the fact that there are no real limitations on depth doping concentrations.

IV. Applications

In this section, we report on some results obtained by using sheet resistance measurements and spreading resistance profiling in order to investigate electrical properties of silicon-implanted samples issuing from two categories:

1. High-dose and high-energy arsenic-implanted samples annealed under various temperatures and durations.
2. Low-energy boron or boron fluorine–implanted layers in preamorphized silicon substrates.

Hereafter, for each study we present experimental details concerning sample preparation and sheet resistance or spreading resistance — or both — measurement conditions. Correlation between the obtained results and other methods of investigation is also given to obtain better understanding of both electrical doping activation and thermal recovery mechanisms occurring in these ion-implanted silicon samples.

1. High-Dose and High-Energy Arsenic-Implanted Silicon Layers

The frequent use of arsenic (As) as an n^+-type dopant in modern technologies for metal-oxide semiconductor field-effect transistor (MOSFET) source–drain areas and in bipolar emitters makes it imperative to properly understand the electrical activation of As as a function of ion implantation and subsequent thermal annealing conditions. Many investigations have been conducted to establish a correlation between electrical activation of implanted dopants and the reordering of lattice damage created by As^+ ion implantation after annealing (Nishi, Sakurai, and Furuya, 1978; Cerofolini *et al.*, 1984). For high-dose implantation, the amount of the induced damage is too large and may lead to the formation of a continuous amorphous layer at the silicon-implanted surface. This induced amorphous layer was shown to regrow epitaxially on the underlying crystalline substrate during annealing at temperatures of around 550°C. The study we report on was conducted in order to examine the relationship between this recrystallization process and As activation at such low-temperature annealing (Boussey-Said *et al.*, 1992).

Samples were prepared by implanting 10 to $20\,\Omega.\mathrm{cm}$ p-type Czochralski-grown Si $\langle 100 \rangle$ single crystal wafers with As at 200 keV, with doses ranging between 10^{13} and 10^{15} atoms/cm^2. All these implantations were done through a 50-nm-thick screen oxide to avoid As ions channeling in the silicon (Si) target, and to prevent out-diffusion of the implanted dopant during further thermal annealing. In addition, care was taken to reduce the implantation temperature to minimize in situ-induced defect annihilation. Isochronal annealing (1 h) was then done in a thermal furnace under flowing nitrogen from 300 to 1100°C. Isothermal treatments were also realized at 400, 420, and 450°C for various durations (15 to 900 min).

After thermal annealing, small pieces (5 × 3 mm) with coating oxide were cut for spreading resistance measurements. Depth profiles of resistivity were obtained using an ASR-100B spreading resistance system from Solid State Measurements, Inc. (U.S.A.) The samples were beveled by a mechanical grinding process using 0.25-μm diamond paste. The spreading resistance profiles were performed with lightweight probes (10 g) for better sensitivity and calibration with well-known standards. Data acquisition and resistivity calculations were conducted using an Apple® computer with Dickey's analysis-based algorithm. In addition, a second set of implanted wafers was realized under the same experimental conditions. Simply patterned 4-level masks were designed to provide both collinear four-point probes and square-shaped (with ohmic contacts at the periphery) configurations for the sheet resistance and Hall effect measurements.

First, cross-section XTEM analysis conducted on the as-implanted samples has shown the existence of an induced amorphous layer at the silicon target when dose implantation was higher than 2×10^{14} As$^+$/cm^2. In addition, the width of this amorphous layer was found to increase as the implantation dose was increased, in very good agreement with experimental data encountered in the literature. Finally, the recrystallization of this layer by solid-phase epitaxy was also confirmed to become significant at temperatures as low as 520°C. The regrowth velocity v, determined by measuring the average distance of the crystalline-amorphous (c − a) interface from the surface after each annealing, was found to satisfy the thermally activated relationship described by (Narayan, 1982)

$$v = v_\infty \exp\left(-\frac{E_a}{kT}\right) \quad (24)$$

with $v_\infty = 3 \times 10^8$ cm/s and $E_a = 2.68$ eV.

Figure 12 shows typical variation of the sheet resistance R_s with the annealing temperature T_a after an isochronal anneal (1 h) performed on As-implanted samples at 200 keV and various doses ranging from 10^{13} to 10^{15} As$^+$/cm^2. Figure 12 clearly shows that a strong modification of the material occurs at about 400°C, when the implanted dose is higher than the

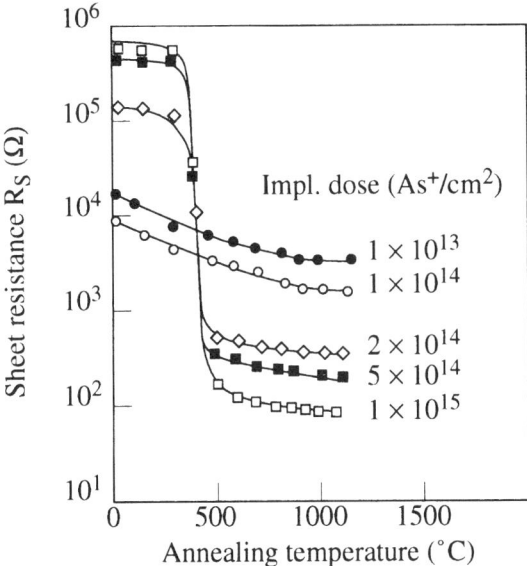

FIG. 12. Variations of sheet resistance as a function of the annealing temperature, as obtained by Van Der Pauw measurements performed on samples implanted with various doses of arsenic (As) at 200 keV and isochronally annealed for 1 h.

threshold amorphization value. The Hall mobility variations deduced from Hall effect measurements conducted on the same samples also present a transition from low to high mobility at temperatures ranging from 400 to 420°C, as depicted in Figure 13. It is worth noting that the annealing temperature at which μ_H and R_s change significantly is well below the amorphous-crystal transition temperature occurring by solid-phase epitaxy at around 520°C, as reported by XTEM analysis. In Figure 14 are represented the variations of the resistivity with depth, as obtained by spreading resistance measurements performed on samples implanted at 2×10^{14} As$^+$/cm^2 and isochronally annealed ($t = 1$ h, $300°C \leq T_a \leq 1100°C$). After annealing, the surface resistivity of the low-temperature annealed sample (400°C) decreases by more than 6 decades, although the surface layer remains amorphous, as evidenced by the corresponding XTEM analysis (Boussey-Said et al., 1992). Furthermore, note that despite the sharpness of the c − a interface revealed by XTEM, no discontinuity in the resistivity profiles appears in this region. Moreover, note that the dopant electrical activation does actually occur in the amorphous overlayer, while recrystallization is not yet completed. The same behavior has been reported for samples

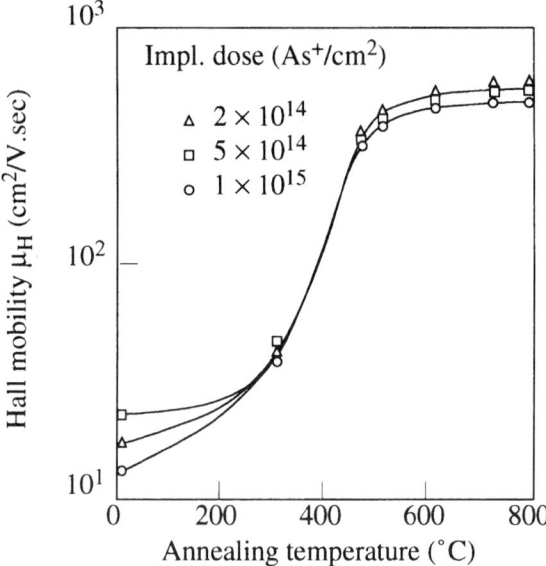

FIG. 13. Variations of the Hall mobility as a function of the annealing temperature, as obtained by Hall effect measurements performed on samples implanted with various doses of arsenic (As) at 200 keV and isochronally annealed for 1 h.

implanted with higher doses. For instance, in Figure 15 we can see the decrease of the surface resistivity occurring—in the amorphous side—just after annealing at 400°C for 1 h.

In order to analyze the effects of the duration of annealing on this mechanism low-temperature isothermal annealing has been conducted at various durations. Figure 16 shows resistivity profiles obtained by spreading resistance measurements from samples implanted at 2×10^{14} As$^+$/cm^2 and annealed at 450°C at various durations. We note that the surface resistivity reduction is efficient only after 15 min of annealing whereas deep diffusion effects are visible for long annealing duration ($t > 900$ min). The same behavior is found by analyzing spreading resistance profiles obtained with higher doses (5×10^{14} As$^+$/cm^2) after isothermal annealing at 420°C and 400°C, as can be seen in Figures 17(a) and 17(b), respectively.

The previous results lead us to conclude that a significant decrease in surface resistivity occurs at an annealing temperature of around 420 to 430°C, mainly because of the electrical activation of the majority of the As atoms within a highly damaged (actually amorphous) surface region (120-nm thick in the case of the samples shown in Figure 14). This suggests that

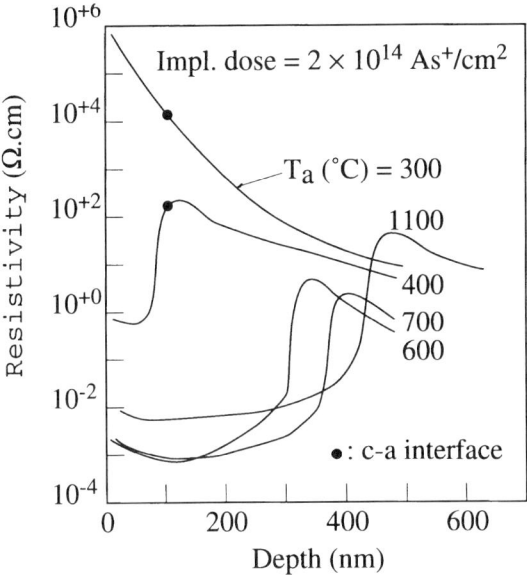

FIG. 14. Resistivity profiles, as obtained by spreading resistance measurements performed on $2 \times 10^{14}\,\mathrm{As^+/cm^2}$, 200 keV implanted samples and isochronally annealed for 1 h at various temperatures. T_a, annealing temperature; As, arsenic; c-a, crystalline-amorphous. (Reprinted from Boussey-Said, et al. (1992). American Institute of Physics, with permission.)

the electrical activation of the impurity takes place in both the crystalline and the amorphous regions via a local reconstruction process in which the material resistivity at a depth x, $\rho(x)$, is described by exponential kinetics as a function of the annealing time t_a (Boussey-Said et al., 1992; Christofides, Ghibaudo, and Jaouen, 1989):

$$\rho(x, t_a) = \rho_f(x) + [\rho_i(x) - \rho_f(x)] \cdot \exp[-t_a/\tau(x)] \quad (25)$$

where $\rho_f(x)$ and $\rho_i(x)$ are, respectively, the final and initial values of $\rho(x)$, and $\tau(x)$ is the relaxation time constant of the local electrical activation process at depth x.

In practice, $\rho_f(x)$ is given by the resistivity profile of a well-annealed sample and $\rho_i(x)$ by that of a nonannealed sample. In this context, $\tau(x)$ can be represented by an expression of the form

$$\tau(x) = \frac{\varepsilon k T_a}{4\pi q^2} \cdot \frac{1}{D(x)N(x)} \quad (26)$$

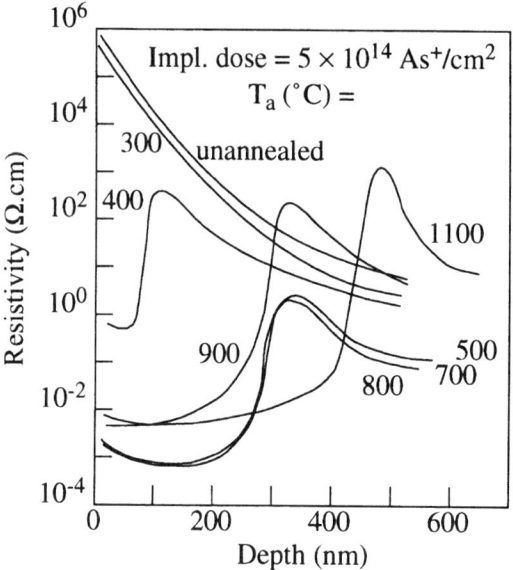

FIG. 15. Resistivity profiles, as obtained by spreading resistance measurements performed on $5 \times 10^{14}\,\mathrm{As^+/cm^2}$, 200 keV implanted samples and isochronally annealed for 1 h at various temperatures. T_a, annealing temperature; As, arsenic.

with

$$D(x) = D_0(x) \exp\left(-\frac{E_a(x)}{kT_a}\right) \quad (27)$$

where $D(x)$ and $N(x)$ are, respectively, the diffusivity and concentration of the point defects (or atom-defect complexes) ensuring the electrical activation of the impurity, ε is the dielectric constant, q is the absolute electron charge, and k is the Boltzmann constant.

The activation energy $E_a(x)$ has been experimentally determined by a proper analysis of the Arrhenius plot of the reduced resistivity defined as

$$Q_\rho(T_a) = \frac{\rho_i(x) - \rho_f(x)}{\rho(x, T_a) - \rho_f(x)} \quad (28)$$

For example, Figure 18 gives the variations of the activation energy and the relaxation time constant, E_a and τ, with depth as obtained by applying this analysis to the samples implanted at $5 \times 10^{14}\,\mathrm{As^+/cm^2}$. E_a is found to be weakly depth-dependent, while always smaller than 0.8 eV and almost

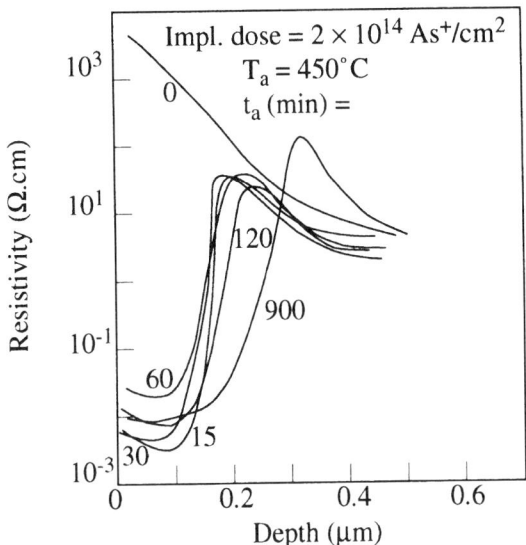

FIG. 16. Resistivity profiles, as obtained by spreading resistance measurements performed on 2×10^{14} As$^+$/cm^2, 200 keV implanted samples and isothermally annealed at 450°C for various durations. T_a, annealing temperature; t_a, annealing time; As, arsenic. (Reprinted from Boussey-Said, J., et al. (1992). American Institute of Physics, with permission.)

independent of annealing duration (Fig. 18(a)), therefore confirming the consistency of the kinetic model of Eqs. (25)–(28).

Likewise, the variation of the relaxation time with depth shows that $\tau(x)$ essentially increases when going from the surface toward the substrate (Fig. 18(b)). This feature is also consistent with the adopted local relaxation process model in which the relaxation time constant is inversely proportional to the diffusivity and concentration of the point defects involved in the dopant electrical activation.

Moreover, because $\tau(x)$ increases experimentally with depth and E_a is almost constant with x, we can conclude from Eq. (26) that (i) the diffusing species is probably the same throughout the implanted layer; (ii) the comparison of the values of the activation energy of Figure 18(a) with the migration of usual point defects (interstitials, vacancies, and their complexes formed with As ions) indicates that the diffusing species at the origin of the activation process should be related to vacancy or diffusion of interstitial atoms, or both; and (iii) the density of these complexes, $N(x)$, decreases with depth x, because D is approximately constant with x. This result confirms that the density of defects decreases with x, as has been proved by X-ray triple crystal diffraction analysis conducted on these samples (Boussey-

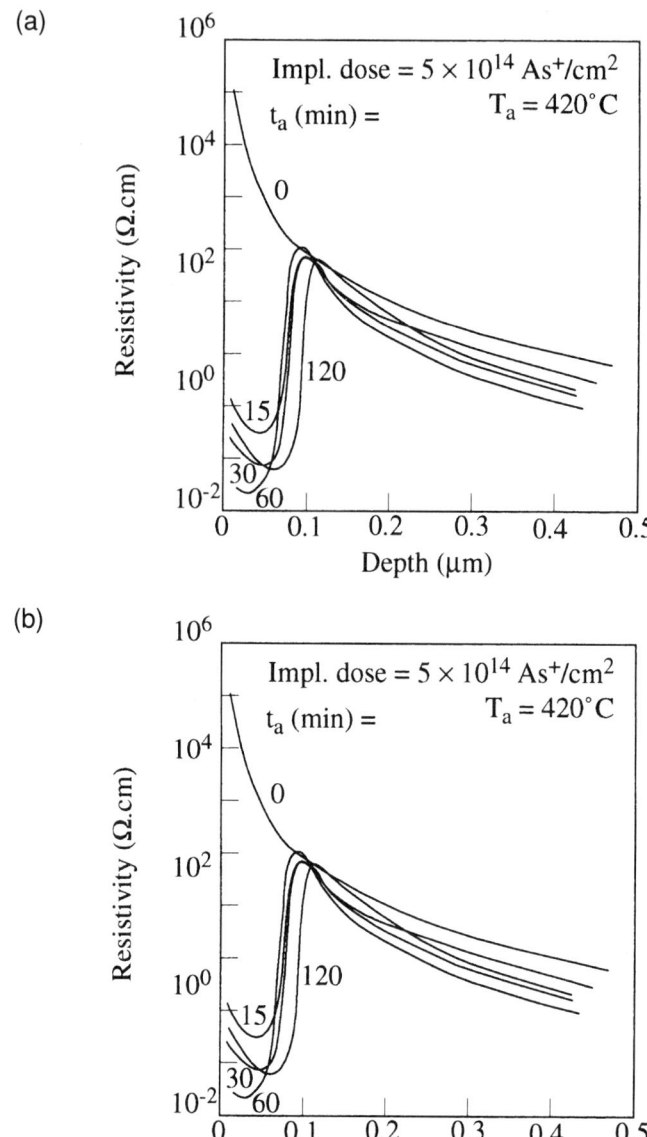

FIG. 17. Resistivity profiles, as obtained by spreading resistance measurements performed on 5×10^{14} As$^+$/cm^2, 200 keV implanted samples and isothermally annealed for various durations at (a) 420°C and (b) 400°C. T_a, annealing temperature; t_a, annealing time; As, arsenic.

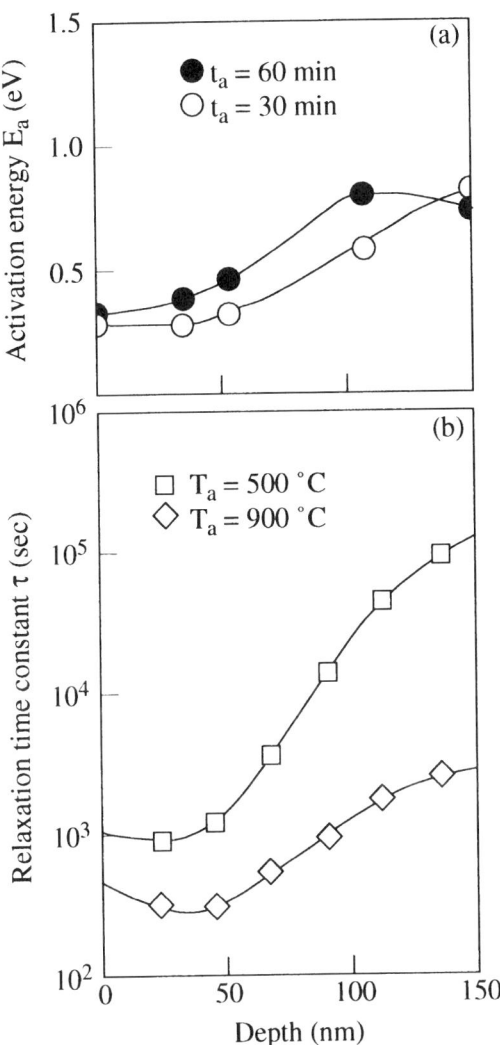

FIG. 18. Variations of (a) activation energy E_a and (b) relaxation time constant τ versus depth, as determined by the local relaxation model for samples implanted with 200 keV, 2×10^{14} As$^+$/cm^2, T_a, annealing temperature; t_a, annealing time; As, arsenic. (Reprinted from Said et al. Nuclear Instruments and Methods in Physics Research B55. (1991). Elsevier-Science, The Netherlands, pp. 576–579, with permission.)

Said, 1992) and predicted by Monte Carlo calculations (Servidori et al., 1987).

In conclusion, the main result of this study is to confirm the existence of a thermally activated electrical activation process taking place within the highly damaged Si layer induced by high-dose ion implantation. This phenomenon is found to be completely independent of the reconstruction mechanism consisting of the propagation of a crystallization front from the substrate toward the surface having a higher activation energy. A proper analysis of resistivity profiles after isochronal and isothermal annealing allowed us to describe this process by a local relaxation model in which the migration of point defects may constitute the prevailing factor of the electrical impurity activation. Finally, this study reinforces the fact that spreading resistance measurements could partially answer crucial questions about the depth distribution of the ion-implantation-induced defects and their behavior under thermal treatment.

2. Low-Energy Boron and Boron Fluoride Implantation in Preamorphized Silicon Substrate

In the fabrication of very large scale integrated (VLSI) circuits, the problem of forming ultrashallow ($<0.1\ \mu m$) and abrupt P^+/N junctions with low-energy boron (B) or boron fluoride (BF_2) ion implantation and conventional furnace or rapid thermal annealing is well known. For P-channel devices used in complementary metal-oxide semiconductor (CMOS) structures — because of their weak nuclear stopping power in Si — implantation of B or BF_2 into source–drain regions is accompanied by a channeling effect, which leads to increased penetration of the implanted profile (Wach and Wittmaack, 1982). In addition, the high diffusivity of B in Si limits the temperature and duration of postimplantation annealing, which makes it difficult to achieve complete electrical activation or complete removal of implantation damage (Cho et al., 1985). It is now well established that the previously mentioned drawbacks can be avoided by amorphizing the Si substrate surface before dopant implantation. This preamorphization stage is easily accomplished by implanting a high dose of Ge ions (an electrically inactive impurity) under well-defined conditions of temperature and energy. The benefits are twofold: (1) B channeling can be suppressed and (2) good electrical activity can be obtained through the solid-phase epitaxial regrowth of the layer during further thermal processing. However, after solid-phase epitaxial regrowth, extended defects are formed beneath the former c–a interface, where the space charge region of the junction may extend, causing very high reverse-bias leakage current.

These defects, usually called end-of-range defects, have been identified as dislocation loops formed by the agglomeration of Si interstitials atoms (due to recoil), left after the bombardment beneath the c–a interface. Furthermore, it has been established that during postimplantation thermal annealing, these dislocation loops release Si interstitials in the doped region, which might influence dopant diffusivity (De Maudit et al., 1994).

The aim of the following experimental investigation is to show how sheet and spreading resistance measurements (well combined with SIMS and XTEM analysis) have been used to illustrate the previously mentioned advantages of Si substrate preamorphization for ultrashallow P^+/N junction formation and to determine the electrical effects of the EOR defects induced by the preamorphization on the concentration of free carriers.

The wafers used in this study are n- and p-type Czochralski-grown silicon $\langle 100 \rangle$ single crystals, with a dopant concentration on the order of $10^{15}\,\text{cm}^{-3}$. The preamorphization steps were performed using Ge^+ ion implantation, through a thin thermal screen oxide (6 nm) with a dose from 2×10^{14} to $2 \times 10^{15}\,Ge^+/\text{cm}^2$ and an energy range between 10 and 150 keV. The thickness of the as-obtained amorphous layers was then measured by either Rutherford Backscattering spectrometry (RBS) or XTEM techniques. Afterward, a p-type dopant was introduced by $^{11}B^+$ or $^{49}BF_2^+$ ion implantation at low energy (3 or 6 keV in the case of B and 15 keV for BF_2) and various doses (from 10^{14} to $10^{15}\,\text{cm}^{-2}$).

The efficiency of the preamorphized surface layer for suppression of channeling tails was then investigated by SIMS profiling as a function of the preamorphization Ge^+ ion energy. For example, this is illustrated in Figure 19 in which are plotted B concentration profiles obtained by 6 keV, $5 \times 10^{14}\,B^+/\text{cm}^2$ implantation in various premorphized substrates. The location of the c–a boundary determined by either the RBS or the XTEM technique is reported in each profile. From this it is easy to see that channeling tails completely disappear with preamorphization energies higher than 50 keV.

Subsequent to this, pieces of all these wafers underwent rapid thermal annealing (RTA) (temperature: 800 to 950°C; duration: 15 sec. to a few minutes under nitrogen ambient) and were analyzed by SIMS, RBS, or XTEM techniques, or by sheet resistance or spreading resistance measurement techniques. Sheet resistance measurements have been conducted, after removal of the screen oxide, with a collinear four-point probe configuration using Prometrix Rs50/e sample mapping equipment. Such an automated investigation was necessary for this study in order to check the homogeneity of the low-energy beam scattering of the B^+ ions on the whole surface of the wafer during implantation. Indeed, when very light ions are weakly accelerated (less than 6 keV in the case of boron), specific precautions must

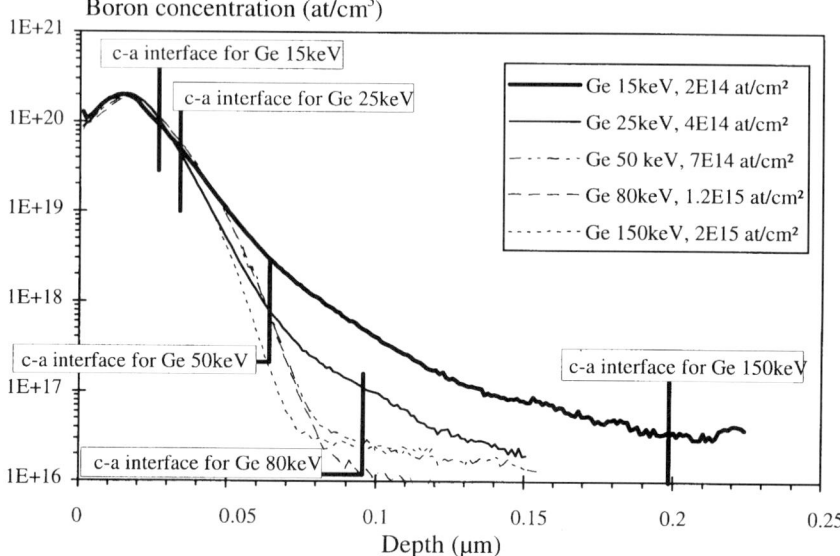

FIG. 19. Secondary ion mass spectroscopy (SIMS) profiles obtained from 6 keV, 5×10^{14} B^+/cm^2 implanted wafers after preamorphization at various doses and energies. The crystalline-amorphous (a-c) interface is reported fore each profile. B, boron; Ge, Germanium. (Reprinted from Minondo, M., et al. (1995). ESSLRC '95. Frontières Editions, pp. 379–382 with permission.)

be taken to keep the scattering beam of the B^+ ions convergent. Finally, the Si wafers used for these resistance mapping measurements were all n-type to minimize the effects of the leakage of current.

Spreading resistance profiles were obtained with the same equipment of the previous study; however, raw spreading resistance data were inverted into resistivity profiles using both Dickey's local slope approach and Schumann and Gardner (1969), multilayer analysis, with a specific additional correction procedure taking into account a bevel effect and the variation of the electrical radius contact with the p-type concentration (Minondo, Roche, and Jaussaud, 1994). There is good agreement between experimental curves obtained by both methods. Finally, spreading resistance measurements were done on a p^+/p-type structure to avoid the carrier spilling effect and to often better knowledge of the depth of the p^+ layer.

a. Electrical Activation of p-Type Dopants

Typical variations of the sheet resistance of 3 keV, $10^{15}/cm^2$ B-implanted samples in various preamorphized Si substrates are reported in Figure 20

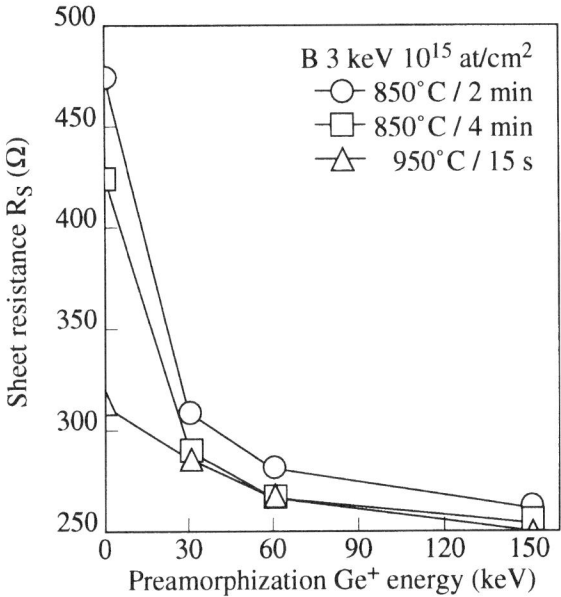

FIG. 20. Sheet resistance variations versus preamorphization Ge$^+$ ion energy, as obtained from 3 keV, B$^+$ 10^{15} at/cm^2 implanted samples that were annealed under several rapid thermal annealing conditions. B, boron.

for three rapid thermal annealing (RTA) experiments. The experimental curves show an important decrease of the sheet resistance as the energy of the Ge ions increases. In addition, when the implanted samples were annealed under equivalent temperature and duration conditions, that is,

$$\sqrt{D_1(T_1 = 850°C) \cdot t_1(=4\,\text{min})} = \sqrt{D_2(T_2 = 950°C) \cdot t_2(=15\,\text{s})}$$

where D is diffusivity, no significant difference between the obtained sheet resistance values was noted, except in the case of the crystalline substrate (Ge$^+$ ion energy = 0) in which the best results are obtained after RTA at 950°C for 15 sec.

If an equivalent amount of B is implanted in the same preamorphized substrate using BF$_2$ molecules accelerated at 15 keV, the sheet resistance also decreases with increasing preamorphized Ge$^+$ ion implantation energy (Fig. 21). However, in comparison with the case of single B implantation, two fundamental differences exist. The first concerns the dopant activation in a nonpreamorphized substrate in which lower sheet resistance values are

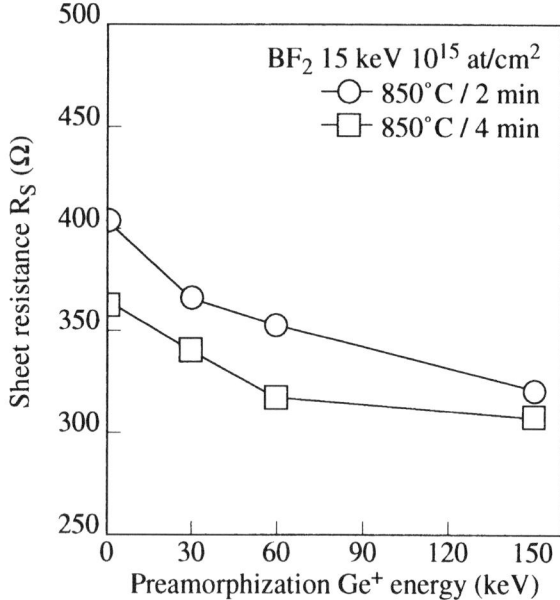

FIG. 21. Sheet resistance variation versus preamorphization Ge$^+$ ion energy, as obtained from 15 keV, 10^{15} BF$_2$/cm^2 implanted samples that were annealed under several rapid thermal annealing conditions. BF$_2$, boron fluoride.

obtained after BF$_2$ implantation. This can be attributed to the fact that BF$_2$ implantation partially amorphizes the Si substrate, reducing — in this manner — the channeling phenomena, in comparison with direct B implantation into completely crystalline substrate. Analysis by XTEM confirms this feature, showing an amorphous layer about 30-nm deep after BF$_2$ implantation in a bare crystalline Si substrate. The second difference is illustrated by better electrical activation, in preamorphized substrates, of B over BF$_2$. This probably is due to the presence of F in the implanted layer, which may influence the doping redistribution during thermal annealing (Fan, Parks, and Jaccodine, 1990).

The activation efficiency, usually expressed by the ratio of the electrically active dose to the implanted one (given by SIMS results) was calculated and plotted as a function of annealing and preamorphization conditions. Using the sheet resistance values shown in Figure 20, the corresponding electrically active dose was extracted from Smith and Stephen (1972) abacus and the electrical activation ratio variations are plotted in Figure 22 for a 3 keV, 10^{15} B$^+$/cm^2 implantation. Enhancement of the electrical activity of B by the preamorphization steps is easily seen. It is worth noting that full

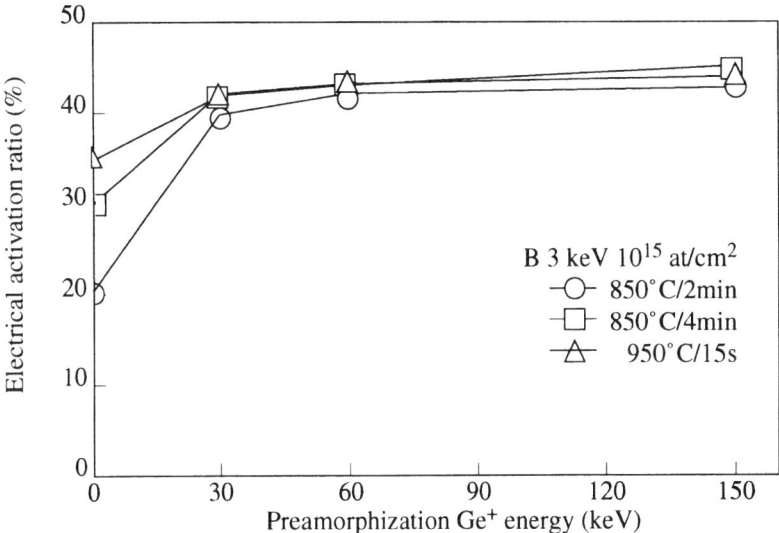

FIG. 22. Activation efficiency (electrical activation ratio), as obtained from annealed samples presented in Figure 20. B, boron.

electrical activation could not be obtained in this case because the B concentration—at the corresponding temperature—was higher than the solid solubility of B in Si. Nevertheless, the exhaustive study (Minondo et al., 1993) conducted on all these samples demonstrates that, in the case of BF_2 implantation, the activation ratio never reach a value of 100%. This probably is because of the behavior of F during thermal annealing, which is not very well understood.

In conclusion, the sheet resistance wafer mapping conducted on low-energy B- or BF_2 implanted samples effectively confirmed a high electrical activation ratio of dopants when the Si substrate is subjected to a preamorphization step before introducing p-type dopants. This conclusion is in agreement with the earlier works of Gibbons (1972), which reported the possibility of reaching 100% of electrically active dopants after ion implantation at doses higher than the amorphization threshold value.

b. *Enhanced diffusion Induced by End-of-Range Defects*

Free carrier profiles obtained by spreading resistance measurements and the corresponding B profiles obtained by SIMS from 3 keV, $10^{15}/cm^2$ B-

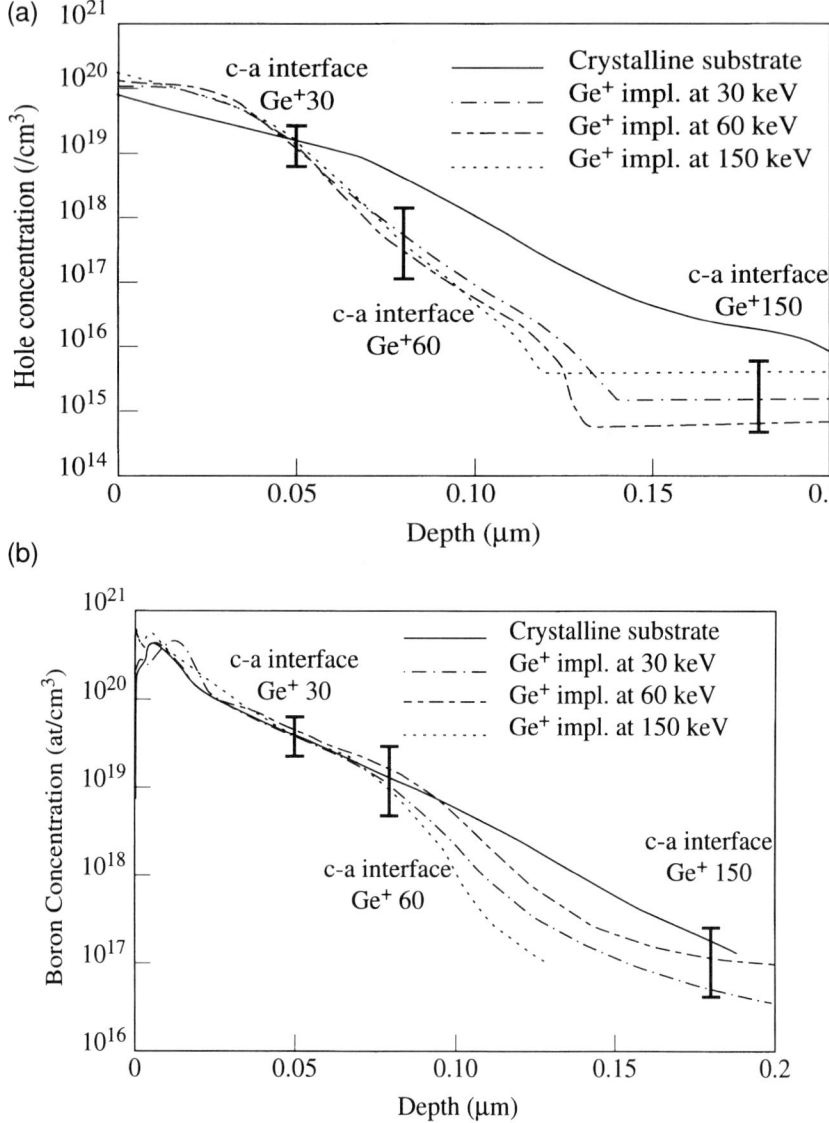

FIG. 23. Hole (a) and boron (b) concentration profiles as obtained by spreading resistance analysis and secondary ion mass spectroscopy (SIMS) profiling, respectively. Samples were doped with a 3 keV, 10^{15} at/cm^2 boron implantation in crystalline and preamorphized substrates with 10^{15} Ge$^+$/cm^2 at various energies and annealed at 850°C for 4 min. Ge, germanium; c-a, crystalline-amorphous.

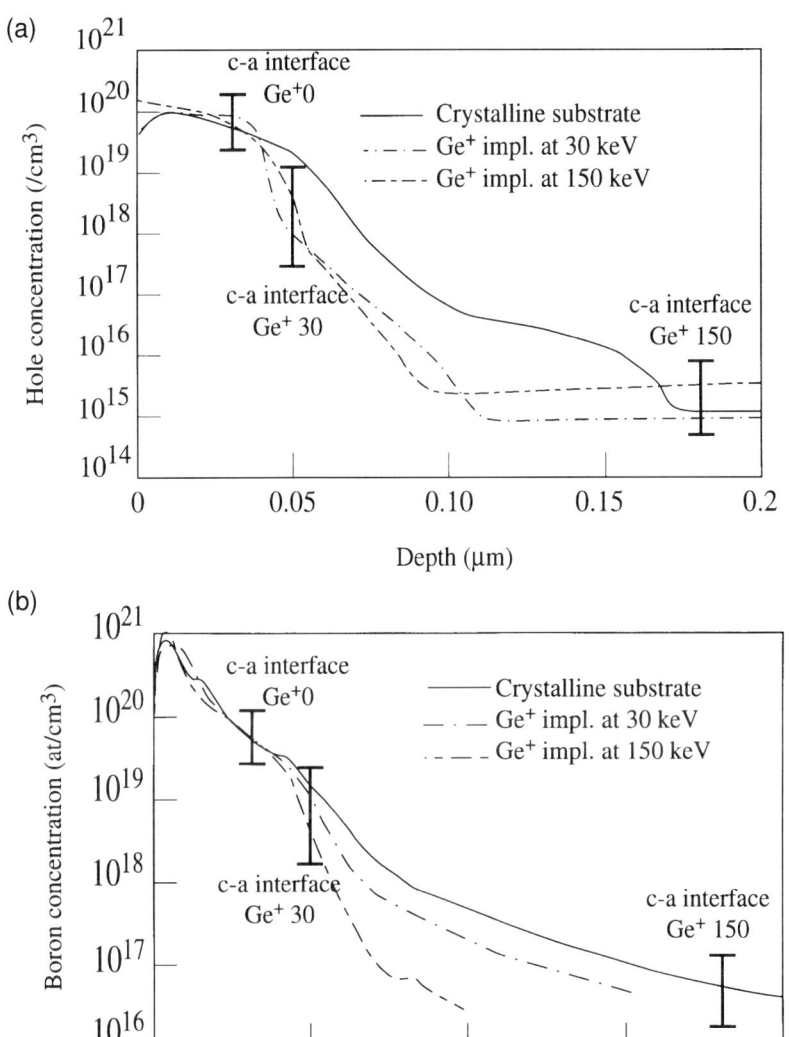

FIG. 24. Hole (a) and boron (b) concentration profiles, as obtained by spreading resistance analysis and secondary ion mass spectroscopy (SIMS) profiling, respectively. Samples were doped with a 15 keV, 10^{15} at/cm^2 boron fluoride implantation in crystalline and preamorphized substrates with 10^{15} Ge$^+$/cm^2 at various energies and annealed at 850°C for 4 min. impl., implanted; c-a, crystalline-amorphous.

implanted samples and annealed at 850°C for 4 min are shown in Figures 23(a) and 23(b), respectively. One can clearly observe the influence of the Ge$^+$-implantation energy with respect to the electrical junction depth (0.14 µm for crystalline substrate as opposed to 95 nm for preamorphized substrates at a concentration value of 10^{17} cm^{-3}). The most important information given by this spreading resistance analysis is the excellent electrical activation of B at the surface. In fact, a maximum hole concentration of only 8×10^{19}/cm^3 is measured in the case of crystalline substrate, whereas preamorphized samples, with similar doping parameters, lead to a hole concentration higher than 10^{20}/cm^3, which is nearly equal to the limit of the solid solubility of B in Si at 850°C. This feature confirms the ability of the preamorphized substrate to offer better electrical activation after recrystallization, as was concluded from the sheet resistance variations study.

In Figure 24, hole and B concentration profiles are plotted, as obtained by spreading resistance measurements and SIMS analysis from BF$_2$-implanted samples at 15 keV, 10^{15}/cm^2, in crystalline and preamorphized substrates, with annealing at 850°C for 4 min. In comparison with the curves of Figure 23, the following three features are pointed out:

1. The reduction of the junction depth is more efficient when the dopants are introduced by BF$_2$ implantation. For example, at a concentration of 10^{18} cm^{-3}, the junction depth is reduced by more than 25% when the doping is performed by BF$_2$ implantation in Si substrate with a 50-nm-thick-induced amorphous layer on its surface.
2. Boron diffusion seems to be enhanced by the former c–a interface only when this boundary is located close to the dopant profile. This is clearly observable in the case of the substrate preamorphized at 60 keV Ge$^+$ ions implantation energy shown in Figure 23(b), where the B profile is significantly shifted from the corresponding c–a interface depth toward the substrate. The same behavior is reported in Figure 24(b) for BF$_2$-implanted samples after a 30 keV, 1×10^{15} Ge$^+$/cm^2 preamorphization step. This enhanced diffusion of B has been studied by several authors (Solmi and Barufaldi, 1991; Marou et al., 1991) and seems to be the consequence of the Si interstitials released during thermal annealing by the dislocation loops located at the former c–a boundary. In this context, the additional information provided by the spreading resistance profiling of such samples (Figures 23(a) and 24(a)) is that the B atoms, which have been enhanced, do not seem to be electrically active. Indeed, the hole concentration profiles shown in Figure 23(a) are almost identical to those of the substrates preamorphized at 150, 60 and 30 keV, Ge$^+$ ion implantation energies, indicating that the EOR defects probably act as recombination centers

during RTA. This assumption has been reinforced by electrical noise analysis conducted on P^+/N shallow diodes elaborated in a preamorphized Si substrate (Minondo, Boussey-Said, and Kamarinos, 1995).
3. Finally, a slight difference is pointed out when the dopant is BF_2, especially in the case of 30 keV, Ge^+ ion implantation energy where the amount of electrically active dopant seems to be reduced inside the former c–a interface region. The presence of F in this case is assumed to be responsible for this effect.

In conclusion, although the spreading resistance analysis reported in this study was very useful in investigating the advantages of the preamorphization stage in terms of channeling depth reduction and electrical activation, combination with other characterization techniques (e.g., SIMS, RBS, Noise, and fluctuations) is still needed to correctly achieve a complete experimental investigation of the annihilation kinetics of such ion-implantation-induced defects.

V. Summary

The techniques of sheet and spreading resistance measurements have been reviewed and several examples of their use for the characterization of ions implanted in Si samples have been reported. First, the principles of the conventional four-point probe method have been examined and both collinear and arbitrarily placed four-point probe sheet resistance measurement techniques have been described.

Next, the spreading resistance profiling technique has been presented with emphasis on the practical precautions to be taken in order to maintain reproducible probe contact parameters. Extraction procedures of resistivity profiles (and free carriers concentration) from spreading resistance data also have been discussed and some inherent profiling artifacts pointed out.

Finally, experimental results based on the use of these electrical characterization methods have been reported. We demonstrated that a good combination of such electrical analysis with other structural characterization methods can offer a highly reliable tool for the determination of the electrical signature of ion-implantation-induced damage. For instance, in the case of high-dose, high-energy As-implanted samples annealed under various temperatures and durations, spreading resistance measurements associated with XTEM investigations have allowed us to point out the existence of a thermally activated low-temperature electrical activation phenomenon occurring independently of the recrystallization by solid-phase epitaxial regrowth. B-enhanced diffusion also has been shown to occur in low-energy

B- or BF_2-implanted samples when the Si substrate is preamorphized, by implantation of Ge ions before dopants are introduced.

ACKNOWLEDGMENTS

The author thanks Dr Gérard Ghibaudo for his helpful assistance in revising this manuscript, and Miss Martine Gri for drawing the figures.

REFERENCES

Albers, J., and Berkowitz, H. L. (1985). "An Alternative Approach to the Calculation of Four Probe Resistances on Nonuniform Structures," *J. Electrochem. Soc.* **132**, 2453–2456.

André, E. C. (1991). "Junction Location Measured by Spreading Resistance Probe," *Jap. J. Appl. Phys.* **30**, 1511–1514.

Battaglia, A., and Campisano, S. U. (1993). "Mechanism of Amorphization in Crystalline Silicon," *J. Appl. Phys.* **74**(10), 6058–6061.

Berkowitz, H. L., and Lux, R. A. (1981). "An Efficient Integration Technique for Use in the Multilayer Analysis of Spreading Resistance Profiles," *J. Electrochem. Soc.* **128**, 1137–1141.

Boussey-Said, J., Ghibaudo, G., Jaoven, H., Stoemenos, Y., and Zaumseil, P. (1992). "Electrical and Structural Properties of Silicon Layers Heavily Damaged by Ion Implantation," *J. Appl. Phys.* **72**(1), 61–68.

Carver, G. P., Kang, S. S., Ehrstein, J. R., and Novotny, D. B. (1987). "Well Defined Contacts Produce Accurate Spreading Resistance Measurements," *J. Electrochem. Soc.* **134**, 2878–2882.

Cerofolini, G. F., Meda, L., Queirolo, G., Armigliato, A., Solmi, I., Nava, F., and Ottaviani, G. (1984). "Damaged and Reconstructed Regions in Silicon after Heavy Arsenic Implantation," *J. Appl. Phys.* **56**(10), 2981–2893.

Cho, K., Allen, W. R., Finstad, T. G., Chu, W. K., and Wortmann, J. J. (1985). "Channeling Effect for Low Energy Ion Implantation in Silicon," *Nucl. Instrum. Meth. Phys. Res.* **B7-8**, 265–272.

Choo, S. C., Leong, M. S., Hong, H. L., Li, L., and Tan, L. S. (1977). "A Multilayer Correction Scheme for Spreading Resistance Measurements," *Solid-State Electron.* **20**, 839–848.

Christofides, C., Ghibaudo, G., and Jaouen, H. (1989). "Electrical Transport Investigation of Arsenic Implanted Silicon. II. Annealing Kinetics of Defects," *J. Appl. Phys.* **65**(12), 4840–4844.

Chwang, R., Smith, B. J., and Crowell, C. R. (1974). "Contact Size Effects on the Van Der Pauw Method for Resistivity and Hall Coefficient Measurements," *Solid-State Electron.* **17**, 1217–1227.

De Maudit, B., Laânab, L., Bergaud, C., Faye, M. M., Martinez, A., and Claverie, A. (1994). "Identification of EOR Defects Due to the Regrowth of Amorphous Layers Created by Ion Bombardment," *Nucl. Instrum. Meth. Phys. Res.* **B84**, 190–194.

Dennis, J. R., and Hale, E. B. (1977). "Crystalline to Amorphous Transformation in Ion Implanted Silicon: A Composite Model," *J. Appl. Phys.* **49**(3), 1119–1127.

Dickey, D. H., and Ehrstein, J. R. (1979). "Semiconductor Measurement Technology: Spreading Resistance Analysis for Silicon Layers with Nonuniform Resistivity," *National Bureau of Standards Special Publication 400-48, PB 296 265*, pp. 1–65.

5 SHEET AND SPREADING RESISTANCE ANALYSIS

Ehrstein, J. R. (1979). "Two-Probe Spreading Resistance Measurements for Evaluation of Semiconductors and Devices." In: *Nondestructive Evaluation of Semiconductor Materials and Devices*, (J. N. Zemel, ed.), Plenum Press, New York, pp. 1–65.

Fan, D., Parks, J. M., and Jaccodine, R. J. (1990). Effects of Fluorine on the Diffusion of Through-Oxide Implanted Boron in Silicon," *Appl. Phys. Lett.* **59**, 1212–1214.

Gibbons, J. F. (1972). "Ion Implantation in Semiconductors—Part II: Damage Production and Annealing," *Proc. IEEE* **60**(9), 1062–1096.

Marou, F., Claverie, A., Salles, P., and Martinez, A. (1991). "The Enhanced Diffusion of Boron in Silicon after High Dose Implantation and during Rapid Thermal Annealing," *Nucl. Instrum. Meth. Phys. Res.* **B55**, 655–660.

Mazure, R. G. (1986). "Doping Profiles by the Spreading Resistance Technique." In *Microelectronics Processing*, (L. A. Casper, ed.), *American Chemical Society Symposium*, **295**, 34–48.

Minondo, M., Boussey-Said, J., Kamarinos, G. (1995). "Preamorphization Induced Defects and their Effects on the Electrical Properties of Shallow P^+/N Junctions," *Proceeding of the 25th European Solid State Device Research Conference, ESSDERC '95*, (H. C. De Graaff, and H. Van Kraneburg, eds.), Editions Frontières, 379–382.

Minondo, M., Jaussaud, C., Roche, D., Van Der Meulen, P., and Mehta, S. (1993). "Comparison of Boron and Boron Fluorine Implantation at Low Energy in Germanium Preamorphized Silicon," *Nucl. Instrum. Meth. Phys. Res.* **B74**, 539–543.

Minondo, M., Roche, D., and Jaussaud, C. (1994). "Characterization of Ultra-Shallow P^+ Profiles by Spreading Resistance Measurements," *Jap. J. Appl. Phys.* **33**(5), 2439–2443.

Narayan, J. (1982). "Interface Structures during Solid Phase Epitaxial Regrowth in Ion Implanted Semiconductors and a Crystallization Model," *J. Appl. Phys.* **53**(12), 8607–8614.

Narayan, J, and Holland, O. W. (1984). "Characteristics of Ion Implanted Damage and Annealing Phenomena in Semiconductors," *J. Electrochem. Soc.* **131**, 2651–2662.

Nishi, H., Sakurai, T., and Furuya, T. (1978). "Electrical Activation of Implanted Arsenic in Silicon during Low Temperature Anneal," *J. Electrochem. Soc.* **125**, 461–466.

Prussin, S., David, I. M., and Tauber, R. N. (1982). "The Nature of Defect Layer Formation for Arsenic Ion Implantation," *J. Appl. Phys.* **54**(5), 2316–2326.

Schumann, P. A., and Gardner, E. E. (1969). "Spreading Resistance Correction Factors," *Solid-State Electro.* **12**, 371–375.

Servidori, M., Zaumseil, P., Winter, U., Cemballi, F., and Mazzone, A. M. (1987). "Defect Distribution in Ion Implanted Silicon: Comparison between Monte Carlo Simulation and Triple Crystal X-Ray Measurements," *Nucl. Instrum. Meth. Phys. Res.* **B22**, 497–499.

Smith, B. J., and Stephen, J. (1972). "Computer Calculations of Sheet Resistance of n- and p-Type Implantations in Silicon," *Radiation Effects* **14**, 181–184.

Solmi, S. and Barufaldi, F. (1991). "Diffusion of Boron in Silicon during Post-Implantation Annealing," *J. Appl. Phys.* **69**, 2135–3258.

Van Der Pauw, L. J. (1958a). "A Method of Measuring Specific Resistivity and Hall Effects of Discs of Arbitrary Shape," *Philips Research Report*, **13**, 1–9.

Van Der Pauw, L. J. (1958b). "A Method of Measuring the Resistivity and Hall Coefficient on Lamellae of Arbitrary Shape," *Philips Tech. Rev.* **20**, 220–224.

Wach, W., and Wittmaack, K. (1982). "Low Energy Range Distribution of Boron in Amorphous and Crystalline Silicon," *Nucl. Instrum. Meth. Phys. Res.* **B194**, 113–116.

Wieder, H. H. (1979). "Four Terminal Nondestructive Electrical and Galvanomagnetic Measurements." In: *Nondestructive Evaluation of Semiconductor Materials and Devices*, (J. N. Zemel, ed.), Plenum Press, New York, pp. 67–104.

Ziegler, J. F. (1992). *Ion Implantation Physics, Handbook of Ion Implantation Technology*, (J. F. Ziegler, ed.), Elsevier Science, Amsterdam, pp. 1–68.

CHAPTER 6

Studies of the Stripping Hall Effect in Ion-Implanted Silicon

M. L. Polignano and G. Queirolo

SGS-THOMSON MICROELECTRONICS
AGRATE BRIANZA, ITALY

I. INTRODUCTION 165
II. SURVEY OF BASIC THEORY 166
  1. *Resistivity Measurements* 168
  2. *Mobility Measurements* 170
  3. *Carrier and Mobility Profiling* 172
III. APPLICATIONS OF THE TECHNIQUE 175
  1. *Experimental Details* 175
  2. *High Fluence Boron Fluoride–Implanted Layers* 175
  3. *Measurements on Shallows, Heavily Arsenic-Doped Layers: Solid Solubility and Mobility Data* 186
  4. *Active Dopant Concentration in Phosphorus-Doped Polycrystalline Layers* . . 189
IV. CONCLUSIONS 192
  References 193

I. Introduction

The basic problem addressed in this chapter is the determination of the carrier depth profile in a doped semiconductor, for instance, in a shallow p-n junction or in the measurement of the active dopant concentration in a thin polycrystalline semiconductor layer. Many analytical techniques can be used to obtain information on the in-depth distribution of both chemical species and charge carries. We will focus our attention on charge carriers, and in particular on the method based on the anodic stripping of silicon, and on the incremental measurements of the sheet resistance and of the sheet Hall coefficient.

We will only touch on two other very popular methods used to obtain carrier depth profiling, the capacitance-voltage ($C-V$) and spreading resistance measurements. $C-V$ methods are limited both in depth and in concentration by the junction breakdown, whereas spreading resistance measurements require knowledge of carrier mobility. Notwithstanding this limitation, these methods are widely used; in particular, spreading resistance is the most useful method for electrical characterization of doped layers, and it probably will be used for many years, due to its ability to perform measurements on small areas without the need for a particularly designed and diffused pattern. The anodic stripping method is more complex. However, it has some unique characteristics and can give direct information on carrier mobility, also making it possible to obtain a reliable carrier depth profile in the presence of scattering centers such as interstitial impurities and lattice defects.

II. Survey of Basic Theory

In an extrinsic (e.g., p-type) semiconductor, the resistivity ρ is expressed as

$$\rho = \frac{1}{\sigma} = \frac{1}{e\mu_p p} \tag{1}$$

where σ is the conductivity of the material, μ_p is hole mobility, p is hole concentration, and e is the positive unity charge.

Mobility usually is macroscopically defined as the proportionality factor between carrier drift veloity v_d and the applied electrical field E at low fields:

$$v_d = \mu E \tag{2}$$

From a microscopic point of view, carrier mobility is determined by the scattering phenomena a carrier experiences during its motion under the action of the electrical field E.

If L_i is the mean free path between two scattering events with two scattering centers of the ith kind, then the mean free path L in the presence of many kinds of centers is

$$\frac{1}{L} = \Sigma_i \frac{1}{L_i}$$

A similar relationship holds for mobility:

$$\frac{1}{\mu} = \Sigma_i \frac{1}{\mu_i}$$

In a real conductor, many scattering centers can exist: phonons, interstitials and vacancies, impurity atoms, precipitates, grain boundaries, and other extended defects such as dislocations and stacking faults. In order to be able to affect carrier motion the size of a scattering center must be comparable with the conduction electron wavelength; at moderate fields and room temperature, $\lambda \approx 100\,\text{Å}$. This means that only isolated defects or small aggregates of both defects and impurities can act as a scattering center. From Eq. (1), it follows that the measurement of resistivity only does not allow evaluation of carrier concentration, unless carrier mobility is known. In many cases, mobility is a unique function of the density of the active dopant, and this allows the density of the carrier to be obtained from conductivity data. However, this is only possible if scattering by ionized dopant atoms is dominant over any other scattering mechanism. This is not necessarily the case, for instance, in the presence of a heavy crystal defectivity, or when, at concentrations in excess of the solid solubility — a very frequent situation in practice — a very small amount of precipitate forms in the diffused layer. Because only substitutional dopants can be electrically active in silicon, the correctness of the data obtained for the very heavily doped layer can be used as a check for the measurement procedure: The maximum carrier concentration should give the solid solubility at the process temperature, provided no precipitation takes place during cooling.

Different expressions have been obtained and are reported in the literature relating carrier mobility to active dopant concentration; however, these expressions disagree with each other at very high doping densities (Masetti, Severi, and Solmi, 1983; Thurber et al., 1980; Baccarani and Ostoja, 1975). For these reasons an accurate determination of carrier density often requires the measurement of both resistivity and mobility of majority carriers. Equation (1) is valid for a homogeneous semiconductor. However, in most cases, it is necessary to determine a concentration of carriers varying with depth, for instance, a junction profile; this is needed not only for the purpose of evaluation, but also to validate the process simulation codes. In the case of a p-type doping profile $p(x)$ in a n-type substrate, Eq. (1) is replaced by

$$\bar{\rho} = \frac{1}{\bar{\sigma}} = 1 \bigg/ \left(\int_0^{x_j} e\mu_p(x) p(x) \, dx / x_j \right) \qquad (3)$$

where x_j is the junction depth and $\bar{\rho}$ and $\bar{\sigma}$ are the average values of resistivity and conductivity. It is evident from Eq. (3) that some differential measurement of conductivity must be carried out in order to determine a junction profile. Besides, in this case, mobility also must be measured as a function of depth.

1. RESISTIVITY MEASUREMENTS

The simplest measurement of the resistivity of a semiconductor sample makes use of the well-known four-point probe method (Fig. 1). Four collinear equally spaced probes make contact at the surface of the material to be measured, current I is forced on the outer probes and voltage drop V is measured between the inner probes. No current must flow through voltage sensing contacts. Therefore, if a voltmeter is used, its input impedance must be high enough to ensure that this condition is always met. The probes are assumed to be located far from the boundary of the sample, which is approximated as an infinite uniform area. If the thickness δ of the layer under consideration is much smaller than the probe spacing (this condition is usually met in semiconductor p-n junctions), the sample can be considered as two-dimensional, and it can be shown that

$$\rho = \frac{\pi}{\ln 2}\frac{V}{I}\delta \qquad (4)$$

It is useful to define the sheet resistance $R_s = (\pi/\ln 2)V/I = \rho/\delta$, and the sheet conductance $G_s = 1/R_s = \sigma\delta$. These quantities are the resistance and the conductance measured between opposite sides of a square of arbitrary dimensions. Indeed, the resistance of a square of side l is

$$R = \frac{\rho l}{\delta l} = R_s$$

Generally, a contact barrier is expected to exist at the probe–silicon interface, with a current-voltage characteristic of the form

$$I = I_0(\exp eV/nK_BT - 1)$$

where I_0 is the saturation current of the barrier, n is its ideality coefficient,

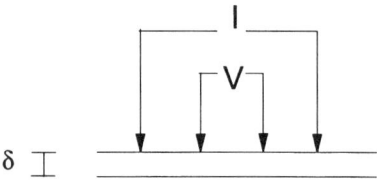

FIG. 1. Scheme of the four-point probe measurement. I, current; V, voltage decrease; δ, layer thickness.

6 STUDIES OF THE STRIPPING HALL EFFECT

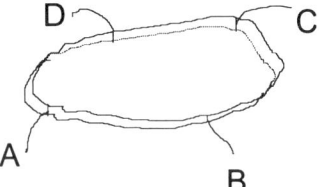

FIG. 2. Uniform plane lamella with four line contacts at its edges (A, B, C, D). This is the geometry to which Van der Pauw theorem applies.

k_B is the Boltzmann constant, and T is the measurement temperature. For this reason, the injected current must be low enough to ensure that $V \ll k_B T/e$, so that the current-voltage characteristic of the barrier is approximately linear.

Provided the previous conditions is met, the contacts between the probes and the semiconductor surface need not be ohmic, so there is no need for sample preparation. This fact makes the method very simple and therefore very commonly used; however, it is unsuitable when the size of the sample is not infinite with respect to the size of the probe. An extension of the four-point probe technique to samples of arbitrary shape was provided by Van der Pauw (1958), who showed that a lamella with four line contacts at its edges can be mapped by conformal transformation into a semi-infinite plane with four contacts on its boundary. The sample (Fig. 2) is assumed to be flat and uniform in thickness but can be of arbitrary shape. Let A, B, C, and D be the contacts at the edge of the lamella, and let

$$R_{ABCD} = \frac{V_{AB}}{I_{CD}}$$

be the resistance obtained by injecting a current I_{CD} between C and D and measuring voltage between A and B (V_{AB}). Van der Pauw showed that the resistivity of the sample can be expressed as

$$\rho = \frac{\pi \delta}{\ln 2} \frac{R_{ABCD} + R_{BCDA}}{2} f\left(\frac{R_{ABCD}}{R_{BCDA}}\right) \quad (5)$$

The function f depends on the ratio R_{ABCD}/R_{BCDA} only, and $f = 1$ if $R_{ABCD} = R_{BCDA}$, so that Eq. (5) reduces to Eq. (4). Equation (5) holds for samples of arbitrary shape. However, if a symmetrical geometry is used, such that the resistances under consideration are equal to one another, a better estimate of the resistance is obtained by measuring eight values (R_{ABCD}, R_{BCDA}, R_{CDAB},

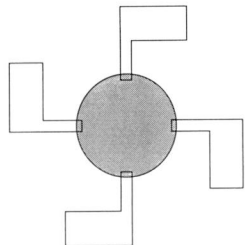

FIG. 3. Cloverleaf structure for sheet resistance and sheet Hall coefficient measurements.

R_{DABC} and those obtained by reversing the sign of the injected current), and taking the average value of these measurements.

A commonly used geometry is the so-called "cloverleaf" structure, sketched in Figure 3. Essentially, this structure consists of a circle with four contacts at its edge. The circle in the center is the layer under investigation. Contacts are symmetrically located with respect to the circle, and are connected to it by narrow arms, thus providing a good approximation to the ideal line contacts required by the Van der Pauw theorem. The outer "leaves" are large areas to be used for assembling the structure. This structure is also convenient for measuring mobility.

2. MOBILITY MEASUREMENTS

Mobility is commonly obtained by measurements of the Hall coefficient. Let us consider for simplicity a p-type uniform bar sample such as the one in Figure 4. A uniform magnetic induction B_x is applied normal to the surface of the sample and a constant current density J_z is forced through the sample. Holes are deflected by magnetic force in a direction normal to their initial average velocity v_z, that is, toward the positive y direction. Under steady-state conditions an electric field E_y is established that opposes the magnetic force, and a voltage decrease (the Hall voltage, V_H) can be measured between opposite sides of the bar. The electric field E_y is proportional to both the magnetic induction B_x and the current density J_z, and the proportionality factor is called the Hall coefficient R_H, defined by

$$-E_y = R_H J_z B_x \qquad (6)$$

By writing $J_z = epv_z$ and by expressing E_y in terms of B_x and v_z, from Eq. (6) it follows that $R_H = 1/ep$. However, this simple expression is based on

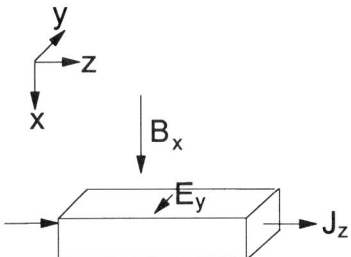

FIG. 4. Hall effect in an uniform bar sample. B_x, uniform magnetic induction; E_y, electric field; J_z, current density.

the rough assumption of a common drift velocity v_z for all carriers. A more correct expression is

$$R_H = \frac{r}{ep} \qquad (7)$$

Here, r is a correction factor called the Hall scattering factor, which can be theoretically evaluated by taking an average of the involved quantities over the energy distribution of carriers.

From Eqs. (7) and (1),

$$\mu = \frac{R_H \sigma}{r} \qquad (8)$$

relating R_H to the drift mobility. It is customary to define the Hall mobility μ_H.

$$\mu_H = R_H \sigma = r\mu \qquad (9)$$

Measuring the Hall coefficient requires a four-contact structure. Current is injected between opposite contacts (rather than between adjacent contacts, as is done for resistivity measurements) in the presence of a magnetic field normal to the surface of the sample, and voltage is measured between the remaining contact. From Eq. (6), for the bar sample in Fig. 2, the Hall coefficient is given by

$$R_H = \frac{V_H \delta}{B_x I_z}$$

where I_z is the injected current. Also in this measurement, it is important to ensure that no current flows through the voltage reading contacts. It is useful to define the sheet Hall coefficient $R_{Hs} = R_H/\delta$ by analogy with the definition of sheet resistance.

For an arbitrarily shaped structure with four contacts at its edge, Van der Pauw (1958) showed that

$$R_H = \frac{\delta}{B_x} \Delta R_{BDCA} \qquad (10)$$

where ΔR_{BDCA} is the variation produced by the magnetic field in the resistance R_{BDCA} measured by injecting current between opposite contacts A and C and measuring voltage between contacts B and D. If the structure is symmetric, like the structure in Figure 1, in the absence of a magnetic field, $R_{BDCA} = 0$. As is done for resistivity measurements, it is customary to reverse the current and the magnetic field and to exchange current and voltage contacts, so that R_H is obtained from the average of eight measurements.

a. Hall Scattering Factor

In order to obtain drift mobility from measurements of the Hall coefficient, the Hall scattering factor r must be known. This factor is related to the energy distribution of majority carriers and to their scattering cross sections. Therefore, r is expected to depend on doping sign as well as on doping density and on temperature. According to available theoretical models (Nagasawa and Zukotynski, 1978; Lin et al., 1981) for what concerns holes at room temperature and doping densities $N_A \geqslant 10^{14}$ cm^{-3}, the Hall scattering factor is expected to be less than 1, $0.7 \leqslant r \leqslant 0.9$. Experimental results support these theoretical models (Andrieu et al., 1988), and several authors (Masetti, Severi, and Solmi, 1983; Thurber et al., 1980) have agreed to use $r = 0.8$ for holes and $r = 1$ for electrons in silicon.

3. CARRIER AND MOBILITY PROFILING

The previous discussion refers to uniform samples. However, because determining carrier and mobility profiles of nonuniform samples is often required, the previously described measurements must be carried out versus depth and related to relevant physical quantities in nonuniform samples.

6 STUDIES OF THE STRIPPING HALL EFFECT

In order to avoid measurements resulting from the parallel connection of the layer under investigation and the substrate, some isolation of the layer is required. This can be obtained by p-n junction, or the layer can be deposited on an insulating substrate. The structure in Figure 1 is very commonly used (see, e.g., (Galloni and Sardo, 1983; Masetti, Severi, and Solmi, 1983; Thurber *et al.*, 1980). Usually, sheet resistance and sheet Hall coefficients are measured versus depth by a sequence of measurements and stripping of thin layers. The method used for layer removal is critical since layer thickness must be carefully controlled. For this reason, a simple chemical etching of silicon is not suitable for this purpose. The growth of a thin oxide by electrochemical methods is more suitable because it allows direct control of the grown oxide (Galloni and Sardo, 1983; Beynon, Bloodworth, and McLeod, 1973) (hence of the consumed silicon) by measuring the voltage decrease through the electrolytic cell. A calibration curve can be determined relating the voltage decrease to the grown oxide. In order to ensure that this curve is independent of the sample to be oxidized, carrier are injected by illuminating the sample during oxidation. Obviously, the contact regions must be protected during the procedure of oxidation and etching. The whole sequence (anodic oxidation, stripping of the grown oxide, R_s and R_{Hs} measurements), of course is performed automatically.

Measurements of sheet resistance and the sheet Hall coefficient can be carried out as a function of the removed thickness. By describing a doping profile as a set of parallel-connected infinitesimal layers, it is easy to see that

$$G_s = \int_0^W \sigma(y) \, dy$$

and

$$R_{Hs}\sigma^2 = \int_0^W R_H(y)\sigma^2(y) \, dy$$

where W is the total thickness of the layer. If G_s and R_{Hs} are known versus the thickness of the removed layer, carrier concentration and drift mobility can be obtained by differentiation of the previous equations with the use of Eqs. (1) and (9):

$$\mu_i = \frac{1}{r} \frac{\Delta(R_{Hs,i} G_{s,i}^2)}{\Delta G_{s,i}} \tag{11}$$

and

$$p_i = -\frac{\Delta G_{s,i}/\Delta x_i}{e\mu_i} \qquad (12)$$

where the equations have been written under finite difference form, since this form is more convenient to elaborate sets of experimental data. Here, Δx_i is the thickness of the ith removed layer, $G_{s,i}$ and $R_{Hs,i}$ are the measured sheet resistance and sheet Hall coefficient at the ith step, respectively, and μ_i and p_i are the mobility and the concentration at a depth $\Sigma_{j=1}^{i} \Delta x_j$, respectively. Equations (11) and (12) allow us to calculate carrier and mobility profiles from a set of sheet resistance and sheet Hall coefficient data. The evaluation of the physical quantities of interest therefore requires a differentiation of the experimental data. For this reason, smoothing methods have been developed and are sometimes used (e.g., Hill, Allen, and Bradley, 1980)) to eliminate fluctuations of raw G_s and R_{Hs} data.

a. Errors and Limitations

In principle, these measurements are simple; however, care must be taken under some conditions:

- As already mentioned, correct measurement of sheet resistance requires that no current flows through voltage-sensing contacts. This point is critical when measuring high resistances but is no longer a major problem with modern digital instruments.
- In order to avoid errors due to photoconductivity, measurements are better performed in the dark.
- If a p-n junction is used for insulating the structure, junction leakage should be checked.
- The thickness of subsequently removed layers must be known accurately. This is especially critical when it is necessary to determine profiles that are rapidly varying with depth, and therefore very thin layers are to be removed.

In addition, it must be taken into account that these measurements provide a *carrier* profile, which may differ from the doping profile for various reasons, for instance, an incomplete activation of dopants or the presence of a depletion layer at the surface or at the p-n junction.

III. Applications of the Technique

In this section we report some results obtained using the anodic oxidation method. In particular, we report some data obtained on

1. Very high fluence boron fluoride–implanted layers annealed at low temperatures
2. Shallow, very heavily arsenic-doped layers
3. Thin doped polycrystalline silicon layers

For each example we stress the advantages of that method with respect to the other analytical methods.

1. EXPERIMENTAL DETAILS

A Van der Pauw structure (Fig. 1) has been used for carrier and mobility profiling in all of the experiments to be discussed. In the experiments concerning ion-implanted layers, isolation of the structure was obtained by a p-n junction, and the perimeter of the structure was defined by local oxidation of silicon. The structure had heavily doped, deeply diffused contacts to ensure reliable contact even when most of the layer to be analyzed has been consumed.

Layer removal was obtained by anodic oxidation of silicon and subsequent etching in a diluted hydrogen fluoride solution. The electrolyte was a solution of 90% ethylene glycol, 10% H_2O, and 0.05% M potassium nitrate (KNO_3). The thickness of the grown oxide was controlled by the voltage drop through the electrolytic cell. The thickness was checked at the end of the measurement cycle by measuring the total removed thickness with a profilometer. Measurements where generally noise-free, therefore smoothing of the experimental data was not necessary.

2. HIGH FLUENCE BORON FLUORIDE–IMPLANTED LAYERS

a. Low-Temperature Furnace Annealing

The solid-phase epitaxial regrowth of the amorphous layer created by ion implantation can be performed at quite low temperatures that is, ranging from 400 to 600°C. The regrown layers exhibit good crystallinity, as judged by Rutherford backscattering spectrometry measurements, and a high conductivity.

A carrier concentration well in excess of the solid solubility at the annealing temperature can be estimated from sheet resistance and spreading resistance measurements, thus showing that the dopant activation during solid-phase epitaxial regrowth is dominated by kinetics. However, the correct measurement of carrier concentration requires evaluation of the carrier mobility. In fact, X-ray multiple-crystal diffraction measurements show a quite high point-defect concentration in these layers, which can act as scattering centers for carriers. In this section we report on some data obtained by the anodic sectioning method on boron-doped layers obtained by low-temperature annealing of high fluence boron fluoride–implanted (BF_2) layers. The carrier depth profiles obtained are compared with those given by spreading resistance measurements, in which the carrier concentration–resistivity relationship valid for perfect silicon (Si) is used. This comparison shows how the spreading resistance method is not able to give meaningful evaluation in these cases.

The low-temperature annealing of BF_2^+-implanted layers was studied (Queirolo et al., 1991) by measurements of active B concentration and hole mobility. Bare Si wafers were implanted in an EATON NOVA (EATON Corporation, Beverly, Massachusetts, USA) 10-160 ion implanter at 60 keV energy with doses ranging from 1 to $5 \cdot 10^{15}$ cm^{-2}. Low-temperature (600°C) and high-temperature (950°C) annealing treatments in N_2 were compared.

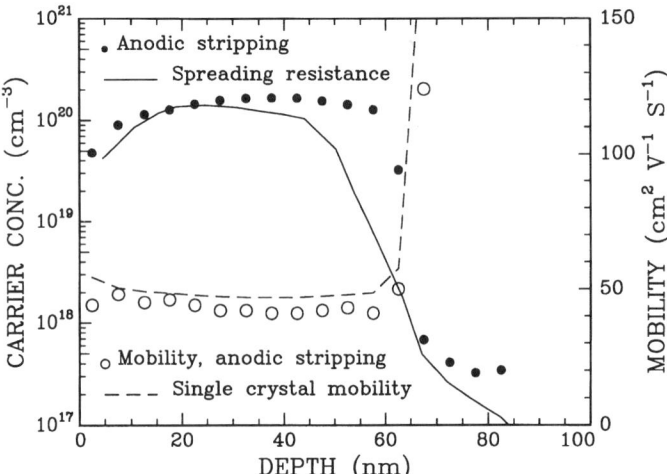

FIG. 5. Carrier and mobility profiles in a sample implanted with 10^{15} cm^{-2} boron fluoride ions and annealed at 600°C for 20 min. The profile of carriers obtained by subsequent sheet resistance and sheet Hall coefficient measurements is compared with the spreading resistance profile, and the measured mobility is compared with the single crystal mobility.

Figures 5 through 7 compare carrier concentration profiles—obtained by subsequent measurements of sheet resistance and the sheet Hall coefficient—to spreading resistance profiles of samples annealed at 600°C for 20 min. These plots also report the experimental mobility versus depth and the mobility expected from the data in the literature (Masetti, Severi, and Solmi, 1983) for single-crystal, defect-free Si having the measured carrier concentration. It is observed that the experimental carrier mobility is gradually degraded with increasing BF_2 dose, indicating the presence of another scattering mechanism, in addition to scattering by ionized dopant atoms. As the experimental mobility departs from the single-crystal value, the spreading resistance profile underestimates carrier concentration.

In all these samples, carrier concentration is much higher than B solid solubility at the annealing temperature (600°C (Armigliato et al., 1977)), showing that a metastable state is obtained by this low-temperature annealing. The highest fluence sample (Fig. 7) shows a strong mobility reduction in a region close to the surface. This sample was also analyzed by secondary ion mass spectroscopy (SIMS) to obtain chemical profiles of B and F. The comparison between chemical and carrier concentration profiles show (Fig. 8) the presence of a large amount of inactive B. In addition, it is observed that F piles up in a region close to the surface, roughly corresponding to the reduced mobility region in Figure 7. Transmisionelectron microscopy

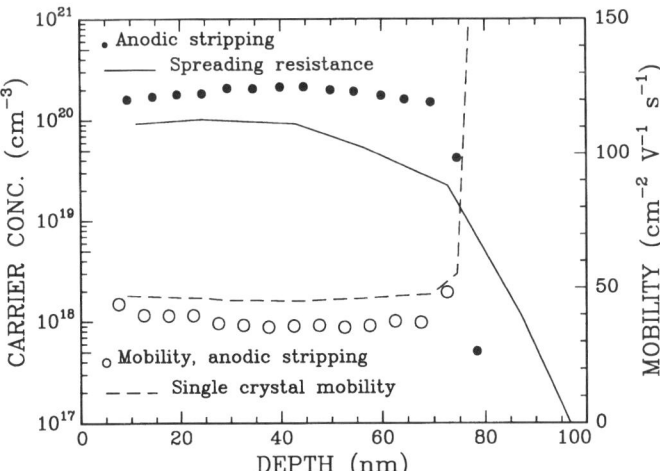

FIG. 6. Carrier and mobility profiles in a sample implanted with $3 \cdot 10^{15}\,cm^{-2}$ boron fluoride (BF_2) ions and annealed at 600°C for 20 min. Carrier profile obtained by subsequent sheet resistance and sheet Hall coefficient measurements is compared with the spreading resistance profile, and the measured mobility is compared with the single crystal mobility.

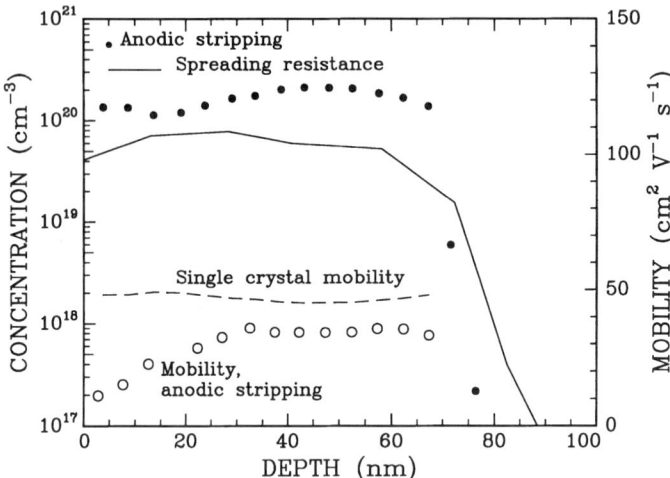

FIG. 7. Carrier and mobility profiles in a sample implanted with $5 \cdot 10^{15}$ cm^{-2} boron fluoride (BF$_2$) ions and annealed at 600°C for 20 min. Carrier profile obtained by subsequent sheet resistance and the sheet Hall coefficient measurements is compared with the spreading resistance profile, and the measured mobility is compared with the single crystal mobility.

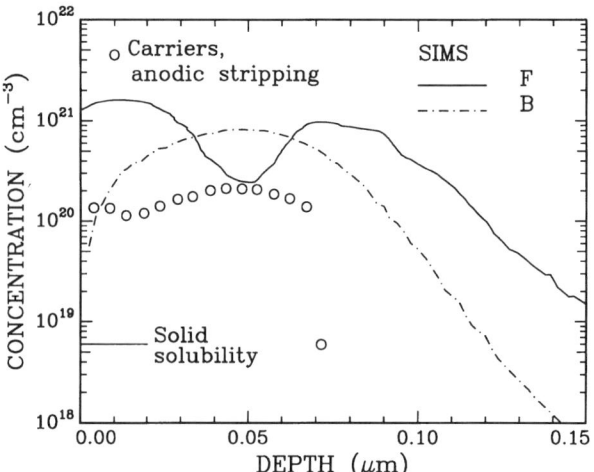

FIG. 8. Carrier and chemical profiles in a sample implanted with $5 \cdot 10^{15}$ cm^{-2} boron fluoride (BF$_2$) ions and annealed at 600°C for 20 min. The solubility limit at the annealing temperature is also shown. SIMS, secondary ion mass spectroscopy.

(TEM) analyses of the highest dose sample (Queirolo et al., 1991) revealed the presence of a heavy crystal defectivity (microtwins) in this region. In addition, a high concentration of intrinsic point defects is likely to be present in these samples. All these factors (high concentration of F and intrinsic point defect, inactive B, crystal defects), in principle, may be responsible for the observed mobility degradation.

b. High-Temperature Furnace Annealing

Similar analyses were carried out in samples receiving an additional thermal treatment at 950°C for 20 min. Figures 9 and 10 show carrier and mobility profiles obtained by anodic stripping for B doses of $3 \cdot 10^{15}$ cm^{-3} and $5 \cdot 10^{15}$ cm^{-3}, respectively. The single-crystal mobility and the chemical profile of B obtained by SIMS are also shown. The comparison between SIMS and carrier concentration profiles shows the presence of a relevant amount of inactive B in the region in which chemical B concentration exceeds solid solubility (Armigliato et al., 1977). TEM inspections suggested (Queirolo et al., 1991) that it is under the form of precipitates. TEM images

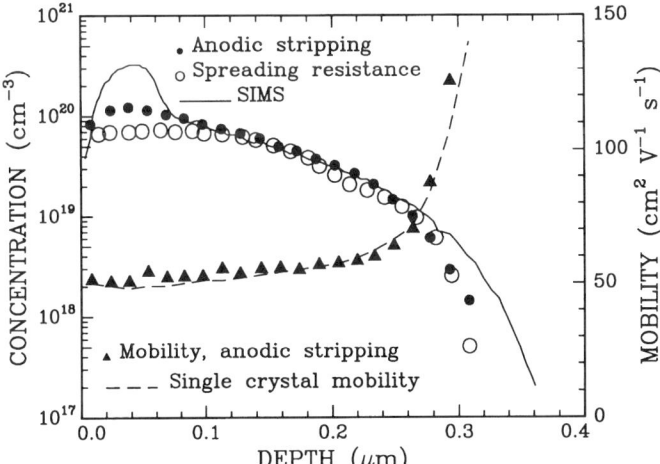

FIG. 9. Chemical concentration of boron (SIMS; secondary ion mass spectroscopy), carrier and mobility profiles in a sample implanted with $3 \cdot 10^{15}$ cm^{-2} boron fluoride (BF$_2$) ions and annealed at 950°C for 20 min. Carrier profile obtained by subsequent sheet resistance and the sheet Hall coefficient measurements is compared with the spreading resistance profile, and the measured mobility is compared with the single crystal mobility.

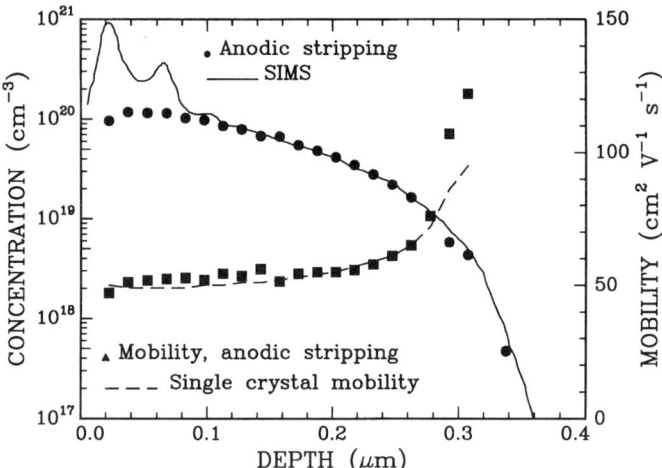

FIG. 10. Chemical concentration of boron (SIMS; secondary ion mass spectroscopy), carrier and mobility profiles in a sample implanted with $5 \cdot 10^{15}$ cm^{-2} boron fluoride (BF$_2$) ions and annealed at 950°C for 20 min. The measured mobility is compared with the single crystal mobility.

of the sample implanted with the highest dose ($5 \cdot 10^{15}$ cm^{-3}) also showed the presence of more extended defects at the surface (twins) and in the bulk (dislocation loops[1]). Despite the heavy crystal defectivity, Figures 9 and 10 show that after the thermal treatment at 950°C, mobility is recovered to the single crystal value in both samples. This fact shows that the observed extended defects do not affect mobility, that is, they do not behave as effective scattering centers. Therefore, point defects (whether intrinsic point defects, F, interstitial B atoms) must be responsible for the degradation of mobility in low-temperature annealed samples.

Figure 9 also compares carrier profiles obtained by anodic stripping and by spreading resistance measurements. These techniques agree quite well up to a carrier concentration of about $8 \cdot 10^{19}$ cm^{-3}. For higher values of the concentration measured by anodic stripping, spreading resistance data are lower than are data from anodic stripping. Since in this sample the measured mobility coincides with single crystal mobility, it is likely that this fact is due to inaccuracy in the mobility data used in the elaboration of the spreading resistance profile and based on the work of Thurber (Thurber et al., 1980).

[1]The depth location of extended defects corresponds to the peaks in chemical boron concentration shown in Fig. 10.

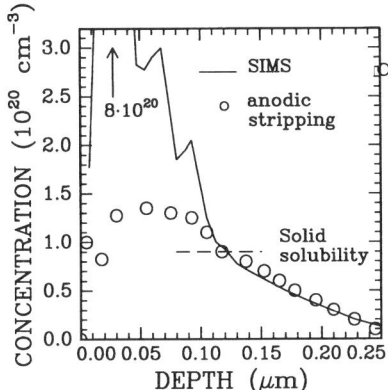

FIG. 11. Chemical concentration of boron (SIMS; secondary ion mass spectroscopy), and carrier profile in a sample implanted with $5 \cdot 10^{15}$ cm^{-2} boron fluoride (BF$_2$) ions and annealed at 950°C for 10 min. The solubility limit at the annealing temperature is also shown.

On close inspection, carrier and chemical profiles of high-dose BF$_2$-implanted samples exhibit a peculiar feature. The sample in Figure 11 was implanted with $5 \cdot 10^{15}$ cm^{-2} BF$_2$ ions and annealed at 950°C for 10 min. The solid solubility of B at the annealing temperature (Armigliato et al., 1977) is also shown. It is observed that the active dopant concentration exceeds solid solubility up to about 50% in the region in which a large amount of inactive B is also found. This fact may be due to the presence of B precipitates. Indeed, inactive B is most likely to be in the form of B precipitates, which may be responsible for an increase in solid solubility (Thomson-Freundlich effect) (Swalin, 1962) depending on their density and size.

c. Rapid Thermal Annealing

Rapid thermal annealing (RTA) of implanted layers is commonly used in device processing, because it provides a very good dopant activation along with negligible dopant diffusion (Miyake et al., 1988; Naem and Calder, 1987). A large excess of point defects is active during rapid thermal treatments and is responsible for a transient enhanced diffusion (Pennycook, Narayan, and Holland, 1985). The question arises if residual point defects may affect carrier mobility in layers obtained by high-dose implantation and RTA. In addition, dopant diffusion during RTA of high-dose implanted layers involves not only point defects created by ion implantation, but also

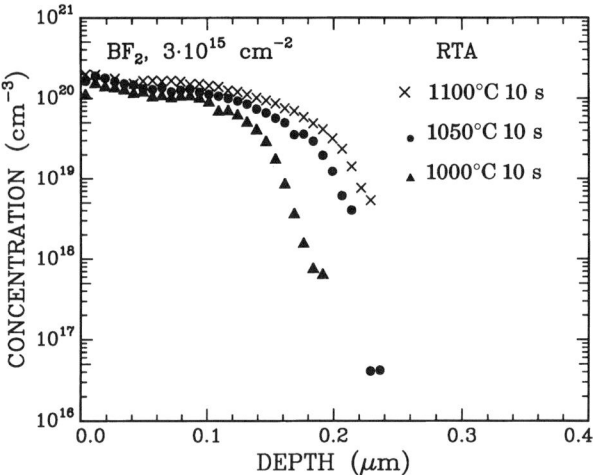

FIG. 12. Carrier profiles obtained by subsequent sheet resistance and the sheet Hall coefficient measurements of boron fluoride (BF_2)-implanted samples annealed in a rapid thermal processor at various temperatures. RTA, rapid thermal annealing.

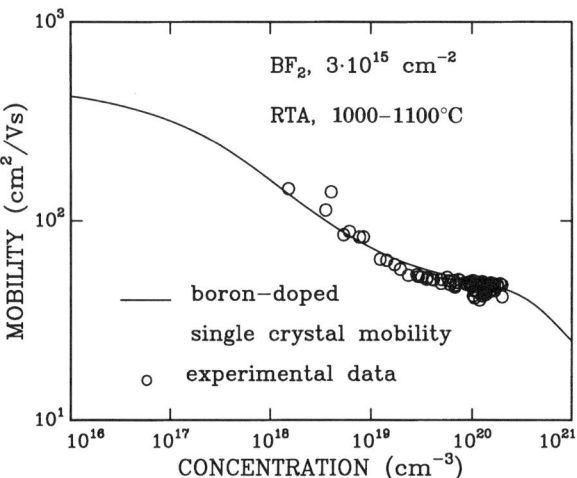

FIG. 13. Experimental mobility data versus carrier concentration in boron fluoride (BF_2) implanted samples annealed in a rapid thermal processor at various temperatures. The single-crystal mobility-carrier curve is also shown.

defects injected by growth and dissolution of precipitates. For this reason, available models for the simulation of dopant diffusion (e.g., SUPREM3 (Ho et al., 1984)) sometimes fail in the simulation of RTA.

In order to investigate some of these problems, RTA of BF_2-implanted layers has been characterized. Implantations of BF_2 were carried out in a EATON NOVA 10/160 implanter through a 100 Å oxide. Implantation energy was 60 keV, and no postacceleration of the beam was used. Implanted samples were annealed in an AG (SAG Associates, San Josè, California, USA) Heatpulse 4100 rapid thermal processor. Annealing temperatures ranging from 950 to 1100°C have been studied.

Figure 12 shows carrier profiles obtained by anodic stripping in wafers implanted with $3 \cdot 10^{15}$ cm^{-2} BF_2 ions and annealed at different temperatures for 10 sec., and Figure 13 collects the corresponding mobility data. It is seen that carrier mobility is recovered to the single-crystal value during the RTA.

Figure 14 compares experimental and calculated carrier profiles of a sample implanted with $3 \cdot 10^{15}$ cm^{-2} BF_2 ions and annealed at 950°C for 10 sec. Calculated profiles have been obtained by SUPREM3 simulation (Ho et al., 1984). As to what concerns dopant activation, the calculation assumes that the dopant is activated up to the solid solubility at the annealing temperature. The experimental profile confirms this hypothesis,

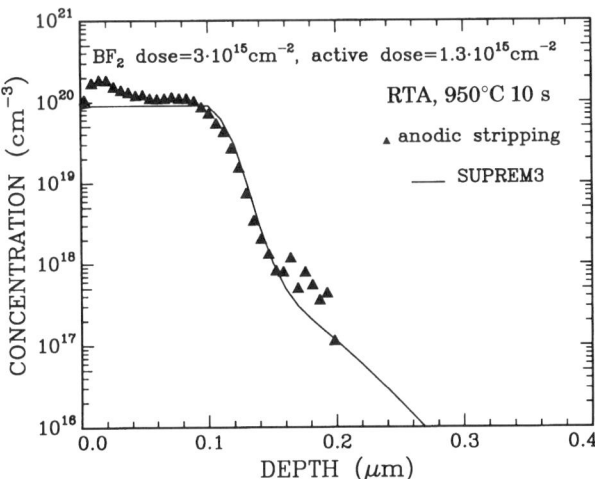

FIG. 14. Experimental and calculated (SUPREM3) carrier profiles of a sample implanted with $3 \cdot 10^{15}$ cm^{-2} boron fluoride (BF_2) ions and annealed at 950°C for 10 sec.

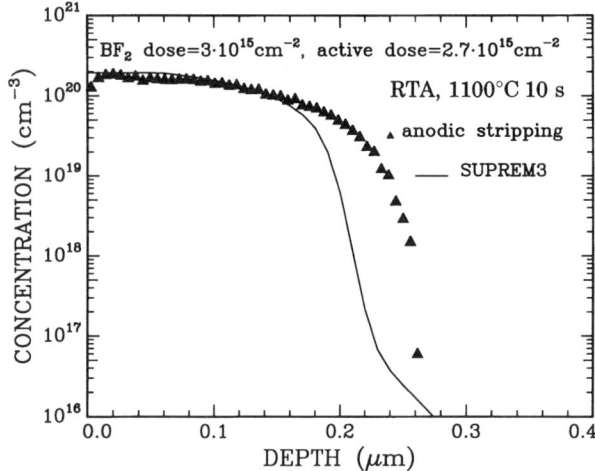

FIG. 15. Experimental and calculated (SUPREM3) carrier profiles of a sample implanted with $3 \cdot 10^{15}$ cm^{-2} boron fluoride (BF$_2$) ions and annealed at 1100°C for 10 sec.

since measured and calculated concentrations coincide in the plateau region of the profile. However, in a region close to the surface (about 200 Å) the concentration increases above the plateau value. This phenomenon is similar to that observed in furnace annealed samples, and it is likely to be due to increased solid solubility because of the presence of precipitates or other extended defects in this region. The presence of precipitates in this region is confirmed by the fact that only about 40% of the implanted dose is electrically active in this sample.

Figure 15 shows experimental and calculated profiles of a sample implanted with $3 \cdot 10^{15}$ cm^{-2} BF$_2$ dose (the same as in Fig. 14) and annealed at 1100°C for 10 sec. In this sample most of the implanted dose (about 90%) is electrically active, so very little or no precipitate should be present at the end of the RTA. In addition, the previously observed concentration peak in excess of solid solubility is not observed, and the concentration in the plateau region equals solid solubility, thus confirming that this peak is related to the presence of precipitates. In addition, in this sample the experimental profile also shows some excess in-depth diffusion with respect to the calculated profile. The calculation code includes models to take into account various different phenomena that concur to determine doping profiles, such as effects related to point-defect excess (e.g., oxidation-enhanced diffusion and transient-enhanced diffusion) and effects of the high concentration on dopant diffusivity. However, the growth and dissolution of

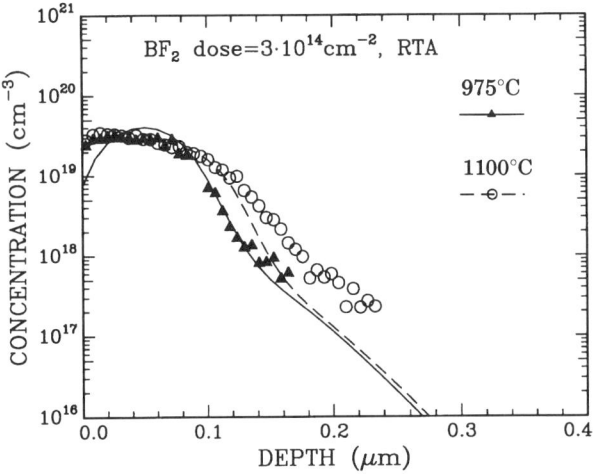

FIG. 16. Experimental (symbols) and calculated (SUPREM3, lines) carrier profiles of samples implanted with $3\cdot10^{14}$ cm^{-2} boron fluoride (BF$_2$) ions and annealed at 975°C and 1100°C for 10 sec.

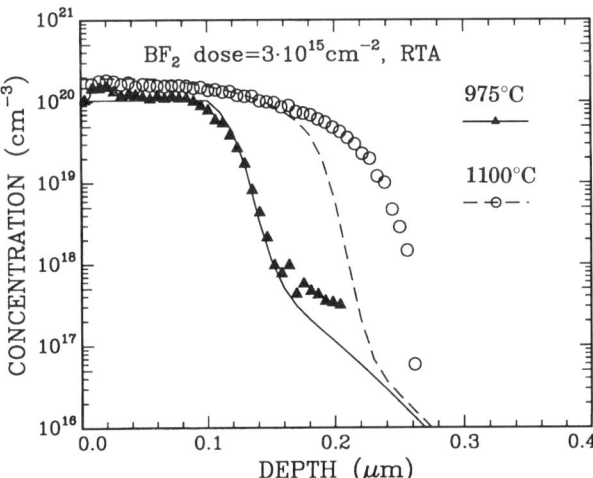

FIG. 17. Experimental (symbols) and calculated (SUPREM3, lines) carrier profiles of samples implanted with $3\cdot10^{15}$ cm^{-2} boron fluoride (BF$_2$) ions and annealed at 975°C and 1100°C for 10 sec.

precipitates, as well as the related point-defect excess, are not taken into account in the present version of the code, although models describing these phenomena have been elaborated on (Solmi, Baruffaldi, and Canteri, 1991). The discrepancy between experimental and calculated profile concerns samples in which dissolution of precipitates and, as a consequence, point-defect injection is likely to take place. For this reason we tentatively ascribe the observed anomalous diffusion to this mechanism.

In order to confirm this interpretation, an investigation has been carried out in samples implanted with moderate doses, such that no precipitation at all is expected. Figures 16 and 17 show carrier profiles obtained at various annealing temperatures for a moderate dose ($3 \cdot 10^{14}$ cm^{-2}) and a high dose ($3 \cdot 10^{15}$ cm^{-2}), respectively. In the absence of precipitation, calculated and experimental profiles (Fig. 16) are in reasonable agreement, showing that in the absence of precipitation current models satisfactorily describe diffusion during RTA. These data confirm the attribution of the observed anomalous diffusion to point defects injected during dissolution of precipitates.

3. MEASUREMENTS ON SHALLOW, HEAVILY ARSENIC-DOPED LAYERS: SOLID SOLUBILITY AND MOBILITY DATA

A comparison between different methods for carrier profiling also has been carried out in arsenic-implanted layers. Arsenic (As) was implanted in p-type wafers at 80 keV energy and $3 \cdot 10^{15}$ cm^{-2} dose in an EATON NOVA 10/160 implanter. Samples were annealed at 900°C for 40 min in dry O_2 and 15 min in N_2.

Figure 18 compares carrier concentration profiles obtained by anodic stripping and by spreading resistance measurements. Mobility data from anodic stripping and single crystal mobility data from the literature (Masetti, Severi, and Solmi, 1983) are also shown. Anodic stripping gives a maximum carrier concentration of about $1.6 \cdot 10^{20}$ cm^{-3}, in good agreement with the solid solubility value at the annealing temperature (Derdour, Nobili, and Solmi, 1991). From these data, the active dose is about $1.9 \cdot 10^{20}$ cm^{-2}, indicating that about 30% of the implanted dose is under inactive form. This fact was confirmed by SIMS measurements (Queirolo and Polignano, 1992).

Spreading resistance data are lower than anodic stripping data by about a factor of two. Experimental and literature mobility data are in very good agreement, showing that no additional scattering centers are present in this sample, that is, inactive As does not affect mobility. Therefore, the discrepancy between anodic stripping and spreading resistance data of carrier concentration must be ascribed to the use of uncorrect mobility values in

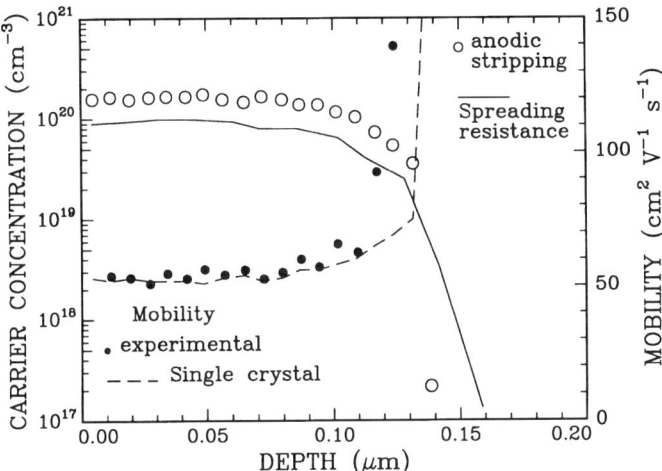

FIG. 18. Carrier and mobility profiles in a sample implanted with $3 \cdot 10^{15}$ cm^{-2} arsenic (As) ions and annealed at 900°C. Carrier profile obtained by subsequent sheet resistance and the sheet Hall coefficient measurements is compared with the spreading resistance profile, and the measured mobility is compared with the single crystal mobility.

converting spreading resistance into carrier concentration profiles (as in the example shown in Figure 9).

The profile in Figure 18 also has been calculated by SUPREM3 simulation. The complete thermal cycle includes temperature ramp up (down) from (to) 800°C. It might be supposed that a correct calculation should include both these ramps. However, it must be taken into account that the models implemented in SUPREM3 at present do not include a kinetic model for dopant deactivation. The dopant is assumed to be at thermal equilibrium instantaneously at each temperature. This assumption may be incorrect, for instance, when the temperature is decreased and the sample is supersaturated because the kinetic of deactivation is not fast enough to ensure thermal equilibrium. This fact may lead to underestimated active dopant concentration as it happens, it is in the example under consideration (Fig. 19). Since a very limited dopant diffusion takes place during temperature ramp-down, a better estimate of the profile is obtained by dropping ramp-down in the calculation. The resulting dopant concentration is calculated in the hypothesis of thermal equilibrium at the oxidation temperature, which is much more realistic, as shown in Figure 19.

Another investigation was carried out concerning dopant activation and mobility recovery in As-implanted samples annealed in a rapid thermal processor. Samples were implanted with 10^{15} cm^{-2} As ions at 150 keV

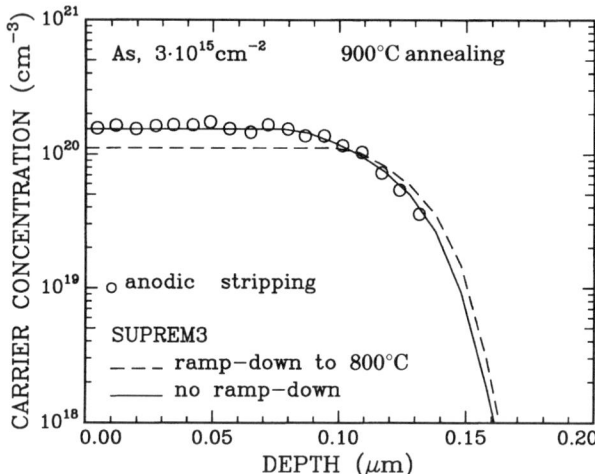

FIG. 19. Experimental and calculated (SUPREM3) carrier profiles of a sample implanted with $3 \cdot 10^{15}$ cm^{-2} arsenic (As) ions and annealed at 900°C.

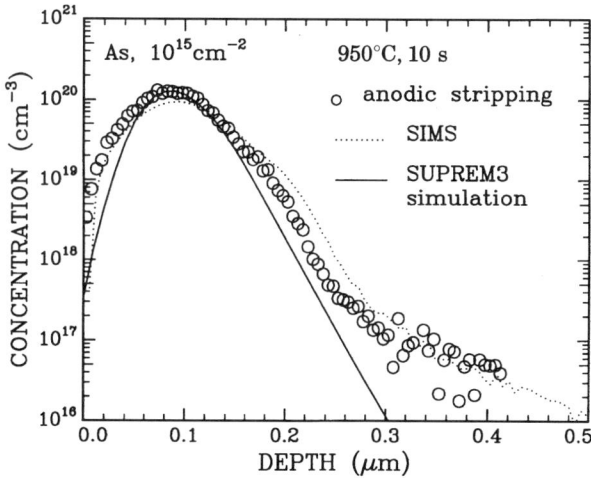

FIG. 20. Experimental carrier profile, chemical concentration of arsenic (As) and calculated (SUPREM3) carrier profile of a sample implanted with 10^{15} cm^{-2} As ions and annealed at 950°C for 10 sec. SIMS, secondary ion mass spectroscopy.

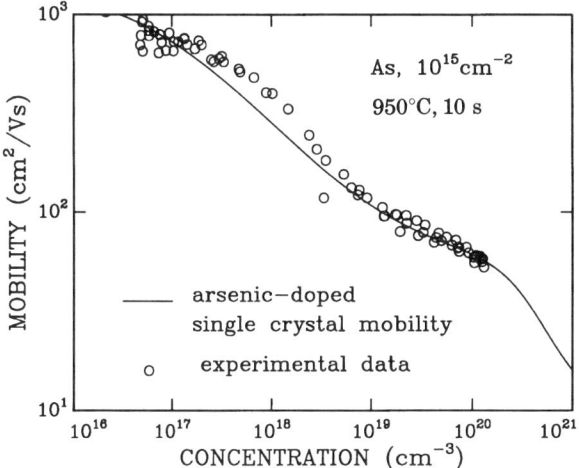

FIG. 21. Experimental mobility data versus carrier concentration in an arsenic (As) implanted sample annealed in a rapid thermal processor at 950°C. The single-crystal mobility-carrier concentration curve is also shown.

energy and 0-degree tilt angle, and annealed at 950°C for 10 sec. Figure 20 compares the carrier profile obtained by anodic stripping and the chemical profile of As measured by SIMS. The as-implanted profile calculated by SUPREM3 simulation is also shown. These data show that complete dopant activation along with negligible dopant diffusion is obtained by the RTA.

Figure 21 compares mobility data measured in this sample with literature data (Masetti, Severi, and Solmi, 1983) for single-crystal As-doped silicon. It is seen that mobility, too, is recovered to the single-crystal value by the RTA.

The experimental profiles in Figure 20 exhibit a channeling tail (expected since the implantation was carried out at 0-degree tilt angle). For As implantations, this phenomenon is not taken into account by the simulation code and does not appear in the calculated profile.

4. ACTIVE DOPANT CONCENTRATION IN PHOSPHORUS-DOPED POLYCRYSTALLINE LAYERS

Doped polycrystalline silicon films are widely used in ultra-large-scale integrated (ULSI) devices for gate interconnection and for floating gates in nonvolatile memories such as electrically erasable programmable read-only memory (EEPROM), electrically programmable read-only memory (EPROM), and Flash EPROM.

In these films, there is a very important scattering mechanism that dominates the usual ones: the scattering against grain boundaries. Due to the grain growth phenomena, this contribution is largely unpredictable; as a consequence, the direct measurement of carrier mobility is necessary when information on the active carrier concentration is needed.

The carrier mobility contribution from grain boundary scattering, in principle, can be evaluated from grain size measurements (Solmi et al., 1982) through the relationship $\mu_{gb} = KD^{\alpha}$, where D is the mean size of the coherently diffracting domains, as evaluated from the measurement of the width of X-ray diffraction peaks, and K and α are constants. This relationship, however, implies that the microcrystals in the polycrystalline films are perfect, and no twins are present in them; on the other hand, the direct measurements of the grain size using TEM measurements is not easily performed. In contrast, the previous relationship can be used to obtain informations on the grain size from mobility measurements (Queirolo et al., 1990), which can be obtained through sheet resistance and sheet Hall coefficient measurements on Van der Pauw patterns obtained on polycrystalline Si films deposited on silicon dioxide. In the same way, active dopant concentration is obtained.

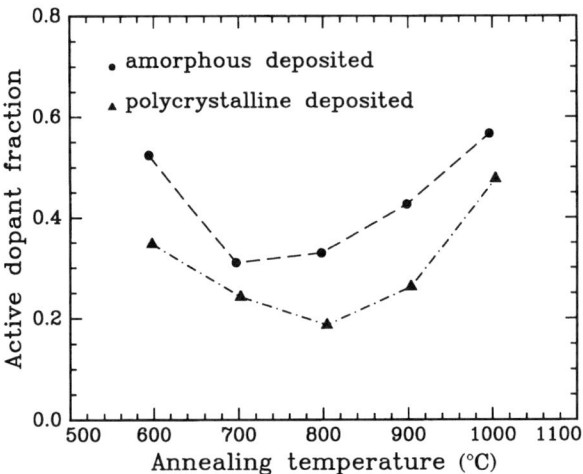

FIG. 22. Fraction of active dopant as a function of the annealing temperature in amorphous-deposited and polycrystalline-deposited silicon films. The films are phosphorus-doped by ion implantation.

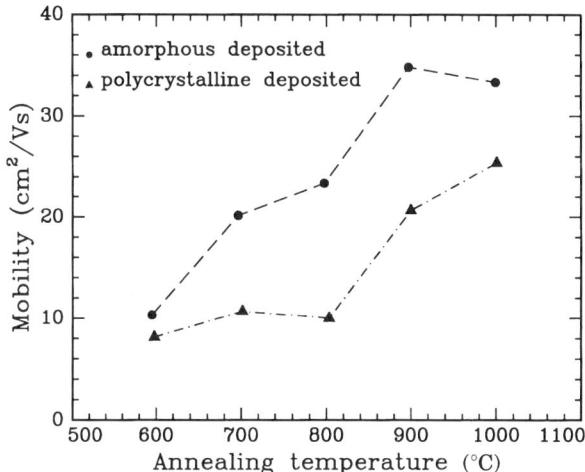

FIG. 23. Carrier mobility as a function of the annealing temperature in amorphous-deposited and polycrystalline-deposited silicon films. The films are phosphorus-doped by ion implantation.

Figure 22 reports the active dopant fraction as a function of the annealing temperature for two polycrystalline Si films, 150-nm thick, deposited by chemical vapour deposition (CVD) methods at two different temperatures (570 and 620°C). After deposition, the two Si films were amorphous and crystalline, respectively. The deposited films were implanted with phosphorus at a dose of $2.5 \cdot 10^{15}$ cm^{-2} and 25 keV energy and annealed at temperatures ranging from 600 to 1000°C for 1 h.

For low annealing temperatures, the active carrier concentration is reduced, due to dopant segregation at the boundaries. For higher annealing temperatures, the carrier concentration increases again. The two films behave in a very different way: The film that was deposited at low temperature and that was amorphous after the deposition step shows much higher dopant activation. This probably is due to the different mean crystal size in the two samples. The mobility measurements reported in Figure 23 support this conclusion: Films deposited at lower temperature exhibit, for any annealing temperature, higher mobilities. For each sample, mobility increases as the annealing temperature is increased, showing a continuous increase in the mean grain size.

Finally, the carrier mobility is plotted for the different annealing temperatures, as a function of active dopant concentration in Figure 24. The carrier mobility increases with the annealing temperature but never reaches the value for the single crystal at the same doping level (continuous line).

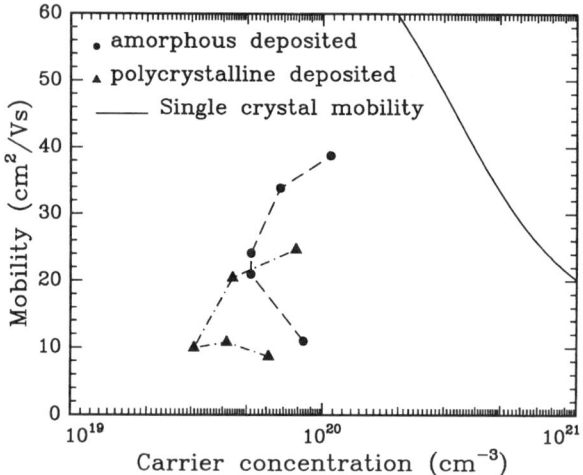

FIG. 24. Carrier mobility versus carrier concentration in amorphous-deposited and polycrystalline-deposited silicon films. The single-crystal mobility–concentration curve is also shown. The films are phosphorus-doped by ion implantation.

IV. Conclusions

The technique of carrier profiling by differential sheet resistance and sheet Hall coefficient measurements obtained by anodic stripping has been reviewed, and a few examples have been examined in order to discuss some specific capabilities of this technique.

The distinctive feature of this technique is the ability to simultaneously measure carrier concentration and mobility profiles. This feature turns out to be essential every time mobility is affected by some other scattering mechanism in addition to scattering by ionized dopant. Examples of this sort have been discussed in Part III, Sections 2(a) and 4. Mechanisms responsible for mobility reduction have been identified. Point defects (whether intrinsic or extrinsic) are found to be much more effective for mobility reduction than are large extended defects in ion-implanted layers, whereas grain boundaries dominate mobility in polysilicon layers.

A comparison between measured mobility and the data in the literature allows us to identify the data in Masetti, Severi, and Solmi (1983) as being the most correct, specifically at high dopant concentration. Available codes for the convertion of spreading resistance into carrier concentration data generally use mobility values that differ from these at high concentrations.

This is probably the reason spreading resistance systematically underestimates carrier concentration at high doping levels. The use of measured values of mobility ensures the correctness of the obtained carrier concentration, so that mechanisms for the activation of dopant in excess of solid solubility can be identified. This has been discussed in Part III, Sections 2 and 3.

REFERENCES

Andrieu, S., Chroboczek, J. A., Campidelli, Y., Andrè, E., and Arnaud d'Avitaya, F. (1988). "Boron Doping of Si MBE Layers: A New High Temperature Effusion Cell," *J. Vac. Sci. Technol.* **6**, 835–841.
Armigliato, A., Nobili, D., Ostoja, P., Servidori, M., and Solmi, S. (1977). "Solubility and Precipitation of Boron in Silicon and Supersaturation Resulting by Thermal Predeposition." (H. R. Huff and E. Sirtl, eds.) *Semiconductor Silicon 1977*, The Electrochemical Society, Princeton, 1977, pp. 638–647.
Baccarani, G., and Ostoja, P. (1975). "Electron Mobility Empirically Related to Phosphorus Concentration," *Solid-State Electron.*, **18**, 579–580.
Beynon, J. D. E., Bloodworth, G. G., and McLeod, I. M. (1973). "The Electrical Properties of Anodically Grown Silicon Dioxide Films," *Solid-State Electron.* **16**, 309–314.
Derdour, M., Nobili, D., and Solmi, S. (1991). "High Temperature Equilibrium Carrier Density of Arsenic-Doped Silicon," *J. Electrochem. Soc.* **138**, 857–860.
Galloni, R., and Sardo, A. (1983). "Fully Automatic Apparatus for the Determination of Doping Profiles in Si by Electrical Measurement and Anodic Stripping," *Rev. Sci. Instrum.* **54**, 369–373.
Hill, A. C., Allen, W. G., and Bradley, R. (1980). "An Automatic Smoothing Algorithm for the Calculation on Impurity Concentration from Sheet Resistivity and Sheet Hall Coefficient Data," *Solid-State Electron.* **23**, 491–496.
Ho, C. P., Plummer, J. D., Hansen, S. E., and Dutton, R. W. (1984). *SUPREM III — A Program for Integrated Circuit Process Modeling and Simulation.* Technical Report SEL84-001, Stanford Electronics Laboratories, Stanford, CA, 1984.
Lin, J. F., Li, S. S., Linares, L. C., and Teng, K. W. (1981). "Theoretical Analysis of Hall Factor and Hall Mobility in p-Type Silicon," *Solid-State Electron.* **24**, 827–833.
Masetti, G., Severi, M., and Solmi, S. (1983). "Modeling of Carrier Mobility Against Carrier Concentration in Arsenic-, Phosphorus- and Boron-Doped Silicon," *IEEE Trans. Electron Devices*, **ED-30**, 764–769.
Miyake, M., Aoyama, S., Hirota, S., and Kobayashi, T. (1988). "Electrical Properties of Preamorphyzed and Rapid Thermal Annealed Shallow p^+-n Junctions," *J. Electrochem. Soc.*, **135**, 2872–2876.
Naem, A. A., and Calder, I. D. (1987). "Formation of Shallow n^+-p Junctions," *J. Appl. Phys.* **62**, 569–575.
Nagasawa, H., and Zukotynski, S. (1978). "Drift Mobility and Hall Coefficient Factor of Holes in Germanium and Silicon," *Can. J. Phys.* **56**, 364–372.
Pennycook, S. J., Narayan, J., and Holland, O. W. (1985). Transient-Enhanced Diffusion during Furnace and Rapid Thermal Annealing of Ion-Implanted Silicon," *J. Electrochem. Soc.* **132**, 1962–1968.

Queirolo, G., Bresolin, C., Robba, D., Anderle, M., Canteri, R., Armigliato, A., Ottaviani, G., and Frabboni, S. (1991). "Low Temperature Dopant Activation of BF_2 Implanted Silicon," *J. Electron. Mat.*, **20**, 373–378.

Queirolo, G., Servida, E., Baldi, L., Pignatel, G., Armigliato, A., Frabboni, S., and Corticelli, F. (1996). "Dopant Activation, Carrier Mobility and TEM Studies in Polycrystalline Silicon Films," *J. Electrochem. Soc.*, **137**, 967–971.

Queirolo, G., and Polignano, M. L. (1992). "Incremental Sheet Resistance and Spreading Resistance: A Comparison." *J. Vac. Sci. Technol.* **B10**, 408–412.

Solmi, S., Baruffaldi, F., and Canteri, R. (1991). "Diffusion of Boron in Silicon during Postimplantation Annealing," *J. Appl. Phys.* **69**, 2135–2142.

Solmi, S., Severi, M., Angelucci, R., Baldi, L., and Bilenchi, R. (1982). "Electrical Properties of Thermal and Laser Annealed Polycrystalline Silicon Films Heavily Doped with Arsenic and Phosphorus," *J. Electrochem. Soc.* **129**, 1811–1818.

Swalin, R. A. (1962). *Thermodynamics of Solids*, John Wiley & Sons, New York, 1962, pp. 143–148.

Thurber, W. R., Mattis, R. L., Liu, R. L., and Filliben, J. J. (1980). "Resistivity-Dopant Density Relationship for Boron-Doped Silicon." *J. Electrochem. Soc.* **127**, 2291–2294.

Van Der Pauw, L. J. (1958). "A Method of Measuring Specific Resistivity and Hall Effect of Discs of Arbitrary Shape," *Philips Res. Reports.* **13**, 1–9.

CHAPTER 7

Transmission Electron Microscopy Analyses

J. Stoemenos

DEPARTMENT OF PHYSICS
ARISTOTLE UNIVERSITY OF THESSALONIKI
THESSALONIKI, GREECE

I. INTRODUCTION 195
II. TRANSMISSION ELECTRON MICROSCOPY 196
   1. *Imaging Ray Path in Transmission Electron Microscopy* 196
   2. *Crystallographic Structure and Chemical Composition* 199
   3. *Specimen Preparation* 199
III. DEFECTS PRODUCED BY ION IMPLANTATION, TRANSMISSION ELECTRON
    MICROSCOPY CHARACTERIZATION 201
   1. *Microelectronics and Microscopy* 201
   2. *Evolution of End-of-Range Defects* 206
   3. *End-of-Range Defects and Electrical Properties* 208
   4. *Kinetics of End-of-Range Defects* 208
IV. IMPLANTATION USING MOLECULAR IONS 210
V. IMPLANTATION CONDITIONS INHIBITING THE FORMATION OF END-OF-RANGE
   DEFECTS 212
VI. MATERIAL MODIFICATION BY ION BEAM SYNTHESIS 216
   1. *Silicon Separation by Implanted Oxygen (SIMOX)* 216
   2. *β-Silicon Carbide Formed by Carbon Implantation into Silicon* 227
VII. ION-BEAM-INDUCED EPITAXIAL CRYSTALLIZATION 230
VIII. CLOSING REMARKS 235
   References 235

I. Introduction

In almost all fields of research in science, engineering, and medicine, transmission electron microscopy (TEM) is indispensable as a method for the direct imaging of submicroscopic structures. In the field of microelectronics, TEM studies generally are applied to two major classes of problems in semiconductor device development: those occurring during device processing and those arising during the development of new processes, such as new patterning procedures and new device materials systems. TEM provides high spatial resolution information on morphology, on crystallographic structure, and with special attachments, on chemical composition of materials. Therefore, a combination of conventional TEM and cross-section transmission electron microscopy (XTEM) can reveal all the essential morphological features of the device.

Doping by ion implantation has many advantages over conventional chemical methods for the formation of the doped layers. Due to radiation damage, however, defects are created by atomic displacements in the host lattice. The depth, nature, and density of the defects depend on the implantation dose, energy, temperature, and current density of the implanted ion. The defects produced by radiation damage are partially removed by subsequent annealing. TEM can give valuable information on the extended defects that are produced during implantation, their distribution, and their evolution during annealing.

II. Transmission Electron Microscopy

1. Imaging Ray Path in Transmission Electron Microscopy

In principle, the imaging ray path in the TEM is analogous to that of the optical microscope, as shown in Figures 1(a) and 1(b). The imaging electrons—typical energy range today is 100 to 400 keV—have a wavelength λ that is short enough (0.0025 nm at 200 keV) to attain subatomic resolution according to the equation

$$\delta = 0.61\,\lambda/D \tag{1}$$

where D is the numerical aperature of the objective lens (Hirsch et al., 1965).

Depending on the material to be investigated, the imaging possibilities are restricted by the fact that electrons with energies ranging from 100 to 200 keV can penetrate specimens only up to a maximum thickness of some tenths of a micrometer. Thus the limiting thickness of silicon (Si) at accelerating voltage 200 keV is about 1 µm. The limiting thickness becomes smaller when specimens for cross-section observations or lattice resolution images are required. For this reason, producing thin foils from bulk material from the area of interest (Part II, Section 3) is one of the main tasks of the TEM preparation technique.

The electron beam incident to the object is modified, depending on the structure of the specimen. On passing through the sample, the beam is scattered by the electrostatic potential variations present, due to the ion cores of the atoms. The scattered beams are refocused by the objective lens

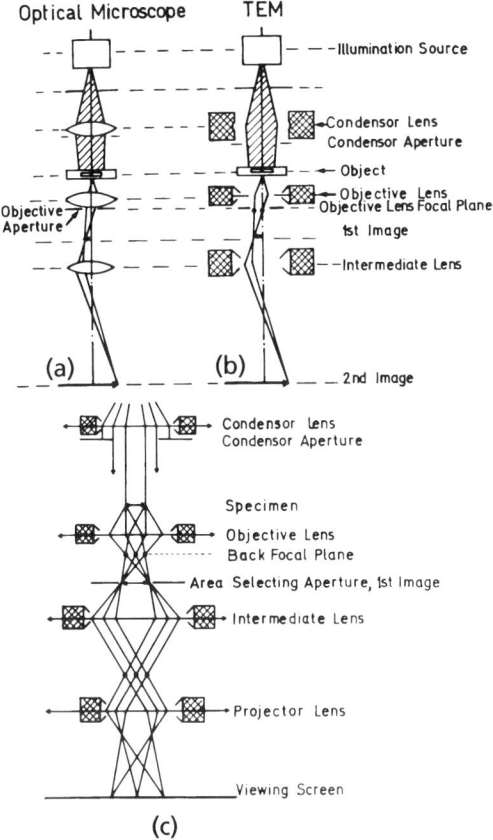

FIG. 1. Simplified ray path: (a) in the optical microscope and (b) in the transmission electron microscope (TEM) for image formation. (c) The ray path in TEM for electron diffraction.

forming the first image. As shown in Figure 2, operation of the objective lens results in formation of the image of the object in the corresponding image plane. At the same time, all those electrons that start from any point of the specimen and have the same direction, are focused in the back focal plane of the objective lens to one point. Therefore, the electrons are diffracted by the specimen, but the beams recombine directly in the instrument without losing the phase information. In this case the Fourier transform of the image is not made by calculation, as with X-rays, but physically, inside the instrument. Thus phase differences present at the exit surface of the specimen are converted into intensity differences as the electrons are recombined to form the image. This is the image due to *phase contrast*, which is important

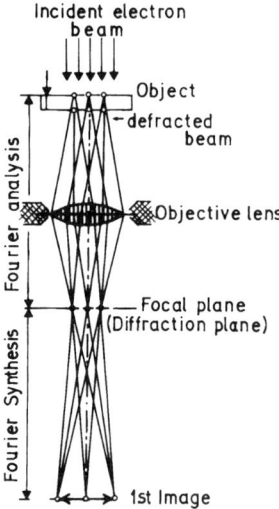

FIG. 2. Schematic representation of the electron optical imaging process.

in lattice resolution studies (Neumann et al., 1987). The resolution obtained by the phase contrast technique is determined by the properties of the electron beam, that is, coherence and stability, the spherical aberration of the objective lens, and the wavelength of the electrons. Today, the resolution of the most advanced TEM with a field emission gun is on the order of 0.1 nm.

Another type of contrast is the *diffraction contrast*. It is possible to eliminate from the image plane all electrons that are scattered and diffracted by inserting an aperture in the focal plane, as shown in Figure 3(a). Thus a bright-field image contrast is obtained. Analogous consideration is valid for dark-field imaging as illustrated in Figure 3(b).

Diffraction contrast is very sensitive in the vicinity of the extended defects. Due to the very small wavelength λ of the electrons, the Bragg condition $\sin/q = \lambda/2d$ is satisfied in TEM, for first-order diffraction, at very small angles $q < 1$ degree, d being the lattice spacing. Therefore, a small distortion of the lattice in the vicinity of the defect results in significant changes in diffraction conditions and, consequently, changes in the contrast, as shown in Figure 3(c) (Edington, 1976).

The diffraction pattern of the illuminated area of the specimen can be observed if the current in the intermediate lens is adjusted so that the back focal plane of the objective is focused on the final screen. If an aperture of diameter D is placed in the first intermediate plane, as shown in Figure 1(c), only diffraction patterns from this area are observed. Diffraction patterns from areas of less than one micron can be observed by this simple method.

7 Transmission Electron Microscopy Analyses

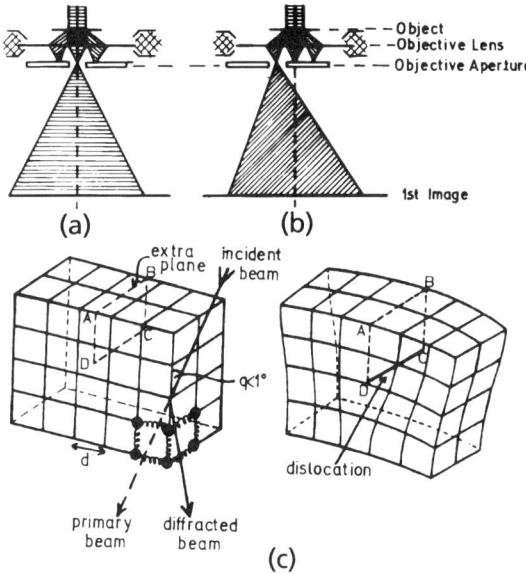

Fig. 3. Ray diagrams of the diffraction imaging: (a) bright-field mode; (b) dark-field mode; (c) change of contrast due to lattice distortion in the vicinity of an extended defect in the lattice.

2. Crystallographic Structure and Chemical Composition

The combined facilities for obtaining TEM images and diffraction patterns from the same specimen offer numerous possibilities for investigating and characterizing the physical and crystallographic nature of the materials. Due to the strong interaction of electrons with matter, very small areas having a diameter of 10 nm can be analyzed.

Chemical information also can be obtained by the use of energy-dispersive X-ray spectroscopy (EDXS). This information can be extracted from regions smaller than 10 nm.

3. Specimen Preparation

The preparation methods for obtaining electron-transparent specimens are some of the most important requirements for practical TEM. The preparation techniques must guarantee that the characteristics of the material are not affected by the preparation method.

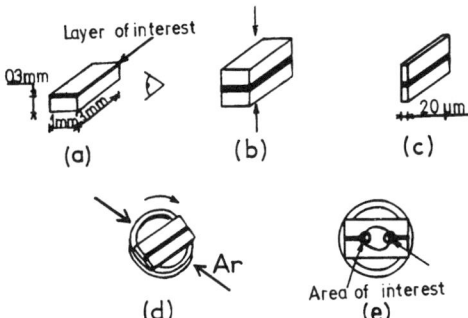

FIG. 4. Steps (a) to (e) for cross-section transmission electron microscopy specimen preparation.

Two types of specimens are used for TEM observations. The first is thin sections of the material parallel to the surface containing the area of interest for plane-view TEM observations. The other method is the cross-section TEM (XTEM) preparation technique, which is very important for the investigation of vertical hetero-systems. This technique enables direct imaging of the vertical structure of epitaxial layers to be made and the ion-implanted layers, interfaces, and device structures.

The cross-section preparation technique includes the following steps, as shown in Figure 4: (a) the material is cut into flakes; (b) the strips are glued face-to-face; (c) the strips are embedded and are ground and polished mechanically down to $30\,\mu$m; and (d) the specimens are glued on a supporting ring and put into an argon (Ar) ion mill for final thinning. Ion milling is accomplished by the impingement of charged Ar ions on a rotating specimen stage. Through this momentum transfer method, material is gradually removed resulting in specimen thickness on the order of a few tenths of a nanometer. Steps (c) and (d) are also applied in the case of plane-view specimens.

The recent advent of highly localized structures in very large scale integrated (VLSI) circuits, with dimensions less than a micron, makes the precise determination of these areas, for XTEM observations difficult using classic methods. For this purpose a new procedure that combines a scanning electron microscope with a focused ion beam for micromachining has been developed. The procedure permits selective thinning in localized areas of $0.2\,\mu$m. Unfortunately, the cost of the apparatus used for this procedure is almost the same as the cost of a modern TEM.

Chemical and electrochemical polishing methods are the most widely used procedures for the final thinning of bulk specimens. Electrochemical methods are based on an electrolytic cell, with the sample as the anode.

Based on these electrolytic preparation methods many different kinds of polishing equipment have developed (Hirsch et al., 1965). The most widely used technique is the jet method in which the electrolyte is directed onto the specimen by a liquid beam generated by a jet. Electrolytic polishing is preferred to chemical thinning, since control of the current in the cell offers an additional parameter for determining the best polishing conditions.

III. Defects Produced by Ion Implantation, Transmission Electron Microscopy Characterization

1. MICROELECTRONICS AND MICROSCOPY

Development of near-micron and submicron VLSI technology calls for better microscope resolution than that provided by optical or scanning electron microscopy (SEM). The optical microscope has played a very important role in the evolution of integrated circuit (IC) technology. The spatial resolution $\delta = 1\,\mu\text{m}$ of the optical microscope was sufficient for the inspection of the IC features of the 1960s.

Concurrent with IC development was a growing interest in the structural characteristics of the material used and the defect formation, during the different processing steps, for large scale integration (LSI). Therefore, TEM was considered useful in exploratory processing projects for LSI circuits. In this scale of integration the essential instrument for the study of the IC device was SEM, where a practical resolution of 10 nm is attainable. The broadening of the incident beam as it penetrates the specimen, however, results in loss of resolution. In addition, the contrast produced by the defects is very poor (Marcus and Sheng, 1983), which is a drawback to the use of SEM for the structural characterization of submicron devices. In VLSI and especially in ultra-large-scale integrated (ULSI) circuits in which the sizes of the devices range from 0.1 to $0.2\,\mu\text{m}$, TEM characterization is necessary in routine process control.

The generation of defects by ion implantation into semiconductor material is a process of primary and secondary collisions of ions with atoms of the crystal. Due to radiation damage, a high concentration of Frenkel pairs, interstitials and vacancies, is generated. The result is the formation of small, highly disordered clusters, finally leading to an amorphous layer for a sufficiently high defect density. Depending on the implanted ion, implant dose, implantation current density, ion energy, and substrate temperature, the amorphous zone can be either completely buried within the host matrix or intersect the upper surface. Because the implanted ions are generally

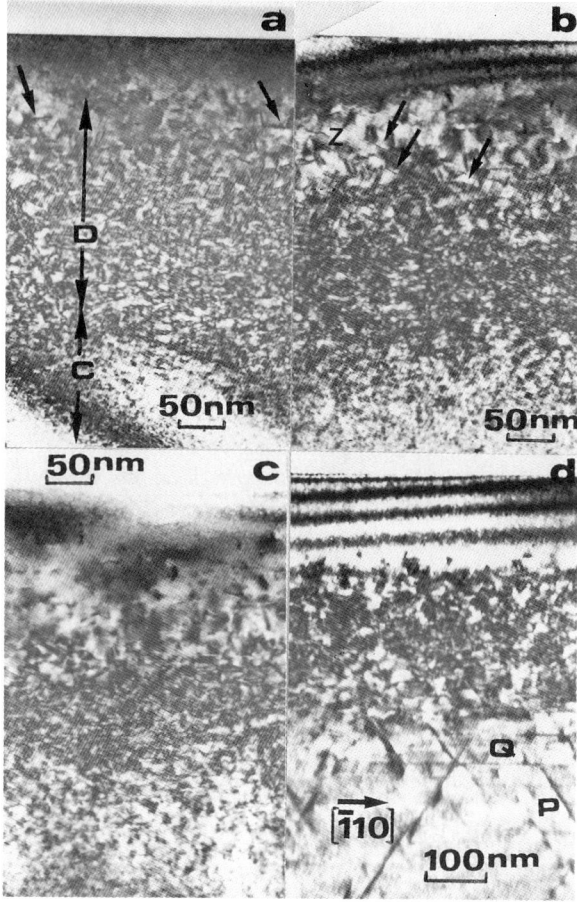

FIG. 5. Cross-section transmission electron micrographs from a (001) silicon (Si) wafer implanted by Si ions at a dose of 10^{14} Si$^+$/cm^{-2} and an energy of 120 keV at room temperature: (a) as-implanted (arrows indicate a zone of high-density small dislocation loop); (b) annealed at 400°C for 1 h, (defects on {113} planes); (c) annealed at 600°C for 1 h; and (d) annealed at 800°C for 1 h. D, high-density defect zone; C, small clusters; Z, zigzag-shaped defects; P, perfect loops; Q, long dislocation dipoles.

located in non–electrically-active sites, postimplantation annealing is applied to recover the damaged zone and to activate the dopant atoms.

For usual implantation conditions the following features should be emphasized. Typically, the defect density increases as the temperature decreases. An exponential dependence between the threshold dose for amorphization and the implantation temperature exists. This is expected since at lower temperatures, vacancies and interstitial atoms have a lower diffusivity, not permitting at least partial recovery of the damage produced during implantation. (Mazey, Nelson, and Barnes, 1968; Vook, 1973). Under constant implantation temperature, as the dose and the ion current density increases, the defect density increases (Vook, 1973).

The extended defects that are produced during ion implantation are visible when the condition of implantation is just below the amorphization threshold, as shown in the cross-section TEM micrograph in Figure 5(a). This is a Si wafer implanted at a dose of 10^{14} Si$^+$/cm^{-2} and energy of 120 keV at room temperature (RT). The wafer was implanted by Si ions in order to avoid the formation of defects due to the presence of impurities in the host lattice. A high-density defect zone was formed, denoted by D, in Figure 5(a). The maximum amount of defects appears at a depth of 200 nm from the surface. This zone consists of a high-density of small dislocation loops, denoted by arrows in Figure 5(a). In the maximum of the defect zone the density of the loops is so high that interaction between the loops occurs, resulting in the formation of dislocations segments. Below zone D some very small clusters are observed, denoted by C. The mean size of the loops is 4 nm and their density is estimated to be 5×10^{11} cm^{-2}.

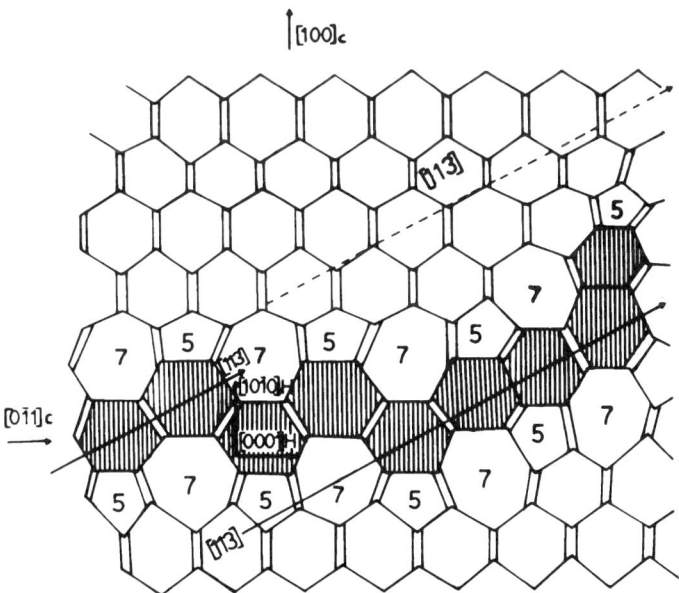

FIG. 6. Schematic representation of the rodlike defect in atomic scale. The hexagonal structure is denoted by the shaded area.

Another commonly observed defect is the $\{11\bar{3}\}$-type defect also called the rodlike defect (RLD). The RLDs are mainly developed after low-temperature annealing at around 400°C for 1 h. They are laid on $\{113\}$ planes, denoted by arrows in Figure 5(b). Very often they take a zigzag shape, switching from the $\bar{1}13$ to the $\bar{1}1\bar{3}$ plane, denoted by Z in Figure 5(b). The RLDs are elongated, directed along the six possible $\langle 110 \rangle$ directions in the diamond lattice (Bartsch, Hoehl and Kastner, 1984).

High-resolution TEM (HRTEM) observation reveals that RLDs are of interstitial type (Bourret, 1987). These defects consists of hexagonal Si layers, as shown by the shaded areas in Figure 6, bounded by five- and seven-membered rings, as those used for describing grain boundary structures (Tan, Foll, and Krakow, 1981). At high temperatures (800°C) this form is no longer stable and, after gliding, it may be transformed into an elongated perfect dislocation loop or a Frank partial interstitial loop (Bourret, 1987). After postimplantation annealing at 600°C for 1 h, a coarsening of the dislocation loops in the uppermost part of the crystal occurs, as shown in Figure 5(c). Zone D shrinks but the high density of defects remains, even after annealing at 800°C for 1 h. This reveals the difficulty for eliminating structure defects once they are formed. Only the uppermost part of the crystal is free of defects, as shown in Figure 5(d). Unexpectedly large defects were formed at the back of zone D, propagating deeper inside the Si substrate. These are perfect loops, denoted by P, and long dislocation dipoles along the $[\bar{1}10]$ direction, denoted by Q in Figure 5(d).

The high density of the defects produced by ion implantation is drastically reduced by amorphization of the implanted area. The amorphous layer is recovered by solid-phase epitaxial (SPE) regrowth at temperatures ranging from 600 to 900°C. The only disadvantage of this technique is the formation of extended defects beneath the former crystalline-amorphous (c-a) interface, which are called end-of-range (EOR) defects.

The evolution versus annealing temperature of the amorphous layer and the EOR defects of a Si specimen implanted with arsenic (As) at a dose of $1 \times 10^{15} \text{As}^+/\text{cm}^{-2}$ at an energy of 120 keV at RT are shown in Figure 7. An amorphous 115-nm-thick layer was formed in the implanted specimen, denoted by A in Figure 7(a), followed by a 35-nm-thick EOR defect zone, denoted by an arrow in Figure 7(a). This zone consists of a high density of small clusters, having a mean diameter of 2 nm. The c-a interface is rough, and very often small crystalline Si islands appear inside the amorphous zone, very close to the c-a interface. Due to the very small size of the clusters in the EOR defect zone, it is difficult to determine their type. It recently has been shown that there are clusters of interstitial type formed during irradiation that co-exist with complexes of vacancy-type agglomerates (Fedina et al., 1995).

7 TRANSMISSION ELECTRON MICROSCOPY ANALYSES

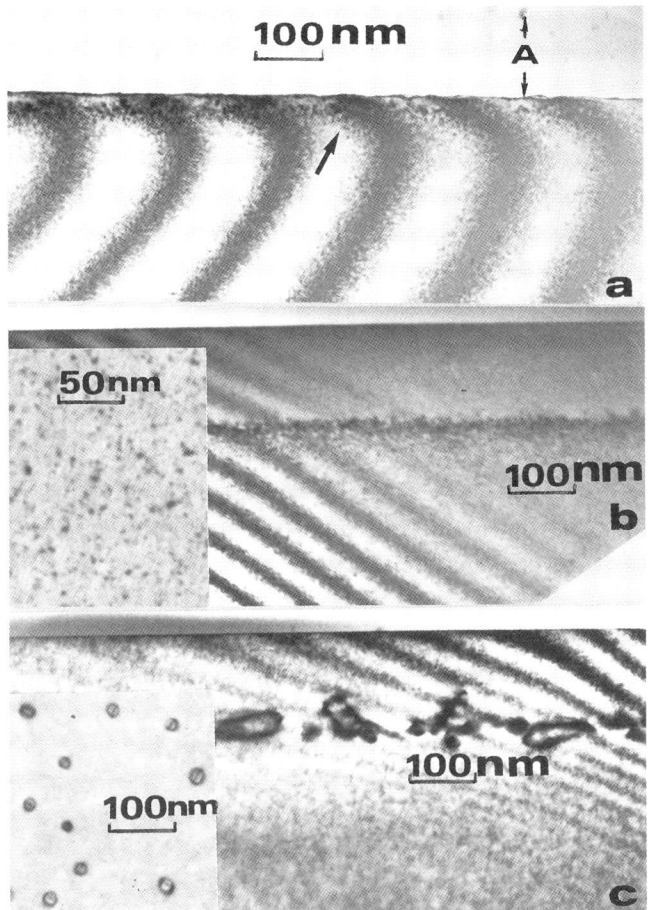

FIG. 7. Evolution of end-of-range (EOR) defects versus annealing time of a (001) silicon wafer implanted with arsenic (As) at a dose of $1 \times 10^{15}\,\mathrm{As}^+/\mathrm{cm}^{-2}$ at an energy of 120 keV: (a) As-implanted (arrow indicates a 35-nm-thick EOR defect zone); (b) Annealed at 600°C for 100 min; and (c) Annealed at 900°C for 100 min. A, implanted specimen.

The amorphous layer is completely recrystallized after annealing at 600°C for 100 min, leaving behind a defect zone located near the c-a interface, as shown in Figure 7(b). The mean size of these defects, which are dislocation loops, remains small—on the order of 5 nm—shown by the plane-view micrograph in the inset of Figure 7(b). Annealing the implanted specimen at 900°C for 100 min results in a significant coarsening of the defected zone, as shown in the cross-section micrograph in Figure 7(c) and in the plane-view micrograph in the figure inset. After high-temperature annealing, only dislocation loops of extrinsic type remain in the defect zone (Mauduit et al., 1994). Identification of these loops reveals that 75% are Frank loops with Burger's vector $b = a/3\langle 111 \rangle$. The remaining 25% are perfect loops with Burger's vector $b = a/2\langle 110 \rangle$ and are located close to $\{111\}$ planes (Mauduit et al., 1994). Therefore, considering the size of the loops and their density, it is possible to estimate the number of the Si-interstitials captured in the loops during ion implantation.

2. Evolution of End-of-Range Defects

End-of-range defects are responsible for the leakage currents, when situated in the space charge region of the p-n function (Tanaka et al., 1990; Hong et al., 1991). Therefore, it is important to study the evolution of these defects during postimplantation annealing. For this purpose, an in situ annealing experiment was performed in order to study the early stage of the solid-phase crystallization (SPC) regrowth in a cross-section TEM specimen. A Si specimen implanted with 5×10^{15} As$^+$/cm^{-2} at 150 keV was used for the situ annealing experiment.

In the nonannealed specimen, an amorphous 90-nm-thick layer, denoted by A, is shown in Figure 8(a). After annealing at 420°C for 25 min the thickness of the amorphous layer did not change, as shown in Figure 8(b), revealing that no SPC occurred. However, small clusters, denoted by arrows in the crystalline substrate, formed well below the c-a interface. Further heating of the same area at 510°C for 70 min resulted in complete recrystallization of the amorphous layer, as shown in Figure 8(c). The recrystallization velocity V was determined by measuring the average distance of the c-a interface from the free surface, which is in agreement with the equation

$$V = 3.07 \times 10^3 \, \text{cm/s} \exp\left(-\frac{2.68 \, \text{eV}}{kT}\right) \qquad (2)$$

proposed by (Narayan and Holland, 1984). The density of the EOR defects was increased after the annealing at 510°C; however, no defects were produced in the recrystallized zone, as shown in Figure 8(c).

The extrinsic character of the EOR defects was also confirmed by lattice strain measurement of the implanted and annealed Si specimens,

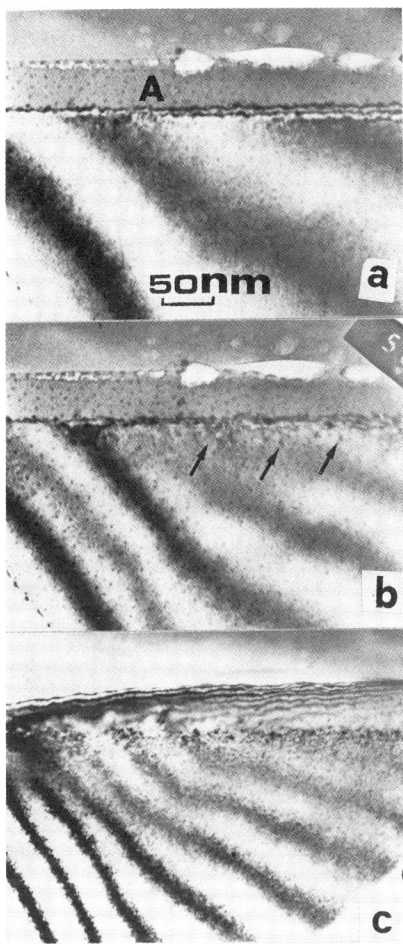

FIG. 8. Cross-section transmission electron micrographs from in situ annealing of a silicon specimen implanted with $2 \times 10^{14}\,As^{+}/cm^{-2}$ at 150 keV: (a) as-implanted (arrows indicate small clusters in the crystalline substrate); (b) annealed at 419°C for 25 min; and (c) annealed at 510°C for 70 min. A, amorphous 90-nm-thick layer; As, arsenic. (Reprinted from Boussey-Said, J., et al. (1992). J. Appl. Phys. **72**, 61, with permission.)

using an X-ray triple-crystal diffractometer (XTCD). The depth profiles of the normal to the surface strain versus annealing temperatures of 400 to 800°C in a sample implanted with $2 \times 10^{14}\,As^{+}/cm^{-2}$ at an energy of 200 keV, is shown in Figure 7 of Chapter 9. The maximum of the strain is in the EOR defect zone and is always positive (expansion strain).

3. END-OF-RANGE DEFECTS AND ELECTRICAL PROERTIES

The structural changes observed in the EOR defect zone at 400°C during the in situ annealing experiment are related to a significant reduction of resistivity, as revealed by spreading resistance measurements (Boussey-Said *et al.*, 1992). In order to make this relation evident, the depth profiles of the spreading resistance measurements are drawn on the XTEM micrographs in the same specimen that was implanted with $2 \times 10^{14}\,\text{As}^+/\text{cm}^{-2}$, 200 keV and annealed for 1 h at various temperatures from 300 to 900°C, as shown in Figure 9. A dramatic change in the resistivity at 400°C is evident in Figure 9(b). This change can be explained by considering the electrical activation of the As atoms at this temperature. It is worth noting that the dopant electrical activation does actually occur in the amorphous overlayer, as revealed by Figures 9(a) and 9(b), respectively. Furthermore, by comparing Figure 9(a) with 9(b), it is evident that no change in the thickness of the amorphous layer occurred during the annealing at 400°C. The only change was observed in the EOR defect zone close to the c-a interface where small clusters appear, which are denoted by E in Figure 9(b). This is in agreement with the observation in the in situ annealing experiment, as shown in Figure 8(b). Moreover, spreading resistance measurements in the same specimen annealed at 450°C for various durations ranging from 15 to 900 min suggest As dopant activation in the first 15 minutes (Boussey-Said *et al.*, 1992). At this low temperature, As dopant diffusion is very low. Only for very long annealing duration (~ 900 min) does As dopant diffusion become significant. It should be noted that after the dopant activation at 400°C, no discontinuity in the resistivity profile appears at the c-a interface of the partially recrystallized amorphous zone during the annealing at 500°C for 1 h, as shown in Figure 9(c). This reveals that dopant activation play the most crucial role in the reduction of resistivity (see Chapter 7 for more).

4. KINETICS OF END-OF-RANGE DEFECTS

Coarsening of the EOR defects at 400°C, before the start of SPC can be explained as follows. The zone beneath the c-a interface is already supersaturated by Si-interstitials due to the recoil of the Si atoms during the implantation (Sadana *et al.*, 1985). As the arsenic atoms are activated by occupying substitutional sites at 400°C, additional self-interstitials are created. The density of the interstitial loops therefore is determined by the competition of the excess interstitials to be diffused to the c-a interface, which is the natural sink for them, and by the nucleation rate of the loops.

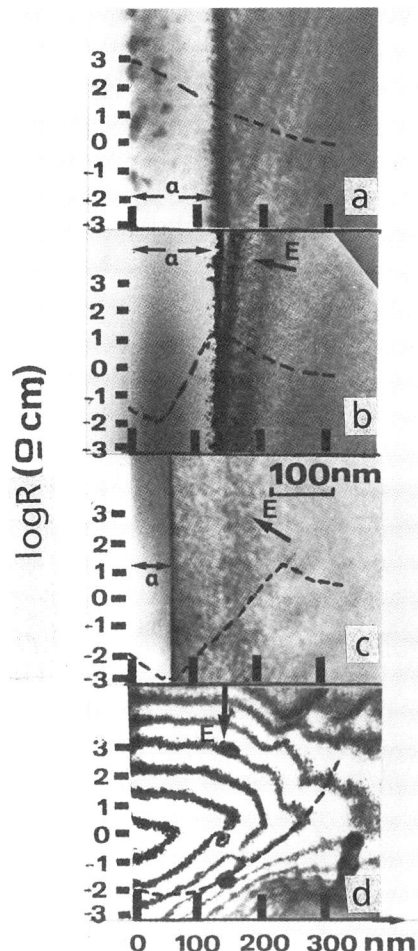

FIG. 9. Spreading resistance measurements depth profiles are combined with cross-section transmission electron micrographs from a specimen implanted with arsenic (As) at a dose of 2×10^{14} As$^+$/cm^{-2} at an energy of 200 keV and annealed for 1 h at (a) 300°C; (b) 400°C; (c) 500°C; and (d) 900°C. a, amorphous; E, change in end-of-range defect zone, close to the crystalline-amorphous (c-a) interface, where small clusters appear. (Adapted from Boussey-Said, J., et al. (1992) J. Appl. Phys. 72, 61, with permission.)

However, the c-a interface does not act as a sink for the interstitials because no recrystallization occurs at 400°C. As evident from Figures 7 and 9, loop density is reduced upon annealing, whereas mean size increases. This behavior is known as "Ostwald ripening." According to the simple nucleation theory, a precipitate has a critical radius below which it is unstable.

The critical radius depends on the annealing temperature and the conditions of supersaturation, (Burke, 1965; Craven, 1981). It recently has been shown that extrinsic loops can emit or capture Si-interstitials, depending on their size and annealing condition (Claverie et al., 1995). In addition to growth kinetics, quantitative analysis shows that the total number of atoms stored in the loops decreases with increasing annealing time at temperatures above 900°C. Thus the loops shrink by emission of interstitials. Therefore, the EOR defects can be considered as reservoirs able to maintain a high supersaturation of self-interstitials, resulting in an enhanced diffusion of substitutional dopants, such as As, phosphorus (P), and boron (B) (Claverie et al., 1995).

IV. Implantation Using Molecular Ions

For the fabrication of shallow junctions in Si, implantation using boron fluoride (BF_2^+) molecules is an important doping technique. This technique has several advantages over B^+ implantation because BF_2^+ is higher than the B^+ beam current and the implantation energy of BF_2^+ is higher than that of B^+ to form identical B range disribution (Nieh and Chen, 1986).

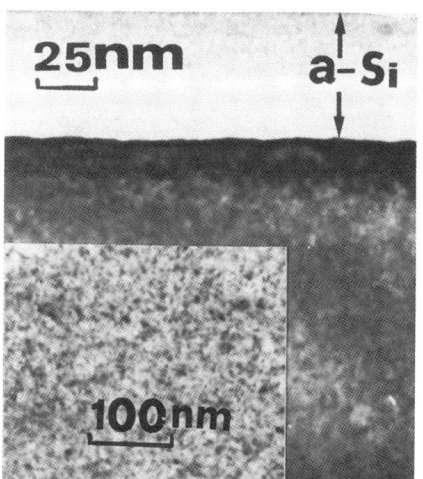

FIG. 10. Cross-section transmission electron micrographs from silicon specimen implanted with 3×10^{15} BF_2^+ cm^{-2} at 40 keV. In the inset a plane view micrograph shows the formation of small clusters in the end-of-range (EOR) defect zone.

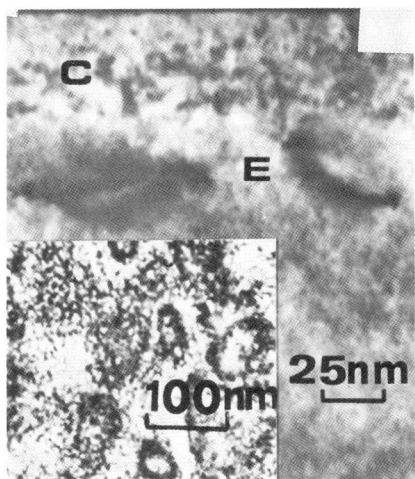

FIG. 11. The same specimen as in Figure 10 after an annealing at 900°C for 100 min. Large loops in the end-of-range (EOR) zone (E) and small fluorine (F) bubbles in the recrystallized layer were formed, denoted by the letter C.

However, due to the presence of fluorine (F), defects such as F bubbles appear after postimplantation annealing.

The evolution of the F bubbles during the postimplantation annealing in a specimen implanted with a dose of $3 \times 10^{15}\,BF_2^+/cm^{-2}$ at 40 keV is shown in Figures 10 and 11. From the XTEM micrograph of the as-implanted specimen in Figure 11, an amorphous zone about 50-nm thick can be seen followed by a defect zone 20-nm thick beneath the c-a interface. The defect zone consists of small clusters, as revealed by the plane-view micrograph in the inset of Figure 10. Annealing at 900°C for 100 min results in the recrystallization of the amorphous zone. Large loops are formed at a depth of 75 nm, which corresponds to the EOR defect zone, denoted by E in Figure 11. However, small black dots with a mean diameter of 7 nm are also observed in the recrystallized zone, having the maximum of their density at a depth of 25 nm from the surface. The black dots, denoted by C in Figure 11, are cavities of F that were not out-diffused to the surface during the postimplantation annealing (Chen, Niek, and Chu, 1988). Superposition of EOR loops and small F cavities are evident in the plane-view micrograph in the inset in Figure 11. In specimens implanted with BF_2^+ and subsequently annealed at less than 1000°C, reverse-bias leakage current has been reported. The leakage current was reduced after longer annealing at higher temperatures by driving the active junction far away from the EOR defect zone (MacIver and Greenstein, 1977).

V. Implantation Conditions Inhibiting the Formation of End-of-Range Defects

The formation of EOR defects can be diminished by different implantation techniques. The three most successful of these techniques are the following:

1. *Multiple-step implantation with subsequent annealing.* The density of the dislocations formed at the EOR defect zone mainly depends on the number of Si atoms displaced during implantation. The critical number of displaced atoms depends on the ion mass; for example, it is 10^{16} Si/cm^2 for B implantation. This critical number occurs at a dose of 10^{14} B$^+$/cm^2 at 200 keV. Below this value no EOR defects are formed. Therefore, multiple-step implantation at the subcritical dose and subsequent annealing after each step prevent the formation of EOR defects (Saris et al., 1992).
2. *Suppression of EOR defects by carbon implantation.* Dopant implantation with subsequent carbon (C) implantation just beneath the c-a interface inhibit the formation of EOR loops during the subsequent postimplantation annealing (Tamura, Ando, and Ohyu, 1991).

The mechanism that prevents the formation of EOR loops is not known exactly. It has been proposed that carbon atoms react with Si-interstitials to form small stable clusters. It is known that carbon in Si can shrink the lattice constant, thereby creating free volumes for carbon–silicon interstitial agglomerates. For a stress-free state of an agglomerate, a 1:1 ratio of carbon and Si atoms is expected. The critical radius r_c of such an agglomerate has been estimated as

$$r_c = \sigma_c \cdot \Omega/k \cdot T \cdot \ln[C_c \cdot C_I/C_c^{eq} \cdot C_I^{eq}] \tag{3}$$

where σ_c is the interface energy, C_c^{eq} is the equilibrium carbon concentration, C_I^{eq} is the equilibrium Si-interstitial concentration, and Ω is the volume per solute carbon atom (Gosele, 1986). From Eq. (3) it is evident that a high supersaturation of Si-interstitials may lead to formation of very small C-Si clusters. An interesting feature is that no extended residual defects are formed after carbon implantation, except for a black-band region located at the R_p depth. No cluster or carbon precipitates can be observed in this case even by high-resolution TEM (Wong et al., 1988).

A direct proof of the preferred silicon carbide (SiC) agglomeration in areas in which Si interstitial supersaturation exists, is the decoration

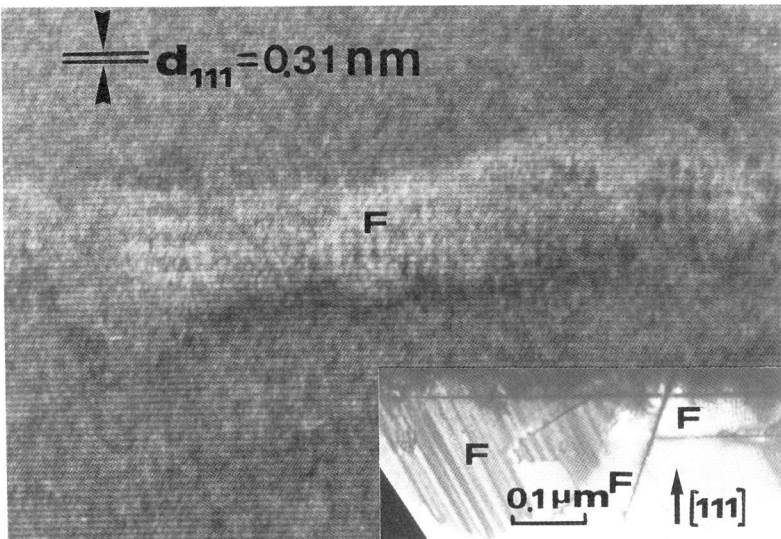

FIG. 12. High-resolution cross-section transmission electron micrograph from a (111) silicon wafer implanted by carbon at a dose of $2 \times 10^{17} \text{C}^+/\text{cm}^{-2}$ at 200 keV at implantation temperature 950°C. Extrinsic stacking faults decorated by small cubic silicon carbide (SiC) precipitates were formed. In the inset the same area at low magnification, stacking faults are denoted by the letter F. (Reprinted from Nejim, A., et al. (1995). App. Phys. Lett. **66**, 15, with permission.)

of extrinsic stacking faults (SF_s) by small 3C-SiC precipitates in Si implanted by carbon at 950°C, as shown in the high-resolution cross-section micrograph in Figure 12 (Frangis et al., 1995). The decorated SF are shown at low magnification in the inset of Figure 12.

Although co-implantation of dopant and carbon suppresses the formation of extended defects, the leakage current of the p-n junction is increased revealing that invisible microdefects degrade the p-n junction (Wijburg et al., 1992).

Carbon precipitates are very efficient for gettering impurities. Thus high-energy carbon implantation in the range of MeV forms a carbon buried layer, which is a very efficient gettering zone for accumulation of gold (Skorupa et al., 1991).

3. *High-temperature implantation.* High-temperature implantation permits in situ annealing of the defects produced by radiation damage.

The specimen can be heated by the beam current or an external heating source. The first case is simple but not well controlled. In general, a high current density is needed to increase the temperature to the range of from 500 to 600°C. The critical dose for amorphization is

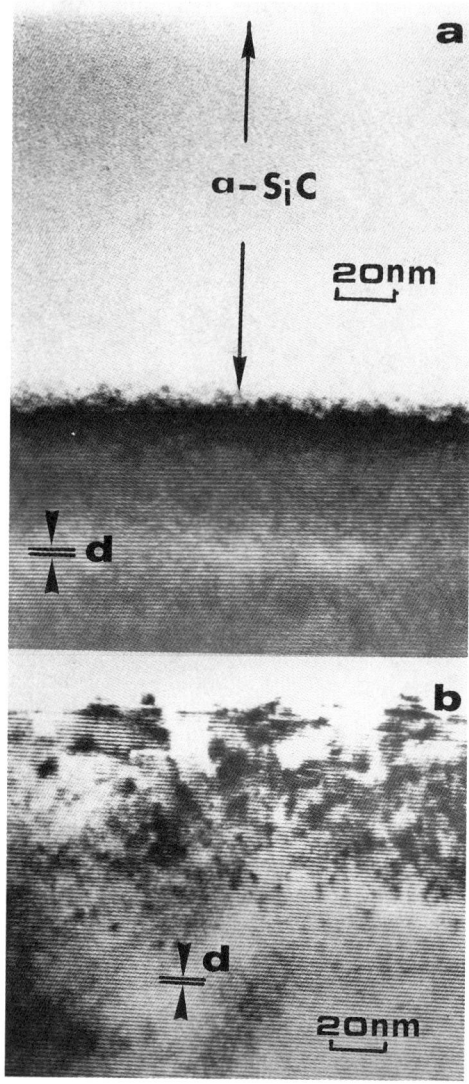

FIG. 13. Cross-section transmission electron micrograph from a hexagonal 6H-SiC polytype implanted by germanium (Ge) at a dose of $1 \times 10^{15}\,\text{Ge}^+/\text{cm}^{-2}$ at 200 keV at room temperature. (a) As-implanted, the amorphous SiC is shown by the arrows, (b) Annealed at 1500°C for 10 min. The amorphous zone was recrystallized. The fringes with periodicity d of 1.5 nm represent the unit-cell length of the 6H polytype along the C axis. Distortion of the fringes at the uppermost part of the recrystallized zone, is denoted by arrows. The off-set of the fringes is associated with the development of strain, which is shown by the strong black and white contrast at these defects. These are growth defects developed during the recrystallization process. A second zone of defects that corresponds to the end-of-range (EOR) defect zone appears deeper in the specimen. It consists of clusters with a diameter ranging from 1.5 to 3 nm, which appear as black dots.

strongly dependent on the implantation temperature, under constant beam-current density. The critical dose decreases as the beam current increases, keeping the temperature constant (Vook, 1973). Therefore, increasing the temperature by increasing the current is beneficial in annealing the radiation damage. However, this advantage is partially eliminated because the high current density increases the density of the collisions.

The use of an external heating stage permits implantation to be carried out under conditions of controlled temperature, unaffected by the beam power. Today, implantation up to 1000°C, using an external heating stage, is feasible from a technological standpoint (Schork et al., 1991).

High-temperature implantation is especially suitable to doped semiconductors, which are recrystallized at high temperatures, for example, in silicon carbide (SiC) which is recrystallized above 1000°C. Even at this temperature the recrystallized layer has a high density of defects. A cross-section micrograph from a hexagonal 6H-SiC polytype amorphized by Ge implantation at a dose of $1 \times 10^{15}\, Ge^+/cm^{-2}$ and

FIG. 14. Implantation of cubic silicon carbide (SiC) by nitrogen (N) at a dose of $5 \times 10^{14}\, N^+/cm^{-2}$ at 50 keV, at a temperature of 800°C. No defects due to ion implantation were formed. The observed defects are stacking faults formed during the epitaxial growth of SiC on Si substrate.

200 keV is shown in Figure 13(a), where an amorphous zone about 122-nm thick was formed. The same specimen was recrystallized after an annealing at 1500°C for 10 min, as shown in Figure 13(b). The uppermost part of the recrystallized zone has a high density of defects, as shown in Figure 13(b) (Pacaud et al., 1996). In contrast, implantation of SiC at high temperatures results in activation of the dopant without defect formation. This is evident in Figure 14, which is a cubic SiC polytype (β-SiC) grown epitaxially on Si substrate implanted with 5×10^{14} N$^+$/cm^{-2} and an energy of 50 keV at 800°C. Figure 14 shows growth defects formed during the epitaxial growth of β-SiC on Si due to 21% misfit. No extended defects due to ion implantation are observed. After high-temperature implantation, nitrogen activation was 12% (Lossy, Obermeier, and Stoemenos, 1995).

VI. Material Modification by Ion Beam Synthesis

In the past, applications of implantation were dominated by the small-beam-current implanters with a beam current on the order of 100 μA. Recently, a new generation of high-current implanters has entered the market, having a current on the order of 85 mA at an energy of 180 keV. These can implant five orders of magnitude faster that can conventional implanters allowing us to implant the large concentration required for compound formation.

Modification by ion beam synthesis is the most advanced part of ion implantation because by carefully tailoring the dose, temperature, and energy of the implant and by subsequent annealing, buried layers of new materials can be formed that leave the host crystal free of defects. Thus buried layers of silicon dioxide (SiO$_2$), Si$_3$N$_4$, and metal disilicides such as C$_0$Si$_2$ are formed (White et al., 1988; Veirman et al., 1990).

1. SILICON SEPARATION BY IMPLANTED OXYGEN (SIMOX)

Silicon separation by implanted oxygen (SIMOX) in order to form buried SiO$_2$ layers is one of the best examples of the successful cooperation of recent technological advancements in the field of ion implantation and materials science. Although SIMOX may not be the most gentle method for the formation of a SiO$_2$ buried layer into Si (Christoloveanu, 1991), it is still the most successful technology for the fabrication of very thin silicon on insulation (SOI) structures for realization of fully depleted metal-oxide semiconductor field-effect transistor (MOSFET) devices (Colinge, 1991).

7 TRANSMISSION ELECTRON MICROSCOPY ANALYSES

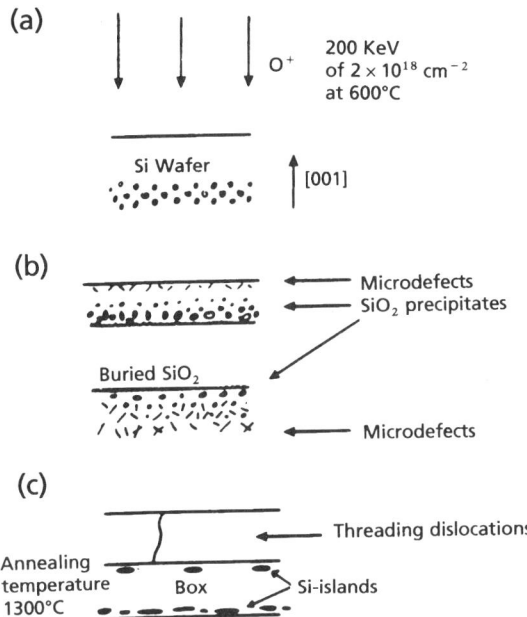

FIG. 15. (a), (b), (c) Schematic representation of the main step for the formation of standard silicon separation by implanted oxygen. SiO_2, silicon dioxide.

The commercially available standard SIMOX wafers are produced in Eaton NV-200 implanters at an energy of 190 keV and a dose on the order of $1.8 \times 10^{18} O^+/cm^{-2}$ and an implantation temperature (T_i) of 600 to 650°C. The wafers are subsequently annealed for 6 h at 1300°C under Ar plus 1% O_2 atmosphere. Wafers subjected to double- or triple-step implantation subsequently annealed at 1320°C for 6 h also are commercially available. The main steps of SIMOX formation are shown in Figure 15. Due to extreme implantation and annealing conditions applied during SIMOX, it is particularly useful to study the evolution of the defects along the different steps of fabrication.

a. Defects in the Silicon Overlayer of Standard SIMOX

The defects in the Si overlayer of standard SIMOX are dislocations and stacking faults.

Dislocations. The dislocation density of standard material is on the order of $5 \times 10^5 \, cm^{-2}$. However, the multistep implantation and annealing pro

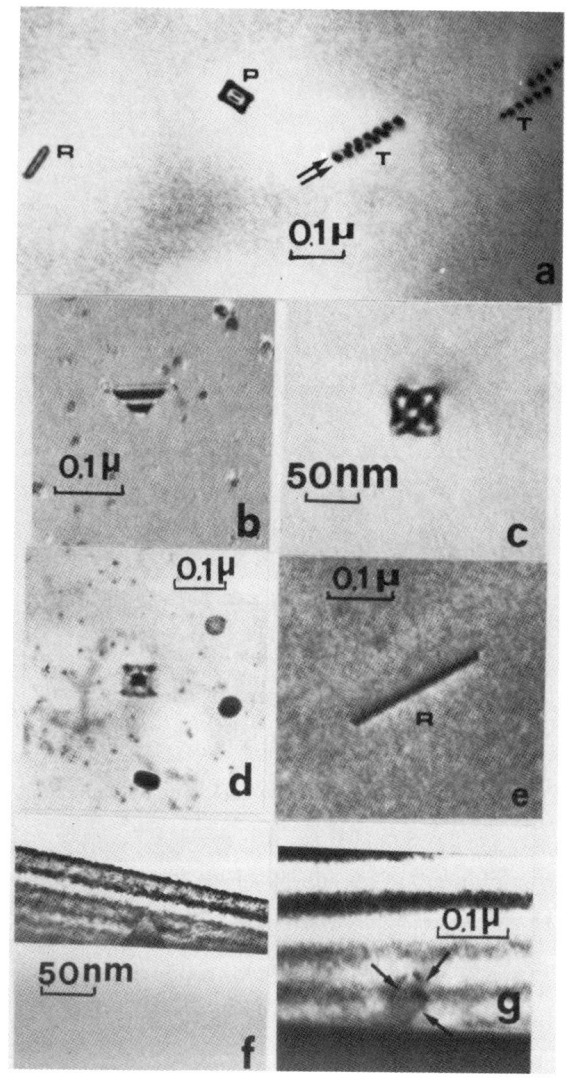

FIG. 16. Defects observed in the silicon (Si) overlayer of the standard silicon separation by implanted oxygen (a) Plane-view micrograph reveals a pair of inclined to the foil dislocations denoted by arrows. (b) Single stacking fault (SF). (c) Tetrahedral stacking faults (TSF) give a very symmetrical contrast under multiple-beam reflection with the electron beam exactly parallel to [001] crystallographic orientation. (d) An SF in the form of a tetragonal pyramid. A similar, slightly distorted tetragonal pyramid is denoted by P in (a). (e) Very shallow prismatic SF denoted by R in (a). (f) Cross-section micrograph of a shallow prismatic SF. The fault is viewed along the long axis. The fault is located on the Si–SiO$_2$ interface. (g) Cross-section micrograph from a complex fault bounded by three single SFs. The fault is viewed along the [110] direction and is located at the silicon–silicon dioxide (Si–SiO$_2$) interface. T, defects. (Reprinted from Stoemenos, J., et al. (1995). J. Electrochem. Soc. **142**, 1248, with permission

cess can reduce the dislocation density to 10^4cm^{-2} (Stoemenos et al., 1995). These defects are edge-type threading dislocations with Burger's vectors $1/2\langle 110\rangle$. These defects appear in pairs and display an oscillating contrast, revealing that they are inclined on the plane of the foil (Margail, Lamure, and Papon, 1992). They are denoted by T in Figure 16(a).

Stacking Faults (SFs). Small SFs having a size less than $0.2\,\mu\text{m}$ and density lower than 10^4cm^{-2} are observed, as shown in Figure 16(b). Small tetrahedral stacking faults (TSF) as well as square-shaped stacking faults are also observed, as shown in Figures 16(c) and 16(d), respectively. Very often the square-shaped SFs are slightly distorted, which is denoted by the letter P in Figure 16(a). Systematic studies by TEM have shown that TSF and square-shaped SFs are located at the back of the Si-overlayer, having a density of about $6 \times 10^4 \text{cm}^{-2}$. However, the most frequently observed fault in this zone is the prismatic SF, denoted by the letter R in Figures 16(a) and 16(e), having a density of 10^6cm^{-2}. The prismatic SFs are shallow and small, not exceeding 50 nm. The configuration of the different SF complexes are shown schematically in Figure 17.

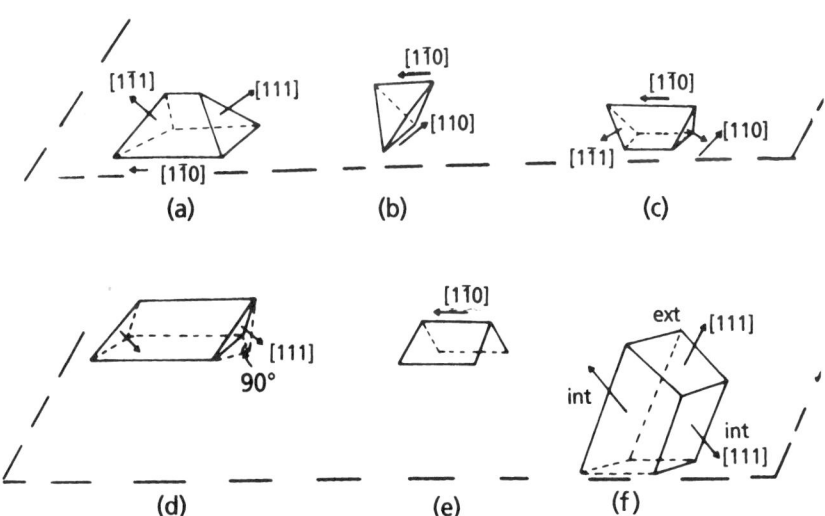

FIG. 17. Schematic showing configurations of stacking fault (SF) complexes at the silicon–silicon dioxide (Si–SiO$_2$) interface. (a) Tetragonal pyramid SF complex. (b) Tetrahedral SF. (c) Not completed tetrahedral SF. (d) SF complex that consists of a pair of parallel and a pair of opposite SFs. (e) Prismatic defect consists of two opposite SFs. (f) Two parallel SFs terminated by a third SF. int., interface. (Reprinted from Stoemenos, J., et al. (1995). *J. Electrochem. Soc.* **142**, 1248, with permission.)

b. Defects in the Buried Oxide (BOX)

The defects in the buried oxide are point defects and Si-islands.

Point defects. The most common point defects are strained $O_3 \equiv Si-Si \equiv O_3$ bonds formed by the trapping of Si atoms in the buried oxide (BOX). The strained Si-Si bonds can relax by capturing a hole and forming an E'_1 center. Point defects are invisible by TEM; however, the E'_1 centers are easily recognized by their characteristic electron spin resonance (ESR) signal (Lelis et al., 1989). Other point defects are the strained Si-O-Si bonds that are formed due to densification of the BOX (Devine and Arndt, 1989).

Si-islands. After high-temperature annealing (HTA), Si-islands with well-developed facets are formed in the BOX close to the back interface. Most have the same orientation as does the Si substrate; however, some are slightly tilted, as shown by the arrows in Figure 18.

In the standard SIMOX the Si islands are located near the back interface. Taking into account the size and density of these islands, it is estimated that they occupy 2% of the volume in the BOX. In the case of multiple-step

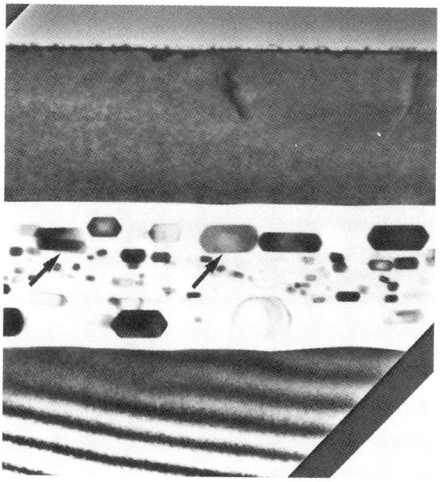

FIG. 18. Cross-section micrograph from a SIMOX specimen implanted with low-dose 1.3×10^{18} O^+/cm^{-2} at 200 kV, at a temperature of 600°C, subsequently annealed at 1300°C for 6h. A very sharp silicon–silicon dioxide (Si–SiO$_2$) interface was formed with a silicon island inside the buried oxide (BOX) having well-developed facets along the {001} and {111} planes. Some of these islands are slightly misoriented (arrows). (Reprinted from Stoemenos, J. et al. (1995). *J. Electrochem. Soc.* **142**, 1248, with permission.)

implantation the percentage of the Si islands in the BOX is reduced to 0.01% only.

c. Formation of the BOX in SIMOX

In order to achieve a better understanding of the generation of the defects in SIMOX, the differences in the formation of thermally grown oxide (TGO) and the BOX are now discussed.

In TGO, the conversion of Si to SiO_2 involves a 2.2-fold increase in molar volume. Since the oxide is constrained to the surface of the wafer, the additional 1.2 molar volume per unit volume of oxidized Si must be obtained in the direction normal to the surface. This extra volume can be accomodated by two different mechanisms:

1. By viscous flow of the oxide, at oxidation temperatures above 950°C, schematically depicted in Figure 19(a). Below this temperature viscous flow is not observed. In this case intrinsic stress is developed (Lewis and Irene, 1986).
2. By emission of Si atoms that become self-interstitial (Si_I) according to the form

$$xSi + O_2 \rightarrow SiO_2 + (x - 1)Si_I \qquad (4)$$

A SiO_2 formation completely free of strain requires that $x = 2.2$, which implies an excess of Si interstitial in the Si matrix. Theoretical and experimental works on Si self-diffusion reveal that silicon interstitials (Si_I)

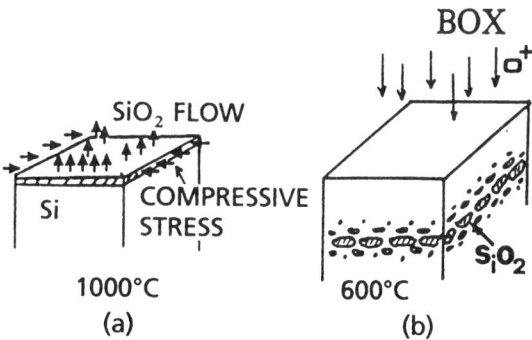

FIG. 19. Schematic showing the differences in the formation of silicon–silicon dioxide ($Si-SiO_2$) interfaces in (a) thermally grown oxide (b) buried oxide. (Reprinted from Stoemenos, J. et al. (1995). J. Electrochem. Soc. **142**, 1248, with permission.)

have a high formation energy and a very low migration energy. The sum of these two activation energies is about 5 eV (Car et al., 1984, 1985; Watkins, 1975; Hu, 1985; Bar-Yan and Joannopoulos, 1984).

Under normal oxidation conditions the formation of SiO_2 by emission of Si_I according to Eq. (4) is insignificant because of the high activation energy of the Si_I formation. Experimental observations and theoretical calculations reveal that under standard oxidation conditions the ratio of the Si interstitials to oxidized Si atoms is less than 10^{-3}.

In SIMOX, where the BOX is formed at 600°C, the oxidation by emission of Si_I, as described in Eq. (4), is dominant because during ion implantation the energy needed to break the Si bonds is provided by the ion beam that creates Si vacancies and self-interstitials. The former participate in the formation of the SiO_2, whereas the latter migrate toward the surface almost athermally by virtue of their high diffusivity (Hu, 1985).

According to Eq. (4), for a free of strain oxidation a flux $0.63F_o$ ($cm^{-2}s^{-1}$) of Si_I out of the BOX is required, where F_o is the flux of the implanted oxygen. Self-interstitial supersaturation can build up a chemical potential that opposes the oxidation reaction unless the Si_I can migrate easily to the surface, which is the natural sink for Si interstitials.

The Si_I supersaturation was estimated by (Hu, 1985) in the case of SiO_2 precipitate formation in Czochralski Si, considering the flux of the migration of the Si_I interstitials to the surface. In a first approximation the same equation is applicable to the case of SIMOX. Thus the Si_I supersaturation C_I/C_I^* is related to the oxygen flux as follows:

$$\frac{C_I}{C_I^*} = 1 + \frac{0.63\,F_o w^2}{8\,D_s C_s} \quad (5)$$

where w is the distance from the free surface, D_s is the Si_I self-diffusivity, C_s is the Si concentration equal to 5.5×10^{22} atoms cm^{-3}, and C_I^* is the equilibrium Si_I concentration. In the case of SIMOX, the supersaturation C_I/C_I^* can be reduced by reducing the oxygen flux F_o, by reducing the implantation depth w or by increasing the Si self-interstitial migration diffusivity D_s.

Molecular dynamics simulations in Si were used to follow low-energy ion-surface interactions, including kinetic energy redistribution and lattice atom trajectories (Kitabake and Greene, 1991). This simulation showed that the diffusivity has a maximum along $\langle 100 \rangle$ directions and a minimum along $\langle 111 \rangle$ directions. It also was shown that the Si_I migration energy decreases toward the surface, which is the sink for Si_I. The migration energy for vacancies is higher than Si_I and there is no tendency for the diffusion of vacancies to proceed preferentially toward the surface. The differences in migration energy of the Si_I and vacancies and the preferential migration of the former toward the surface readily explain the formation of voids that were observed in the top of the Si-overlayer (Maszara, 1988).

The three sources of the defects in the Si-overlayer are the following:

1. Defects due to Si interstitial supersaturation. The formation of defects in the Si-overlayer is related to the inability of Si_I to be epitaxially incorporated into the surface. The growth defects at the uppermost part of the Si-overlayer are mainly small dislocation loops of extrinsic type, semiloops, and segments of dislocations, as shown by the cross-section and plane-view micrographs in Figures 20(a) and 20(b). Most of these semiloops escape to the surface during HTA; however, some are extended downward to the Si-overlayer in the form of semiloops, as shown in Figures 20(c) and 20(d) (Stoemenos et al., 1991). These loops are pinned at the Si–BOX interface, resulting in the formation of pairs of threading dislocations denoted by T in Figure 16(a).
2. Defects due to radiation damage and strain. These defects are located at the back of the Si-overlayer near the BOX interface. This is the most defective zone because the radiation damage is at its maximum there and stacking faults are created due to strain development in the BOX (Visitserngtrakul et al., 1989).

Defects in the Si-overlayer related to the strain development in the BOX are the multiple fault defects (MFDs) appearing near the Si–BOX interface, as shown in Figure 21. These defects consist of overlapping SFs that are extrinsic and intrinsic in character. These SFs are randomly spaced between 2 and 8 atomic layers apart (Visitserngtrakul et al., 1989). The coexistence of extrinsic and intrinsic SFs in MFDs reveals that the SFs are created by Si lattice deformation at the $Si-SiO_2$ interface. Most of the existing defects in this zone are eliminated during HTA because they are pinned by the growing SiO_2 precipitates and finally annihilated at the Si–BOX interface (Stoemenos et al., 1991). The role of the defects near the Si–BOX interface becomes decisive when this zone comes very close to the surface, because in this case the defects easily can be extended to the surface. Processes that bring this defect zone closer to the surface are: (i) low-energy implantation ($<160\,keV$); (ii) Implantation at $190\,keV$ followed by sacrificial oxidation before the HTA; and (iii) Implantation through a 100-nm thick SiO_2 capping layer. In all these cases a highly defective Si-overlayer is formed after HTA.
3. Defect formation due to dissolution of the SiO_2 precipitates. During HTA significant reconstruction occurs in the Si overlayer, mainly in the

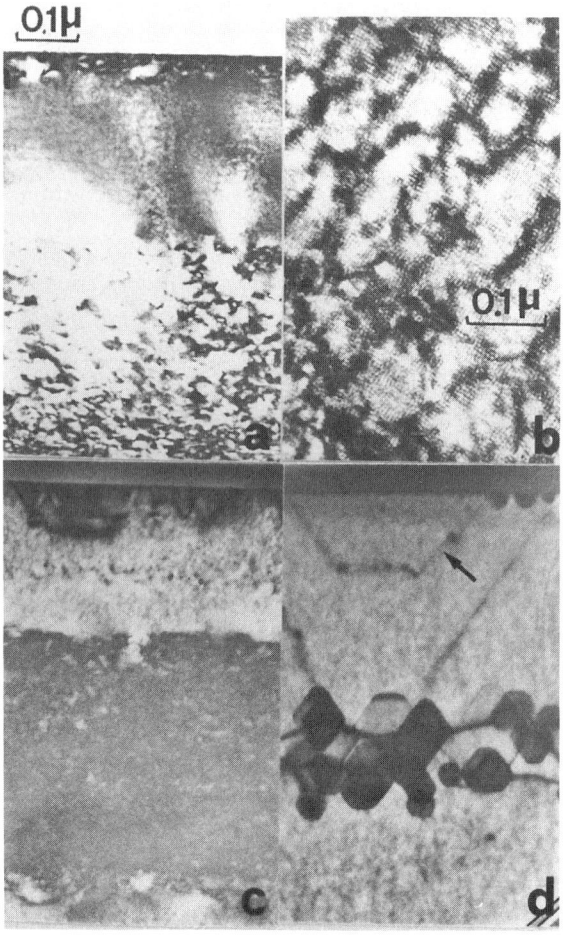

FIG. 20. Defects at the uppermost part of the silicon (Si) overlayer. (a) Cross-section micrograph from an as-implanted specimen with a low dose of 0.15×10^{18} O^+/cm^{-2} at 200 kV and 520°C. Segments of dislocations are evident at the uppermost part. (b) Plane-view micrograph from the uppermost part of the as-implanted specimen with a dose of 1.6×10^{18} O^+/cm^{-2} at 200 keV. The net of dislocations is clearly evident. The small dots are SiO_2 precipitates that form a three-dimensional net along the $\langle 100 \rangle$ crystallographic directions. (c) The same specimen as in (a) annealed at 1000°C for 1 h. Some of the defects at the surface were grown. (d) The same specimen annealed at 1405°C for 30 min. A semiloop denoted by an arrow is extended downward to the Si-overlayer. Large SiO_2 precipitates were formed at a depth at which the implanted oxygen has a maximum. Dislocations were pinned between the SiO_2 precipitates. All the cross-section micrographs are at the same magnification. (Reprinted from Stoemenos, J., et al. (1991). J. Appl. Phys. **69**, 793, with permission.)

7 TRANSMISSION ELECTRON MICROSCOPY ANALYSES   225

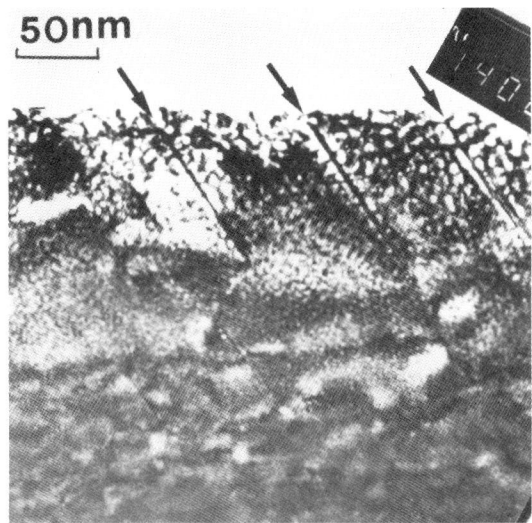

FIG. 21. Cross-section micrograph from a specimen implanted with a dose of 1.8×10^{18} O^+/cm^2 at 200 keV at 600°C, revealing the formation of multiple stacking faults in the silicon (Si) overlayer (arrows), due to stress development at the silicon–silicon dioxide (Si–SiO$_2$) interface. (Reprinted from Stoemenos, J. et al. (1995) *J. Electrochem. Soc.* **142**, 1248, with permission.)

zone near the Si–SiO$_2$ interface (Stoemenos and Margail, 1986). The reduction of the Si-overlayer width is expected due to the dissolution of the SiO$_2$ precipitates. Consequently, defects are created in the Si-overlayer near the Si–SiO$_2$ interface during HTA where three-dimensional reconstruction occurs.

The dissolution of the SiO$_2$ precipitates during HTA is described by the reverse process of Eq. (4), which implies the absorption of 1.2 Si$_I$ per dissolved SiO$_2$ molecule due to available molar volume. The extra 1.2 Si$_I$'s are provided by the formation of SiO$_2$ at the Si–SiO$_2$ interface according to the reaction in Eq. (4). Thus perfect balance is struck between the absorbed and emitted Si$_I$'s.

Dissolution of SiO$_2$ precipitates and absorption of Si$_I$ comprise a three-dimensional coalescence process that implies dislocation formation for accomodating the translation and rotation displacements between the agglomerating Si areas.

The influence of the angular misorientation of the neighboring Si sites in the dislocation formation can be understood on the basis of the arguments put forward by Read and Shockley (1950) to explain the mosaic structure of bulk crystals. It has been shown that a lattice misorientation by a small angle Q along a length L introduces a

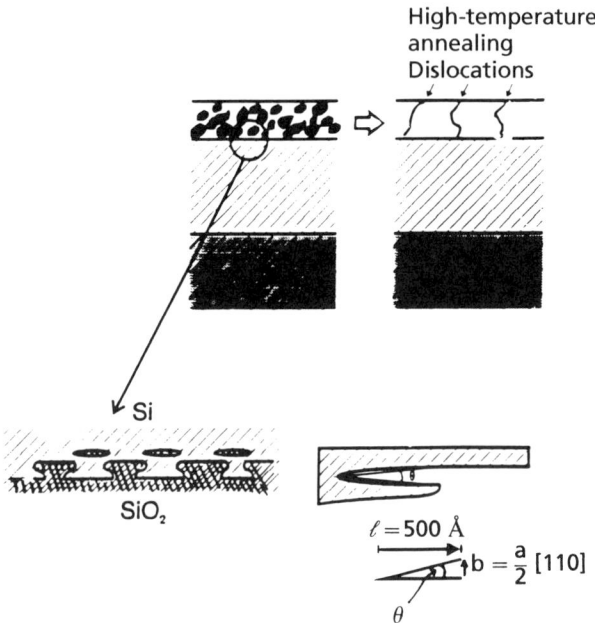

FIG. 22. Schematic showing the generation of dislocations in the silicon (Si) overlayer during high-temperature annealing (HTA) due to the small Si matrix misorientation near the silicon–silicon dioxide (Si–SiO$_2$) interface. The dislocations are introduced during the dissolution of the SiO$_2$ precipitates. The three-dimensional coalescence of the agglomerating Si areas introduce dislocations in order to accommodate small translation and rotation displacements that were developed by radiation damage and stress imposed by the BOX layer. Most of these dislocations are pinned at the Si–SiO$_2$ interface and finally are annihilated there. Only a few dislocations are extended to the surface, resulting in the formation of threading dislocations. The phenomenon is enhanced in the case of shallow implantation in which the defect zone is formed very close to the surface so that the probability of the dislocation to propagate to the surface is very high. The black dots represent precipitates. The generation of a perfected dislocation with Burger's vector $b = (a/2)$ [110] due to misorientation by an angle θ of the agglomerated Si areas during HTA is shown. l, length. (Reprinted from Stoemenos, J., et al. (1995). J. Electrochem. Soc. **142**, 1248, with permission.)

dislocation so that the condition $L\theta = b$ is satisfied, where b is Burger's vector of the dislocation, as schematically shown in Figure 22. Another consequences of the translational displacement between two agglomerating Si areas is the formation of small SFs. It is energetically favorable that adjacent areas having a relative displacement close to a displacement vector $a/6$[112] be accommodated by a SF. In this case, small single SFs can be formed (Fig. 16(b)). This process occurs mainly in the last 60 nm. The prismatic defects observed at the back of the

Si-overlayer in SIMOX are similar to the small prismatic faults observed in epitaxial Si layers grown on (111) Si substrates, suggesting that we are dealing with growth defects. This configuration also was observed in autoepitaxial growth of Si (Booker and Stickler, 1962).

Despite the crude conditions of the BOX formation, the defect density in the Si-overlayer remains low thanks to the excellent flexibility of the Si-SiO$_2$ system.

2. β-Silicon Carbide Formed by Carbon
   Implantation into Silicon

Another example of material modification by ion beam synthesis is β-SiC formed by carbon implantation into (100) and (111) Si wafers at high temperatures (850° to 950°C) at doses ranging between 0.2×10^{18} and $1 \times 10^{18}\,\text{cm}^{-2}$ and 200 keV (Nejim, Hemment, and Stoemenos, 1995). In these cases, a buried β-SiC layer is formed having the same orientation as the Si matrix. Cross-section TEM observations reveal that high-density β-SiC precipitates were formed in a 200-nm-thick zone. Above this zone a perfect 380-nm-thick Si-overlayer is evident, as shown in Figure 23(a). No dislocations or other defects are observed in the overlayer. In the buried β-SiC layer the precipitates are perfectly aligned with the Si matrix. The SiC precipitates are shown at higher magnification in Figure 23(b). Due to double diffraction from the Si-matrix and the SiC precipitates, perfectly aligned satellite diffraction spots are formed, as shown in Fig. 23(b).

The structure of the β-SiC precipitates in the buried layer has been studied by cross-section HRTEM, as shown in Figure 24 (Frangis et al., 1995). In the β-SiC precipitates, displacement type (111) moiré fringes were formed because the electron beam originated from the superposition of the two lattices, denoted by D_{111} in Figure 24.

The mean spacing of the moiré pattern along the [111] direction was found to be $D_{111} = 1.29\,\text{nm}$. This value was compared with the theoretical value given by

$$D^m_{111} = \frac{d_{Si}\,d_{SiC}}{d_{Si} - d_{SiC}} \quad (6)$$

where d_{Si} and d_{SiC} are the d-lattice spacing for Si and β-SiC, respectively, when the operating reflections is (111). From Eq. (6) the theoretical value of the D^m_{111} spacing of the moiré pattern is 1.254 nm. Comparing the

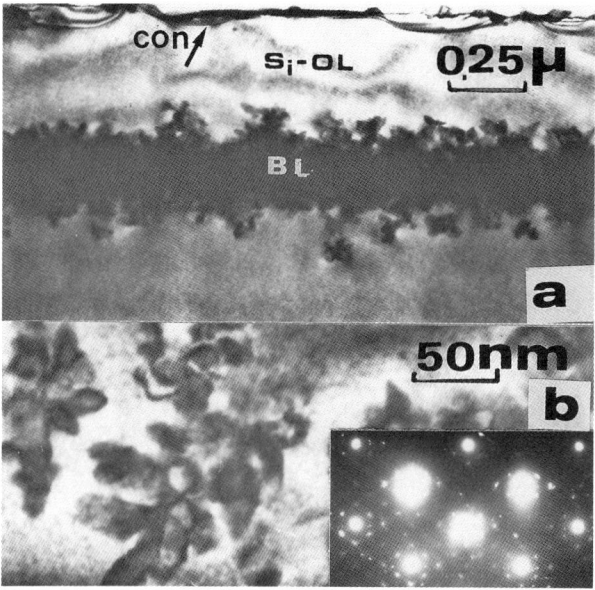

FIG. 23. (a) Cross-section micrographs from a (001) specimen implanted at $T_i = 950°C$ with a dose of $2 \times 10^{17} C^+/cm^2$. The Si-overlayer (Si = OL) is free of extended defects. Contamination at the surface is denoted by the arrow labeled (con). (b) β-SiC precipitates at higher magnification. The inset shows the related diffraction pattern from the β-SiC buried layer. Due to double diffraction from the Si-matrix and the SiC precipitates, perfectly aligned satellite diffraction spots are formed. T_i, implantation temperature. (Reprinted from Nejim, J. et al. (1995). App. Phys. Lett. **66**, 15, with permission.)

experimental and theoretical values of the moiré pattern, we can deduce the remaining misfit due to the residual strain of the lattices δ_{111}; which in this case was estimated to be $\delta_{111} = 0.5\%$ along the [111] crystallographic direction. The actual misfit in the system β-SiC–Si is 22%. Therefore, it is evident that the buried β-SiC is almost relaxed. The possible mechanisms for this relaxation will be discussed subsequently.

Rotations of the moiré pattern up to 4 degrees also have been found. They are denoted by R in Figure 24. The actual misorientation Q of the lattice can be estimated from the misorientation R of the moiré pattern according to

$$Q = R \times \delta \qquad (7)$$

where δ is the misfit. Actual misorientations of the lattice up to one degree have been observed. Local strain variations result in the observed fringe irregularities.

7 TRANSMISSION ELECTRON MICROSCOPY ANALYSES   229

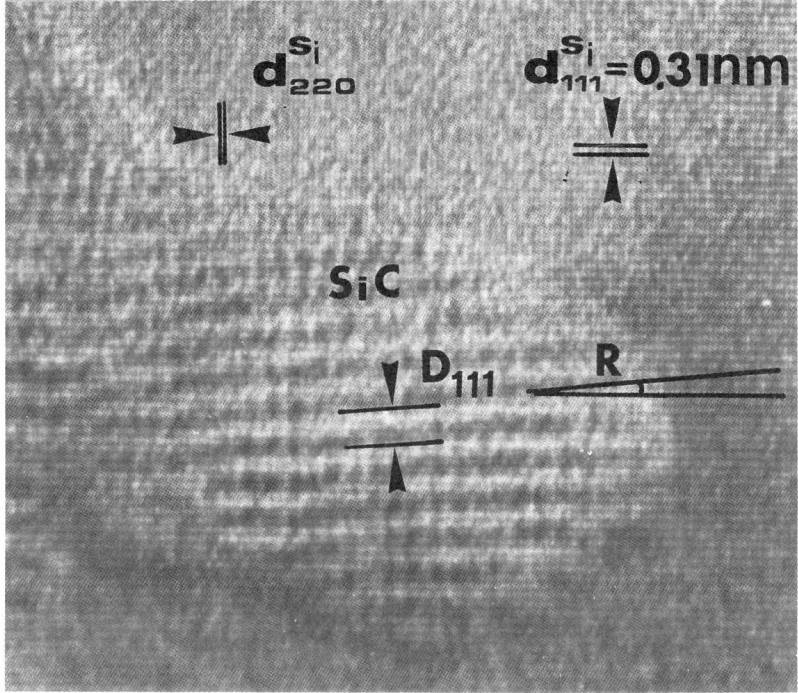

FIG. 24. Cross-section high-resolution from a (111) wafer implanted at 950°C at a dose of $4 \times 10^{17} \, C^+/cm^{-2}$. The electron beam was parallel to the [112] crystallographic direction so that the (220) and (111) lattice planes are visible. Inside the β-SiC precipitate moiré patterns of the displacement type are formed along the [111] direction, denoted by D_{111}. R, Rotations of the moiré pattern.

The almost-perfect alignment of the β-SiC precipitates with the Si matrix can be explained by considering that carbon occupies substitutional sites in the Si lattice, which is an example of topotactic transformation. The topotactic transformation is a solid-state transformation in which the product is structurally and orientationally related to the starting material (Wilson, 1987). Since the β-SiC lattice constant is 22% smaller than Si, a 48% local volume contraction is expected if a Si unit cell is replaced by a β-SiC cell. This volume reduction also implies the generation of 4 Si interstitials (Si_I) per unit cell, which are replaced by 4 carbon atoms in the unit cell. In this case, a constrained strain of about 0.12 is expected, which is very high. According to (Ashby and Johnson, 1969) when the value of strain exceeds 0.05, dislocation loops are formed even for precipitates with a diameter smaller than 1 nm. However, during implantation β-SiC precipitates with a diameter of 45 nm are formed without the generation of dislocations. Therefore, a different mechanism for the growth of β-SiC precipitates must be considered. The most probable is that interstitial Si generated during ion bombardment reacts with the implanted carbon, forming β-SiC in the available 48% empty space per unit cell. Therefore, each Si atom is substituted by an SiC molecule, namely, a Si unit cell produces two β-SiC unit cells, which results in a volume increase of 3.25%. In this case, although the constrained strain is only 7×10^{-3}, which is very low, the mismatch between the β-SiC and Si lattices remains high — about 20%. Therefore, if these precipitates are coherent, the generation of misfit dislocations represents the most likely mechanism to accomodate such mismatch. Misfit dislocations are formed during the epitaxial growth of β-SiC on Si having a periodicity of 1.6 nm (Carter, Davis, and Nutt 1986; Becourt et al., 1993; Stoemenos et al., 1995). No periodic misfit dislocations were observed in the case of β-SiC precipitates. A possible explanation for this discrepancy is to take into account the small spacing and the complexity of the net of misfit dislocations in the case of β-SiC precipitates, which is surrounded by a very dense three-dimensional net of dislocations. This entanglement is very difficult to resolve because of the complexity of the TEM image contrast.

VII. Ion-Beam-Induced Epitaxial Crystallization

Ion-implanted amorphous layers in Si recrystallize at temperatures above 500°C by solid-phase epitaxial growth. However, ion beam bombardment can result in recrystallization of amorphous Si layers at temperatures as low as 300°C (Holmen, Linnros, and Svensson, 1984). This process was characterized as ion-beam-induced epitaxial crystallization (IBIEC). The IBIEC mechanism is attributed to the mobile point defects that are produced during the collision cascade. Thus vacancies and interstitials generated by the beam become mobile due to the elevated temperature. The c-a interface acts as a sink for these migrating point defects that, when they arrive at the interface, contribute to recrystallization. An activation energy of about 0.3 eV was found in this process, which is considerably below the thermal activation energy of 2.5 eV (Linnros, Holmen, and Svensson, 1985).

IBIEC is particularly useful in annealing radiation damage produced during ion implantation into SiC. Ion-beam-amorphized SiC is very stable against thermal annealing, therefore temperatures higher than 1450°C are necessary for its recrystallization (McHardue and Williams, 1993). Such extremely high temperatures are not suitable for most device fabrication processes. This problem can be overcome by avoiding amorphization or, if amorphization is inevitable, by the IBIEC process, which permits recrystallization at lower temperatures.

The structural characteristics of the 6H-SiC samples, implanted with germanium (Ge) and subsequently recrystallized by IBIEC, were studied by combined cross-section and plane-view TEM observations.

Germanium implantation with a dose 10^{15} Ge$^+$/cm^2 (200 keV) results in the formation of a 166-nm-thick amorphous zone followed by a 30-nm-thick zone with many defects as shown in Figure 25(a). The c-a interface is very rough, and in some cases small crystalline grains are observed inside the amorphous zone located close to the c-a interface, denoted by arrows in Figure 26(a).

Annealing of this specimen at 480°C for 4 h results in a slight reduction of the defected zone to a width of about 22 nm. No recrystallization of the amorphous layer occurred during this low-temperature annealing (see the XTEM micrograph in Figure 26(b). However, if the same specimen is subjected to an IBIEC process with a dose of 3×10^{17} Si$^+$/cm^{-2} (300 keV) at 480°C, significant recrystallization occurs, as revealed by the XTEM micrograph in Figure 25(b). The IBIEC process resulted in the formation of a polycrystalline zone about 86-nm thick denoted by P and a single crystalline zone of about 80 nm denoted by B (Fig. 25(b)). Zone B can be divided into subzones B_1 and B_2, as shown in the high-magnification

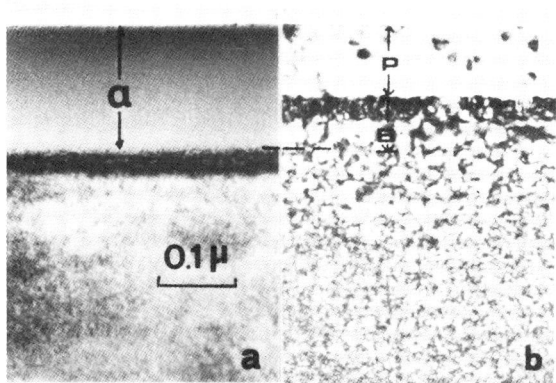

FIG. 25. (a) Cross-section micrograph after amorphization at a dose of 10^{15} Ge$^+$/cm^{-2} at 200 keV; the amorphous layer is denoted by the letter α. (b) The same specimen after the IBIEC process at a dose of 3×10^{17} Si$^+$/cm^{-2} (300 KeV) at 480°C. The polycrystalline and the single crystalline layer after recrystallization are denoted by the letters P and B, respectively. (Reprinted from Heera, V. et al. (1995) J. Appl. Phys. 77, 2999, with permission.).

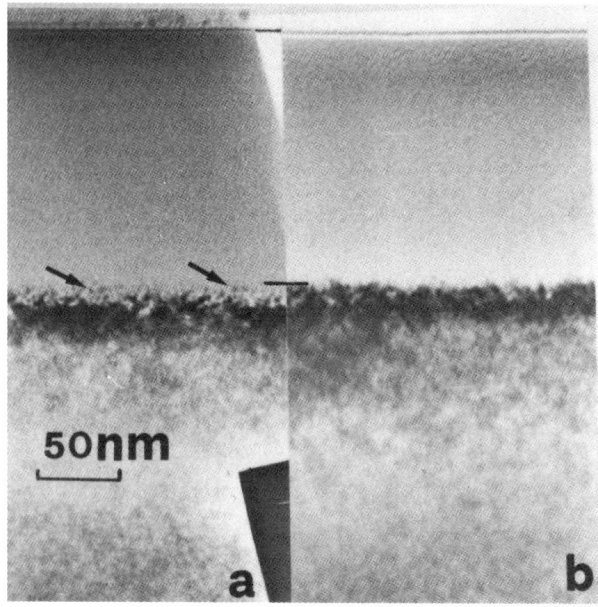

FIG. 26. Cross-section micrographs from the crystalline-amorphous (c-a) interface at high magnification. (a) Implanted at a dose of $10^{15}\,Ge^+/cm^{-2}$ (200 KeV). Small crystallites are observed in the amorphous layer close to the interface. (arrows). (b) The same specimen after annealing at 480°C for 17 min. The amorphous zone close to the interface where the small crystalline islands were located is recrystallized. Ge, germanium. (Reprinted from Heera, V., et al. (1995). J. Appl. Phys. 77, 2999, with permission.)

micrograph in Figure 27. Zone B_1, which is close to the 6H-SiC substrate, has a low defect density. Zone B_2 has many defects and exhibits a columnar structure. Below these layers a 480-nm wide zone with a high density of defects was formed due to the Si implantation denoted by C in Figure 25(b). Comparing Figure 25(a) with Figure 25(b), it is evident that the sum of zones A and B in Figure 25(b) is equal to the thickness of the amorphous layer. From Figure 27, which is a cross-section micrograph in $(10\bar{1}0)$ section, it is deduced that the defects in zone B_1 are mainly small loops and segments of dislocations, with a density on the order of $10^8\,cm^{-2}$. The defects in zone B_2 are mainly low-angle grain boundaries that are developed along the direction of recrystallization. The columnar growth is evident by the high-resolution micrograph in Figure 28. The 0.25-nm periodicity of the individual tetrahedral layers is observed. The lattice planes in zone B_2 are interrupted by perpendicular walls of amorphous material, denoted by arrows in Figure 28. The amorphous walls become wider as the growth advances and until loss of coherence occurs, resulting in a very rough

7 TRANSMISSION ELECTRON MICROSCOPY ANALYSES   233

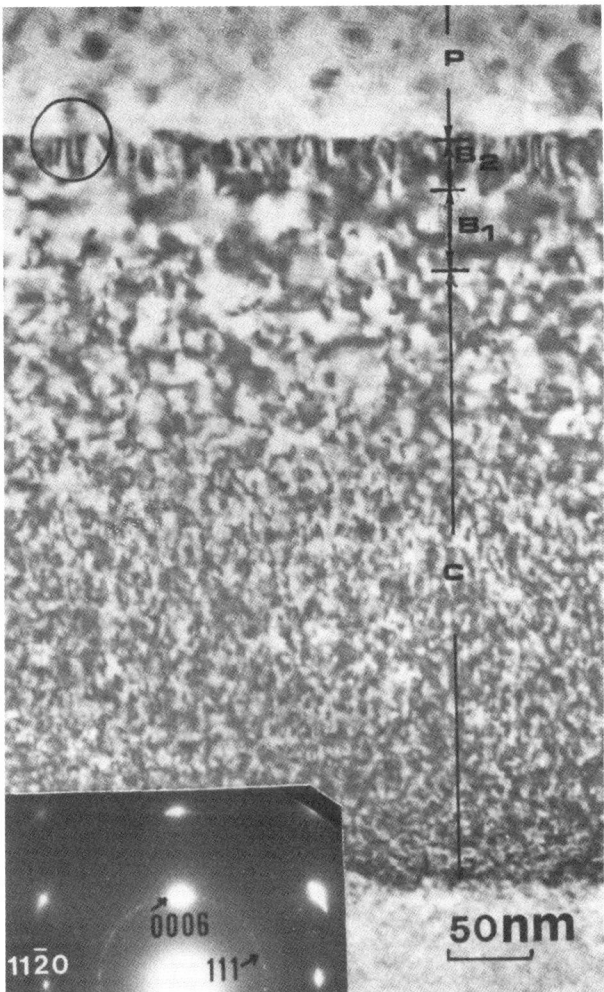

FIG. 27. High-magnification micrograph from the specimen of Figure 25(b). The related diffraction pattern from the crystalline-amorphous (c-a) interface is shown in the inset. The 111 diffraction ring is from the polycrystalline zone, denoted by P. The arcs in the diffraction spots reveal a small misorientation of the recrystallized zone B_2. (Reprinted from Heera, V., et al. (1995). J. Appl. Phys. 77, 2999, with permission.)

interface. Above this interface lattice, randomly oriented fringes with a periodicity of 0.25 nm are evident, revealing the formation of polycrystalline SiC. Cross-section TEM observations show that the 6H-polytype prevails in the single-crystalline recrystallized zone. However, in the uppermost part of this zone other polytypes are formed, with the cubic ones prevailing.

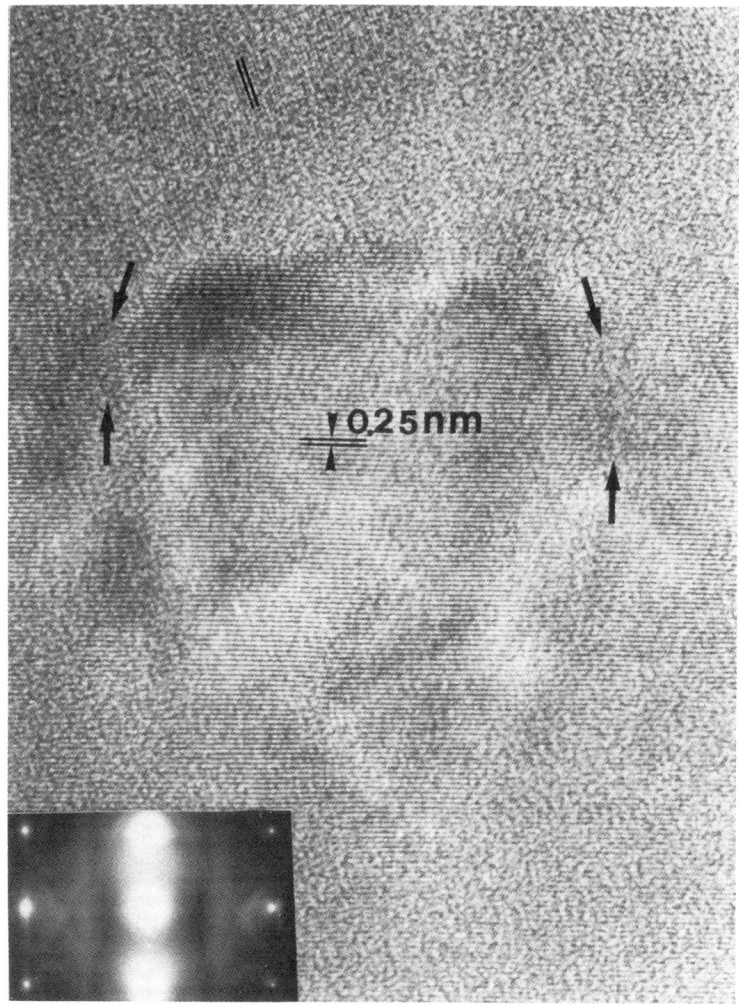

FIG. 28. High-resolution cross-section micrograph taken from the $000\bar{6}$, 0000, and 0006 reflections; see the related diffraction pattern in the inset. Arrows denote perpendicular walls of amorphous material. (Reprinted from Heera, V., et al. (1995). J. Appl. Phys. 77, 2999, with permission.)

The IBIEC process can result in single crystalline SiC layer at temperatures as low as 480°C. The thickness of this layer increases with the IBIEC dose; however, the quality is better for lower IBIEC doses. The mechanism that is responsible for the loss of coherence in the uppermost part of the recrystallized zone is not known.

VIII. Closing Remarks

The structural characteristics of implanted and annealed semiconductors have been studied. It has been shown that TEM is the most elegant technique yet applied to these types of studies.

It is widely accepted that Si will continue to dominate other semiconductors. It is expected that Si will own more than 98% of the semiconductor market for the next years, whereas all the other semiconductors will have less than a 2% share (Dunn, 1994; Courtois, 1993). Therefore, the observations have focused on Si and Si-related semiconductors.

The various electron microscopy techniques permit a uniquely detailed analysis of the implanted semiconductors highlighting the underling mechanisms of the defect formation during the early stage of implantation below the amorphization threshold.

The role of EOR defects as reservoirs able to emit self-interstitials has been shown. The emission of interstitials due to the decrease of EOR defects during annealing and its influence on the diffusion of substitutional dopants has been discussed.

ACKNOWLEDGMENT

The author is indebted to Drs. C. Jaussaud and J. Margail at LETI Grenoble, Professor P. L. F. Hemment, at the University of Surrey, and D. Tsoukalas at NCSR Democritos, for their helpful discussions.

REFERENCES

Ashby, M. F., and Johnson, L. (1969). "On the Generation of Dislocations at Misfitting Particles in a Ductile Matrix," *Philos. Mag.* **20**, 1009–1022.
Bar-Yan, Y., and Joannopoulos, J. D. (1984). "Barrier to Migration of the Silicon Self-Interstitial," *Phys. Rev. Lett.* **52**, 1129–1132.
Bartsch, H., Hoehl, D., and Kastner, G. (1984). "Radiation-Induced Rodlike Defects in Silicon and Germanium," *Phys. Stat. Sol. (a)* **83**, 543–551.
Becourt, N., Ponthenier, J. L., Papon, A. M., and Jaussaud, C. (1993). "Influence of Temperature on the Formation by Reactive CVD of a Silicon Carbide Buffer Layer on Silicon," *Phys. B* **185**, 79–84.
Booker, G. R., and Stickler, R. (1962). "Crystallographic Imperfections in Epitaxially Grown Silicon," *J. Appl. Phys.* **33**, 3281–3290.
Bourret, A. (1987). "Defects Induced by Oxygen Precipitation in Silicon: A New Hypothesis Involving Hexagonal Silicon," *Inst. Phys. Conference Series* **87**, Institute of Physics, Bristol, pp. 39–48.

Boussey-Said, J., Ghibaudo, G., Stoemenos, J., and Zaumseil, P. (1992). "Electrical and Structural Properties of Silicon Layers Heavily Damaged by Ions Implantation," *J. Appl. Phys.* **72**, 61–68.

Burke, J. (1965). *The Kinetics of Phase Transformation in Metals*, Pergamon Press, Oxford, Chapts V, VII.

Car, R., Kelly, P. J., Oshiyama, A., and Pantelides, S. T. (1984). "Microscopic Theory of Atomic Diffusion Mechanisms in Silicon," *Phys. Rev. Lett.* **52**, 1814–1817.

Car, R., Kelly, P. J., Oshiyama, A., and Pantelides, S. T. (1985). "Microscopic Theory of Impurity-Defect Reactions and Impurity Diffusion in Silicon," *Phys. Rev. Lett.* **54**, 360–363.

Carter, C. H., Davis, Jr., R. F., and Nutt, S. R. (1986). "Transmission Electron Microscopy of Process-Induced Defects in β-SiC Thin Films," *J. Material. Res.* **1**, 811–819.

Chen, L. J., Nieh, C. W., and Chu, C. H. (1988). "Cross-Sectional Transmission Electron Microscope Study of BF_2^+ Implanted (001) and (111) Silicon," *Solid State Phenomena* **182**, 45–58.

Claverie, A., Laanab, L., Bonatos, C., Bergaud, C., Martinez, A., and Mathiot, D. (1995). "On the Relation Between Dopant Anomalous Diffusion in Si and End-of-Range Defects," *Nucl. Instrum. Meth. Phys. Res.* **B96**, 202–209.

Colinge, J. P. (1991). *Silicon-on-Insulator Technology: Materials to VLSI*. Kluwer, London.

Courtois, G. (1993). "Where Are We Going?," *CAD and Testing of IC's and Systems*, Editions TIMA/CNRS; INP, Grenoble, May 1993; see also Dataquest Inc., 1993.

Craven, R. A. (1981). "Oxygen Precipitation in Czochralski Silicon." In: *Semiconductor Silicon*, Vol. 81-5, Electrochemical Society, Pennington, NJ, pp. 254–271.

Cristoloveanu, S. (1991). "A Review of the Electrical Properties of SIMOX Substrates and Their Impact on Device Performance," *J. Elect. Soc.* **138**, 3131–3139.

Devine, R. A. B., Arndt, J. (1989). "Correlated Defect Creation and Dose-Dependent Radiation Sensitivity in Amorphous SiO_2," *Phys. Rev. B* **39**, 5132–5138.

Dunn, P. (1994). "The Importance of European Silicon VLSI" (Invited), *Proc. 24th European Solid State Dev. Reas. Conference* (C. Hill and P. Ashburn, eds.), p. 521, Editions Frontieres, Gilf-sur-Yvette, France.

Edington, J. W. (1976). *Practical Electron Microscopy in Materials Science*, MacMillan Philips Technical Library, Part 3.2.

Fedina, L., Van Landuyt, J., Vanhellemont, J., and Aseev, A. L., (1995). "Observation of Vacancy Clustering in Fz-Si Crystals during in situ Electron Irradiation in a High Voltage Electron Microscope," *E-MRS Spring Meeting, Strasbourg, May 22–26, Symposium C*, Strasbourg.

Frangis, N., Nejim, A., Hemment, P. L. F., Stoemenos, J., and Van Landuyt, J. (1995). "Ion Beam Synthesis of β-SiC at 950°C and Structural Characterization," *E-MRS Spring Meeting, May 22–26, Symposium C*, Strasbourg.

Gosele, U. (1986). "The Role of Carbon and Point Defects in Silicon," *Mat. Res. Soc. Symp. Proc.* **59**, 419–431.

Heera, V. Stoemenos, J., Kogler, R., and Skorupa, R. (1995). "Amorphization and Recrystallization of 6H-SiC by Ion Beam Irradiation," *J. Appl. Phys.* **77**, 2999.

Hirsch, P. B., Howie, A., Nicholson, R. B., Pashley, D. W., and Whelen, M. J. (1965). *Electron Microscopy of Thin Crystals*, Butterworths, London, Chapt. 1.

Holmen, G., Linnros, J., and Svensson, B. (1984). "Influence of Energy Transfer in Nuclear Collisions on the Ion Beam Annealing of Amorphous Layers in Silicon," *Appl. Phys. Lett.* **45**, 1116–1118.

Hong, S. N., Ruggles, G. A., Wortman, J. J., Myers, E. R., and Hren, J. J. (1991). *IEEE Trans. Electron Devices* **38**.

Hu, S. M. (1985). "Oxygen Precipitation in Silicon," *Mat. Res. Soc. Symp. Proc.* **59**, 249–267.
Kitabake, M., and Greene, J. E. (1991). "Simulations of Low-Energy Ion/Surface Interaction Effects during Epitaxial Film Growth," *Mat. Res. Soc. Symp. Proc.* **223**, 9–20.
Lelis, A. J., Oldham, T. R., Boesch, H. E., and McLean, J. F. B. (1989). "The Nature of the Trapped Hole Annealing Process," *IEEE Trans. Nucl. Soc.* **36**, 1808–1815.
Lewis, E. A., and Irene, E. A., (1986). "Models for the Oxidation of Silicon," *J. Vac. Sci. Technol.* **A4**, 916–925.
Linnros, J. Holmen, G., and Svensson, B. (1985). "Proportionality Between Ion-Beam-Induced Epitaxial Regrowth in Silicon and Nuclear Energy Deposition," *Phys. Rev. B* **32**, 2770–2776.
Lossy, R., Obermeier, E., Stoemenos, J. (1995). "High Temperature Implantation of β-SiC and Its Characterization," *Internat. Conference on Silicon Carbide and Related Materials–1995, September 18–21*, Kyoto Research Park, Kyoto, Japan.
MacIver, B. A., and Greenstein, ?. ?. (1977). "Damage Effects in Boron and BF_2 Ion-Implanted p^+-n Junctions in Silicon," *J. Electrochem. Soc.* **124**, 273–275.
Margail, J., Lamure, M., and Papon, A. M. (1992). "Defects in SIMOX Structures: Some Process Dependence," *Materials Sci. Engrg.* **B12**, 27–36.
Marcus, R. B., and Sheng, T. T. (1983). *Transmission Electron Microscopy on Silicon VLSI Circuits and Structures*. John Wiley & Sons, New York, pp. 3–6.
Maszara, W. P. (1988). "Oxygen Bubbles along Individual Ion Tracks in O^+ Implanted silicon," *J. Appl. Phys.* **64**, 123–128.
Mauduit, B., Laanab, L., Bergaud, C., Faye, M. M., Martinez, A., and Claverie, A. (1994). "Identification of EOR Defects due to the Regrowth of Amorphous Layers Created by Ion Bombardment," *Nucl. Instrum. Meth.* **B84**, 190–194.
Mazey, D. J., Nelson, R. S., and Barnes, R. S. (1968). "Observation of Ion Bombardment Damage in Silicon," *Philos. Mag.* **17**, 1145–1161.
McHardue, C. J., and Williams, J. M. (1993). "Ion Implantation Effects in Silicon Carbide," *Nucl. Instrum. Meth.* **B80/81**, 889–894.
Myers, E., Hren, J. J., Hong, S. N., and Ruggles, G. A. (1989). "*Ion Beam Processing of Advanced Electronic Materials Symposium, San Diego,*" April 25–27, 1989 Material Research Society, Pittsburgh, PA, pp. 27–32.
Narayan, J., and Holland, O. W. (1984). "Characteristics of Ion-Implantation Damage and Annealing Phenomena in Semiconductors," *J. Electrochem. Soc.* **131**, 2651–2662.
Nejim, A., Hemment, P. L. F., and Stoemenos, J. (1995). "SiC Buried Layer Formation by Ion Beam Synthesis at 950°C," *Appl. Phys. Lett.* **66**, 2646–2648.
Neumann, W., Hillebrand, R., and Werner, P. (1987). "High Resolution Electron Microscopy." In: *Electron Microscopy in Solid State Physics*, (H. Bethge and J. Heydenreichs, eds.), Elsevier, Amsterdam, Chapt. 4.
Nieh, C. W., and Chen, L. J. (1986). "Formation of Bubbles in BF_2^+-Implanted Silicon," *Appl. Phys. Lett.* **48**, 1528–1530.
Pacaud, Y., Weishart, W., Voelskow, M., Skorupa, W., Perz-Rodriguez, A., Stoemenos, J., and Brauer, G. (1996). "An Annealing Study of Ge Implanted 6H-SiC," *MRS Spring Meeting 96. April 8–12, 1996 San Francisco, Symposium E: III-Nitride, SiC and Diamond Materials for Electronic Devices*, San Francisco, CA.
Read, W. T., and Shockley, W. (1950). "Dislocation Models of Crystal Grain Boundaries," *Phys. Rev.* **78**, 275–289.
Sadana, D. K., Snds, T., Maszara, W., and Rozgonyi, G. A. (1985). "Transmission Electron Microscopy of Preamorphized, Shallow Implanted and Rapid Thermally Annealed Silicon," *Microscopy of Semiconducting Materials, Institute Physics Conference Series* **76**, 93–97.

Saris, F. W., Custer, J. S., Schreutelkamp, R. J., Liefting, R. J., Wijburg, R., and Wallinga, H. (1992). "Avoiding Dislocations in Ion-Implanted Silicon," *Microelect. Engrg.* **19**, 357–362.
Schork, R., Pichler, P., Kluge, A., and Ryssel, H. (1991). "Radiation-Enhanced Diffusion during High Temperature Ion Implantation," *Nucl. Instrum. Meth. Phys. Res.* **B59/60**, 499–503.
Skorupa, W., and Kogler, R., Schmalz, K., Bartch, H. (1991). "Proximity Gettering by MeV-Implantation of Carbon: Microstructure and Carrier Lifetime Measurements," *Nucl. Instrum. Math. Phys. Res.* **B55**, 224–229.
Stoemenos, J., Dezauzier, C., Arnaud, G., Contreras, S., Camassel, J., Pascual, J., and Robert, J. L. (1995). "Structural, Optical and Electrical Properties of State-of-the-Art Cubic SiC Films," *Materials Sci. Engrg.* **B29**, 160–164.
Stoemenos, J., Garcia, A., Aspar, B., and Margail, J. (1995). "Silicon on Insulator Obtained by High Dose Oxygen Implantation, Microstructure and Formation Mechanism," *J. Electrochem. Soc.* **142**, 1248–1260.
Stoemenos, J., and Margail, J. (1986). "Nucleation and Growth of Oxide Precipitates in Si Implanted with Oxygen," *Thin Solid Films* **135**, 115.
Stoemenos, J., Reeson, K. J., Robinson, A. K., and Hemment, P. L. F. (1991). "Dislocation Formation Related with High Dose Oxygen Implantation on Silicon," *J. Appl. Phys.* **69**, 793–802.
Tamura, M., Ando, T., and Ohyu, K. (1991). "MeV-Ion-Induced Damage in Si and Its Annealing," *Nucl. Instrum. Meth.* **B59/60**, 572–583.
Tan, T. Y., Foll, H., and Krakow, W. (1981). "Detection of Extended Interstitial Chains in Ion-Damaged Silicon," *Appl. Phys. Lett.* **37**, 1102.
Tanaka, A., Yamaji, T., Uchiyama, A., Hayashi, T., Iwabuchi, T., and Nishikawa, S. (1990). "Optimization of the Amorphous Layer Thickness and the Junction Depth in the Preamorphization Method for Shallow-Junction Formation," *Jpn. J. Appl. Phys.* **292**, L191–L194.
Veirman, De A., Landuyt, Van J., Reeson, K. J., Gwilliam, R., Jeynes, C., and Sealy, B. J. (1990). "Identification of CoSi Inclusions within Buried $CoSi_2$ Layers Formed by Ion Implantation," *J. Appl. Phys.* **68**, 3792–3794.
Visitserngtrakul, S., Jung, C. O., Ravi, T. S., Cordts, B., Burke, D. E., and Krause, S. J. (1989). *Institute of Physics Conference Series* **100**, Institute of Physics, Bristol, 1989, p. 557.
Vook, F. L. (1973). *Radiation Damage and Defects in Semiconductors* (J. E. Whitehouse, ed.) Institute of Physics, London, 60.
Watkins, G. D. (1975). Institute of Physics Conference Series **23**, Institute of Physics, London.
White, A. E., Short, R. C., Dynes, R. C., Gibson, J. M., and Hull, R. (1988). "Synthesis of Buried Silicon Compounds using Ion Implantation," *Material Res. Symposium* **107**, pp. 3–15.
Wijburg, R. C. M., Lietting, J. R., Custer, J. S., Wallinga, H., Saris, F. W. (1992). "Improvement of Device Characteristics by Multiple Step Implants or Introducing a C Gettering Layer," *Microelect. Engrg.* **19**, 543–546.
Wilson, I. H. (1987). *Ion Beam Modification of Insulators* (P. Mazzoldi and G. Arnold, eds.), Elsevier Science, Amsterdam, Chapt. 7.
Wong, H., Cheung, N. W., Chu, P. K., Liu, J., and Mayer, J. W. (1988). "Proximity Gettering with Mega-Electron-Volt Carbon and Oxygen Implantations." *Appl. Phys. Lett.* **52**, 1023–1025.

CHAPTER 8

Rutherford Backscattering Studies of Ion Implanted Semiconductors

Roberta Nipoti and Marco Servidori

CONSIGLIO NAZIONALE DELLE RICERCHE
ISTITUTO LAMEL
BOLOGNA, ITALY

I. INTRODUCTION 239
II. MEASUREMENT OF DISORDER DEPTH PROFILES BY RBS-CHANNELING 240
III. TYPICAL RESULTS FOR SILICON AND SILICON CARBIDE 248
    1. *Silicon* 248
    2. *Silicon Carbide* 255
    References 258

I. Introduction

Rutherford backscattering spectrometry (RBS) is based on the elastic backscattering process that energetic ions can experience when impinging on a solid target. The energy value of the backscattered ions depends on the ion, the mass of the target atoms, and the depth at which the collisions occur. The RBS technique is a mass dispersive spectrometry that allows depth characterization of the sample. In particular, for crystalline samples the measurement of the backscattering efficiency as a function of the relative orientation between the probe beam and the sample contains information about the crystal structure. In fact, lattice atomic rows and planes can "steer" the energetic ions by means of a correlated series of gentle, small-angle collisions with subsequent reduction in the close encounter probability between ions and crystal atoms. This action of the crystal lattice on the energetic ion trajectories is called channeling and the RBS analysis in such a geometry is called RBS-channeling analysis. In a typical RBS experiment, the ratio between the backscattering events and the ion fluence is on the order of magnitude of 10^{-6}. This value is reduced up to two orders of magnitude in the channeling geometry.

The elastic scattering phenomenon at the basis of the RBS technique is the same as that responsible for the crystal damage in implanted layers. For the characterization of structural disorder, beam artifacts must be avoided. The probe ions have such mass and energy that they undergo a low number of scattering events with a low amount of transferred energy on penetrating a near-surface layer. The thickness of this layer is the sample depth that can be characterized. Light ions (hydrogen or helium) with MeV energies are generally used to analyze up to a few micrometers of material thickness.

The reader is referred to the fundamental books by Chu, Mayer, and Nicolet (1978) and Feldman, Mayer, and Picraux (1982) for a broad description of the RBS-channeling technique. In this chapter, we only discuss the information that can be extracted from an RBS-channeling spectrum and the aspects related to quantitative measurement of defect profiles in semiconductor crystals by RBS-channeling.

The study of the annealing kinetics of defects in ion-implanted semiconductors is generally concerned with how point-defect profiles evolve to extended defect profiles. In fact, in as-implanted samples, clusters of small defects are mostly present, whereas thermal annealing produces a secondary damage composed of extended defects. The density of the implantation damage, also called primary damage, determines the types and the depth distributions of the residual damage. A classification scheme of the different forms of this secondary damage as a function of the primary damage has been presented by Jones, Prussin, and Weber (1988) for the Si case. This chapter deals with cases in which quantitative and qualitative RBS-channeling measurements are possible. In addition, some examples of kinetics studies are reported that concern implanted silicon and silicon carbide.

II. Measurement of Disorder Depth Profiles by RBS-Channeling

An example of RBS-channeling characterization of an implanted semiconductor is presented to show which qualitative information is contained in the RBS spectra and when quantitative disorder profiles can be extracted from them. Figure 1 shows an example of random and aligned spectra of a virgin and an implanted silicon carbide (SiC) crystal obtained by a 2 MeV He^+ (helium) beam backscattered at an angle of 165.3 degrees. The implanted ions were $^{52}Cr^+$ at 260 keV and $1 \times 10^{16} cm^{-2}$ fluence. Every RBS-random or RBS-channeling spectrum is composed of the independent contribution of the backscattering events from the elements present in the sample: carbon (C), silicon (Si), and chromium (Cr). Even the implanted element Cr gives a non-negligible contribution due to its high scattering

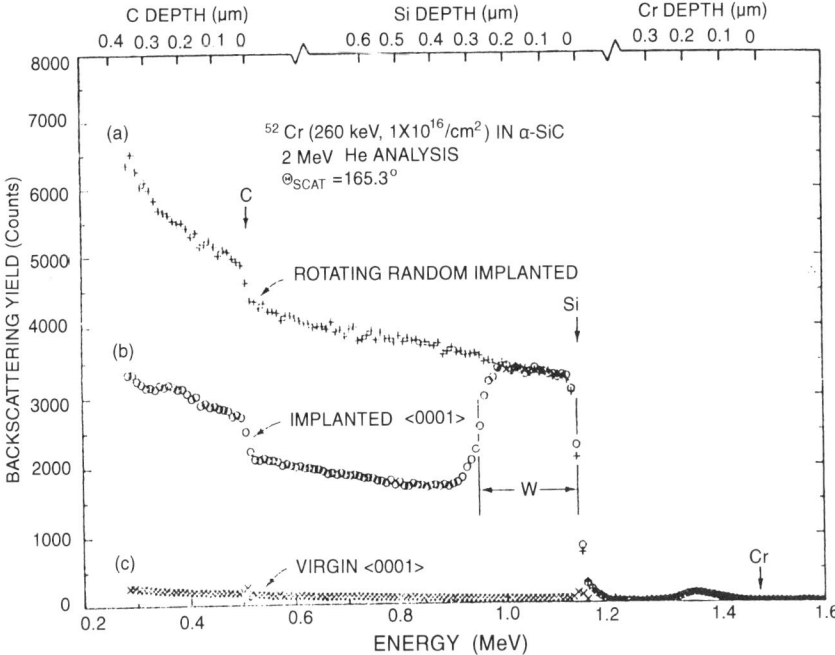

FIG. 1. Random (a) and ⟨0001⟩-aligned (b), (c) Rutherford backscattering spectra for a virgin and an implanted silicon carbide (SiC) sample. The random depth scales for the ion species present in the sample are drawn on the upper horizontal axis. He, helium; C, carbon; Si, silicon; Cr, chromium. (Adapted from Fig. 2 of Bohn, Williams, McHargue, and Begun (1987). "Recrystallization of Ion-Implanted α-SiC," *J. Material Res.* **2**, 107–116.)

cross section. The yields of the spectra depend on the relative orientation between the beam and crystal. The backscattering efficiency of the aligned spectra (Fig. 1, spectra (b) and (c)) is much lower than that of the random spectrum (Fig. 1, spectrum (a)) because in the first two cases the direction of the He beam is parallel to the ⟨0001⟩-SiC axis. The backscattering efficiency of the implanted crystal (Figure 1, spectrum (b)) is higher than that of the virgin crystal (Fig. 1, spectrum (c)), because the ion implantation process has completely destroyed the order of the crystal in a near-surface layer. The backscattering yield from this damaged layer is equal to that of a crystal in the random geometry. The yield due to the backscattering events from the perfect crystal below the damaged layer is higher than that of spectrum (c) because the damaged layer produces a larger He-beam divergence than does an equivalent thickness of perfect crystal. This effect is called dechanneling by the defective region. The backscattering efficiency of a damaged crystal in the channeling geometry is then higher than that of a perfect crystal because of the direct backscattering events from the atoms displaced off the crystal axis or plane and because of the dechanneling effect due to the displaced atoms themselves. The yield differences between damaged and undamaged crystals can be converted into disorder profiles, once the magnitude of the direct scattering and of the dechanneling factor are known as a function of the sample depth. When compound semiconductors are studied, as in the example of Figure 1, the signal from the element with the highest kinematics factor is generally used to compute the damage profile if the backscattering cross section is sufficient for good statistics and the energy of the backscattering events from the damaged layer falls in the range between the signal edge of the chosen element and that of the next one. The damage produced by the implantation process in SiC (Fig. 1) is generally characterized by studying only the Si signal. This approach is typical of RBS analysis of compound semiconductors, because of the difficulty of obtaining good count statistics from light elements or of drawing a correct dechanneling yield for overlapping signals. Such an approach gives reliable results if the damage states of all species comprising the crystal lattice are equal. This is certainly true in the case of damage above the amorphization threshold, but must be demonstrated for very low damage. To our knowledge this has not been done yet.

The energy scale for every element in an RBS spectrum is converted to a depth scale, taking into account the kinematics of the collisions and the beam energy losses along the inward and outward paths. Taking the Si signal as an example, the energy versus depth relation is

$$\Delta E_{Si} = [\bar{\varepsilon}]_{Si}^{SiC} N^{SiC} \Delta x$$

where ΔE_{Si} (dimension = $energy$) is the spectrum width on the Si signal; $[\bar{\varepsilon}]_{Si}^{SiC}$ (dimension = $energy \times unit\ area$) is the He stopping cross section in SiC for collision with Si atoms (channeling and random trajectories have different $[\bar{\varepsilon}]_{Si}^{SiC}$ values); N^{SiC} (dimension = $number/unit\ volume$) is the atomic density of SiC; and Δx (dimension = $length$) is the sample width corresponding to the energy width ΔE_{Si}. As $[\bar{\varepsilon}]_{Si}^{SiC}$ depends on the kinematics factor and the energy values, the relation between ΔE_{Si} and Δx is not linear. The non-linearity and the mass dependence can be better appreciated when the analyzed layer is almost some micrometers thick, that is, one order of magnitude larger than in this case. Furthermore, the difference between the channeling and random $[\bar{\varepsilon}]_{Si}^{SiC}$ determines different depth-scale factors for channeling and random spectra, respectively, that cannot be neglected at micrometric depths. The scales at the top of Figure 1 were computed using the values of the random stopping power (Ziegler, 1977). This also can be correctly adopted for measuring the thickness of the damaged layer from the

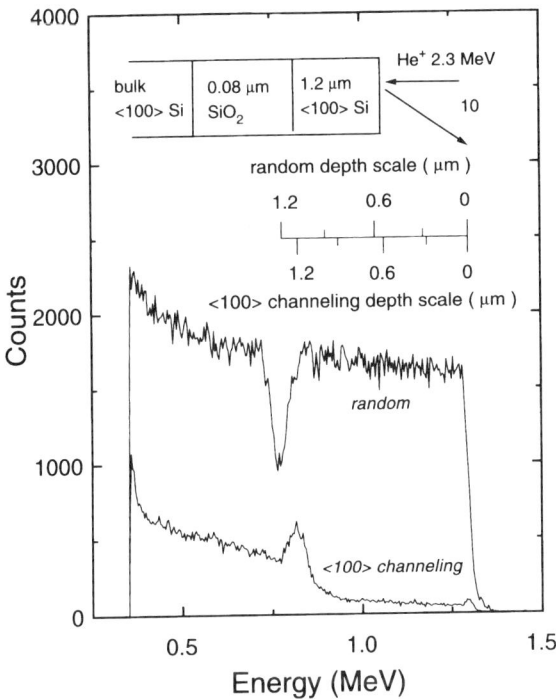

FIG. 2. Random and ⟨100⟩-aligned Rutherford backscattering spectra of a ⟨100⟩ Si crystal with a 0.08-μm silicon dioxide (SiO$_2$) layer buried below a 1.2-μm-thick Si layer. The energy shift in the position of the Si signal from the oxide layer is due to the different values of stopping power of channeling and random trajectories. The random and channeling depth scales corresponding to the spectra energy scale are shown. He, helium.

channeling spectrum (b) in Figure 1, because the damaged layer is amorphous and placed at the sample surface. In any case, the known reduction in the density of the amorphized SiC with respect to the crystal phase must be taken into account. This density has been determined by Heera et al. (1994) to be equal to 2.9 gr/cm^3 as opposed to 3.2 gr/cm^3 of the perfect crystal.

The differences in the depth scale of an aligned and a random spectrum are evident in Figure 2, which shows the random and the axially aligned spectra of a ⟨100⟩ Si crystal with a 0.08-μm SiO$_2$ layer buried below 1.2-μm-thick Si layer. The low-energy peak of the aligned spectrum is due to the direct scattering events between the well-channeled He ions and the Si atoms of the buried oxide. The spectrum yield increases because the Si atoms in the oxide occupy uncorrelated positions with respect to those in the top crystal layer. The dip in the random spectrum also corresponds to scattering events from the Si in the oxide. The yield reduction with respect to the crystal is due to the lower Si concentration in the oxide than in the Si crystal even if, due to events occurring at the same depth, the peak and dip have different energy positions because of the different energy loss of the He ions along the inward channeled and random paths in the perfect crystal (see the quantitative data by Santos et al., 1995). In particular, the peak is located at higher energy than is the dip because the channeling energy loss is lower than the random energy. For a correct computation of both the concentration and the depth scale of a damage profile from an RBS spectrum, the channeling stopping power or an average between the channeling and the random values must be used depending on the damage level itself. Generally, only the most refined simulator codes take into account the energy loss reduction of channeled ions and the dependence of the channeling stopping power on the crystal disorder. Nevertheless, this effect is not negligible for correct computation of the damage profile and becomes more and more important when medium-to-high damage levels are produced at buried depths, as in the case of sub-MeV- or MeV-implanted samples.

Although a detailed description of the yield in an RBS-channeling spectrum is outside the purpose of this chapter, a simple description can be given from a conceptual point of view. Let us imagine that the analyzing beam has the only two random and channeled components. In this case, the normalized yield $\chi_s(z)$ from the sample depth z can be written as

$$\chi_s(z) = \chi_{\text{dech}}(z) + [1 - \chi_{\text{dech}}(z)] \cdot f \cdot N_D(z)$$

where $\chi_{\text{dech}}(z)$ is the dechanneling contribution from both perfect and damaged crystals and $[1 - \chi_{\text{dech}}(z)] \cdot f \cdot N_D(z)$ is the direct scattering contribution proportional to the function of the still channeled beam $1 - \chi_{\text{dech}}(z)$ and to the normalized defect density $N_D(z)$ corrected by a factor f that represents the effective number of scattering centers per defect type.

The magnitude of the factor f depends on the geometry of the defect. It ranges from one for isolated randomly displaced atoms to approximately zero for dislocation loops. The presence or absence of a large direct scattering peak in the aligned spectra often allows us to determine if the scattering factor is significant for the defect under study.

The dechanneling factor $\chi_{\text{dech}}(z)$ contains the deflecting action beyond the critical angle for channeling for defect type and determines the spectrum yield within and beyond the defective region. Extended defects produce distortions or curvatures of the channel walls that can be quite significative with respect to the critical angle for channeling, thereby inducing a dechanneling action that can be much stronger than isolated point defects within the channel. Theoretical expressions for defect families have been proposed that satisfactorily described the trend of the experimental spectra, when a single type of defect is present (see the book Chu, Mayer, and Nicolet, 1978). However, description of the dechanneling when more than one extended defect type is present is a difficult problem.

To compute a defect profile from an RBS-channeling spectrum, a hypothesis must be made for f and $\chi_{dech}(z)$, that is for the defect types present in the sample. The energy dependence of $\chi_{dech}(z)$ beyond the defective layer can help to recognize the defect family; however, this is rarely done, because the needed energy range might overcome that of the MV accelerators used for ion beam analysis (see, e.g., Bentini, Bianconi, and Servidori, 1987). Furthermore, such an identification is useful for recognizing the defect family but is not sufficient to univocally identify the defect type. The use of other structural techniques is preferable to identify the defect types, whereas RBS can be used to compute the defect profile when a single defect family of known f and $\chi_{dech}(z)$ is dominant in the sample.

As far as point-defect profiles are concerned, the most widely used simulators of RBS-channeling spectra assume a scattering factor equal to one and a density of defects equal to the fraction of displaced atoms. Different approaches have been developed for the description of the dechanneling factor (Sigmund and Winterbon, 1974; Ziegler, 1971; Gärtner, Hehl, and Scholtzhauer, 1983, 1984a, 1984b; Bianconi, et al., 1994). A common feature of all these theories is their inability to reproduce the RBS-channeling spectrum of any disorder profile; however, the most suitable model can be chosen according to the defect distribution under consideration. A good criterion for this choice is the ability of the model to reproduce the spectrum yield for the undamaged crystal beyond the defective region, with a damage profile tail going to zero. In Si crystals, displaced atom distributions along submicrometric or micrometric depths can be determined with a depth resolution of tens of nanometers and a sensitivity as low as 1% for the fraction of displaced atoms. Figure 3 shows four RBS-channeling spectra of Si samples implanted with As$^+$ at 800 keV at different fluences. The random and channeling spectra of a virgin crystal are shown for comparison. A ^{4}He$^+$ beam at 2 MeV and 170-degree scattering angle was used. The aligned spectra were reproduced with the simulator code by Bianconi et al. (1994), assuming the damage profiles shown in Figure 4. These are typical results for as-implanted samples and the quality of the reproduction of the RBS spectra is generally considered satisfactory. This code takes into account the dependence of the stopping power on the channeling direction and uses a semi-empirical dechanneling function that depends on the damage level crystal depth and beam energy. A comparison with more refined codes was

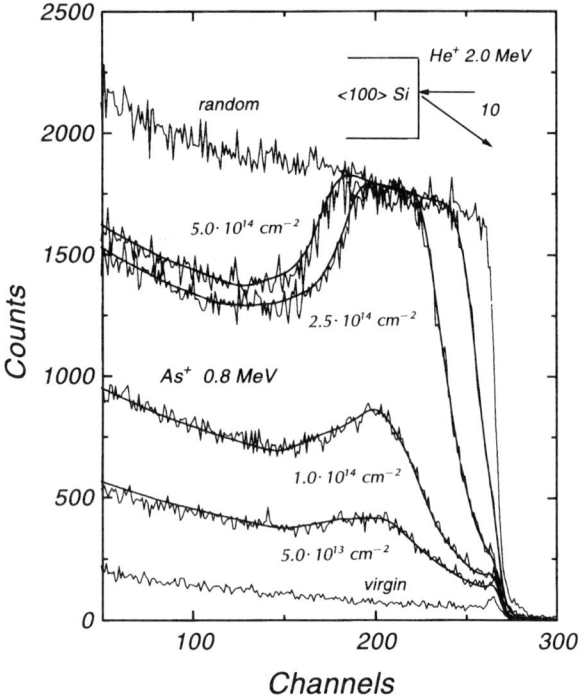

FIG. 3. ⟨100⟩-axial Rutherford backscattering spectrometry–channeling spectra (light lines) of silicon (Si) samples implanted with As$^+$ (arsenic ions) at 0.8 MeV and different fluences between $5 \cdot 10^{14}$ and $5 \cdot 10^{13}$ As$^+$/cm^{-2}. The random and channeling spectra of a virgin Si are shown for comparison. By the simulator of Bianconi et al. (1994), the spectra of the as-implanted samples were reproduced (heavy lines), assuming the damage profiles shown in Figure 4. He, helium.

made by Albertazzi et al. (1996) and showed that the code by Bianconi et al. (1994) gives very good results. A further improvement is expected if the dependence of the channeling stopping power on the damage level is introduced.

As far as extended defect distributions are concerned, although the computation of a damage profile is possible in principle (see the book by Feldman, Mayer, and Picraux, 1982), and was actually done in particular cases such as dislocation loops by Picraux et al. (1980), Gärtner, Hehl, and Schlotzhauer (1984a, 1984b) or Bentini, Bianconi, and Servidori (1987), the technique has never reached systematic use in the quantitative analysis of the residual damage in implanted and annealed semiconductors. This might be due both to the difficulty of handling the RBS analysis over the necessary

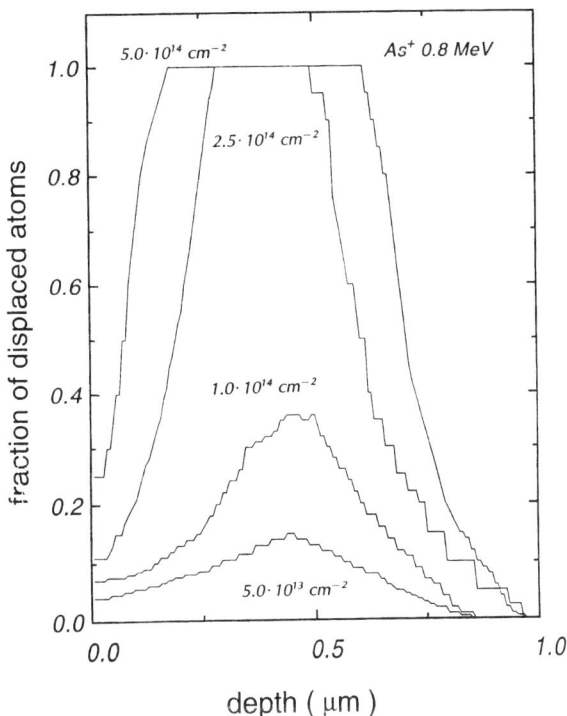

FIG. 4. Profiles of the function of the displaced atoms used to simulate the channeling spectra of Figure 3. As, arsenic.

energy range and to the often concomitant presence of more than one defect type in implanted and annealed semiconductors (for Si see the review article by Jones, Prussin, and Weber (1988). Finally, the available experimental data of the dechanneling factors for many types of extended defects in semiconductors are very few. The advent of microbeams in ion beam analysis (Breese et al., 1995; King et al., 1995a, 1995b), or the availability of samples with only one type of extended defect (such as in the heterostructures described by Mazzer, Drigo, Romanato (1992) or in the experiment by Picraux et al. (1980) will probably compensate for this.

In conclusion, the RBS-channeling technique is systematically used for a quantitative measurement of isolated scattering center distribution and for a qualitative description of extended defect distribution. From the point of view of implantation technology, the RBS-channeling technique is useful to study the evolution of the disorder as a function of irradiation conditions and post-irradiation thermal treatments. A typical example is given by the study of the solid-phase epitaxial regrowth of an amorphous layer from the crystalline substrate. To our knowledge, the first measurement of the activation energy of such a process was made by the RBS-channeling technique (Csepregi et al., 1977). The value of 2.7 eV derived by those first studies, which were done with self-implanted Si, was only slightly corrected by more recent measurements performed by the time-resolved reflectivity technique (Olson and Roth, 1988). In the case of samples implanted below the amorphization threshold, RBS-channeling is generally usefully employed for a qualitative screening of the sample versus the annealing parameters, with the aim of choosing the most significant specimens. These specimens can then be studied in more detail by other diagnostic techniques.

III. Typical Results for Silicon and Silicon Carbide

We have chosen silicon (Si) and silicon carbide (SiC) to give some examples of RBS-channeling studies of ion-implanted and annealed semiconductors. Today, ion implantation is widely used for microelectronic device fabrication in Si technology and also seems promising for applications involving SiC. For electronic device applications the annealing kinetics of ion-induced defects are studied with particular interest because of the influence of the residual damage on device performance.

1. SILICON

As a first example, a series of RBS spectra is presented that shows the annealing evolution of the implantation damage versus temperature. Then the determination of the minimum number of displaced Si atoms necessary to the formation of extended defects during the annealing process is reported on. Finally, the measurements of the solid-phase regrowth rate at the two crystalline-amorphous (c-a) interfaces of a buried amorphous layer are reviewed. All these examples refer to sub-MeV- or MeV-implanted samples. They were chosen because the advent of high-energy ion implantation technology brought some new insights about the mechanism of recovery of the ion damage in annealed Si. First, it was demonstrated that, independently of the implantation energy, a minimum amount of displaced Si atoms is needed for the formation of extended defects during annealing. Second, an enhanced solid-phase regrowth process seems to be active at the two c-a interfaces of a buried amorphous layer.

It is commonly known that ion damage recovery in Si is accompanied by formation of secondary damage in the forms classified by Jones, Prussin, and Weber, (1988). Five classes of extended defects have been recognized, depending on the damage density in the as-implanted crystal:

Category I. Rod-like defects, {113} stacking faults or dislocation loops are formed at the implanted dopant tail for ion damage below the amorphization threshold, depending on the annealing temperature.

Category II. Dislocation loops nucleate and grow at the initial c-a interface during solid-phase epitaxial regrowth of a surface amorphous layer at any annealing temperature.

Category III. Mostly hairpin dislocations and microtwins are formed due to imperfect regrowth of an amorphous layer.

Category IV. Clamshell defects form at the depth at which the two c-a interfaces of a buried amorphous layer meet.

Category V. Precipitates of the implanted impurity in Si.

This classification scheme is the result of transmission electron microscopy (TEM) studies. A quantitative evaluation of the depth profiles for each of these types by RBS-channeling is not possible for the reasons explained in Part II. To our knowledge, among those listed, point defects and dislocations loops are the only ones for which a quantitative analysis of the corresponding RBS spectra can be done. (Sigmund and Winterbon, 1974; Ziegler, 1971; Gärtner, Hehl, and Schlotzhauer, 1983, 1984a, 1984b). Generally, as far as the RBS-channeling spectra can show the possible presence of extended defects, TEM is used for unique identification of defect type. The presentation of such cases would be a set of RBS spectra of similar shape that corresponds to different TEM images. This does not seem significant to us in the framework of this chapter.

In Figure 5 a set of RBS-channeling spectra shows the evolution of low-density ion damage versus annealing temperature. The random and aligned spectra of a virgin crystal also are shown for comparison. Si wafers with $\langle 100 \rangle$ orientation were implanted at a dose rate of $1.4 \cdot 10^{12}$ Si$^+$/cm^2/sec above room temperature (91°C) at an energy of 0.54 MeV and a fluence of $2 \cdot 10^{15}$ ions/cm^2), to produce a slightly buried disordered layer. The samples were submitted to isochronal annealing treatment for 20 min at temperatures between 400 and 950°C. The interest in this complete set of RBS-channeling data resides in the fact that these authors compared all the RBS results with those of the positron annihilation spectroscopy made on the same samples and gave a new interpretation of the damage annealing evolution at the lower temperatures. In fact, although RBS-channeling is more sensitive to interstitial-type defects, positron annihilation spectroscopy is sensitive to both interstitial- and vacancy-type defects. The yield reduction

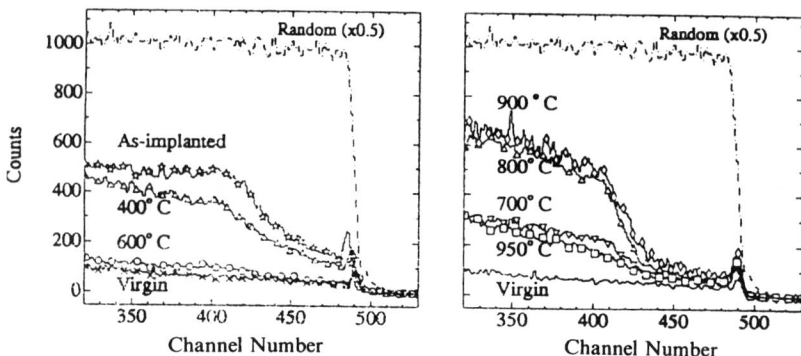

FIG. 5. ⟨100⟩-axial Rutherford backscattering spectrometry–channeling spectra of as-implanted and annealed silicon (Si) crystals. The implantation parameters were 0.54 MeV, 2×10^{15} As$^+$/cm^{-2} fluence, and 91°C substrate temperature. The annealing treatments were isochronal (20 min) for increasing temperatures between 400 and 950°C. As$^+$, arsenic ions. (Adapted from Figure 1 of Goldberg, Simpson, Mitchel, and Schultz (1994). "Studies of Dislocation Formation in Annealed Self-Ion-Irradiated Silicon," *Materials Research Society Symposium Proceedings* **316**, 39–44.)

in the spectra of the samples treated at increasing temperatures from 400 to 600°C is commonly interpreted to be due to a reduction in the number of point scattering centers, that is, of interstitial Si atoms. Nevertheless, Goldbert et al. (1994), detected a structural change of a defect type growing rich with vacancies with increasing temperatures in the same range. In this case, the yield reduction in the RBS spectra might be explained by a lower backscattering cross section of the new defect structure compared with that of the as-implanted defects, which are randomly displaced interstitial-type defects with a direct scattering factor equal to one. If this were to be confirmed by further investigations, the traditional way of analyzing RBS spectra should be revised. The slight dechanneling at 600°C with respect to the virgin crystal may be due both to very small loops and to small point-defect clusters. The dechanneling behavior between 600 and 950°C is typical of the presence of dislocation loops placed at a certain depth, the diameters of which increase at the expense of their density with an increase in annealing temperature. From TEM studies we know that this hypothesis is correct (Bentini, Bianconi, and Servidori, 1987; Zaumseil et al., 1987). The results of positron annihilation spectroscopy show a progressive dissolution of vacancy-type defects for increasing annealing temperatures.

The determination of the critical parameter for the formation of Category I defects in implanted and annealed Si samples is still under discussion. Schreutelkamp et al. (1991) performed a systematic study by varying the ion

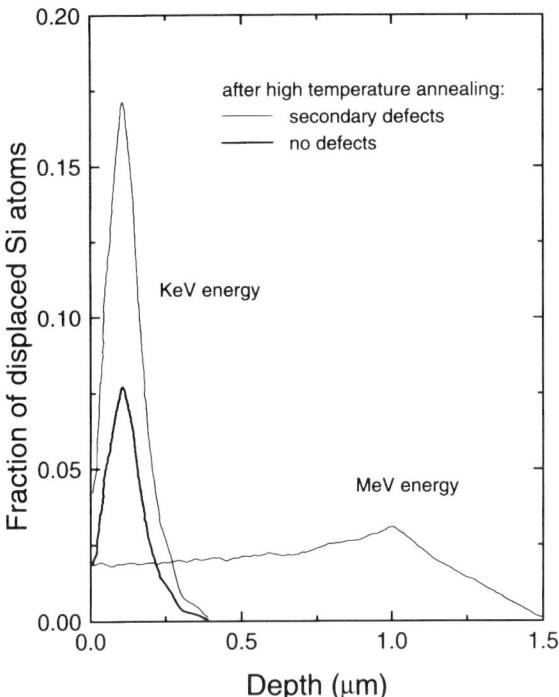

FIG. 6. Typical damage distribution for keV or MeV energy implantation processes in silicon (Si). Light lines show damage profiles with the same integrated number of displaced Si atoms, but different implantation energies. Both profiles correspond to the presence of extended defects after high-temperature annealing. The maximum low-energy damage profile for which any defects are formed is shown by a heavy line. (Adapted from Fig. 12 of Schreutelkamp, Custer, Liefting, Lu, and Saris (1991). "Preamorphization Damage in Ion-Implanted Silicon," *Material Science Reports* **6**, 275–366.)

mass between 11 and 121, and the ion energy between tens and thousands of keV. They used the RBS-channeling technique to measure the fraction of displaced atoms in the implanted samples and TEM to verify the presence or absence of extended defects after annealing at a high temperature (950°C for 15 min). Figure 6 schematically illustrates three significant primary damage profiles, as measured by RBS-channeling, due to the same ion species implanted at low and high energies with different fluence values. The profiles indicated by the light line at low and high energies have the same integral, and the corresponding samples contain extended defects and dislocation loops when annealed. The sample with the primary damage

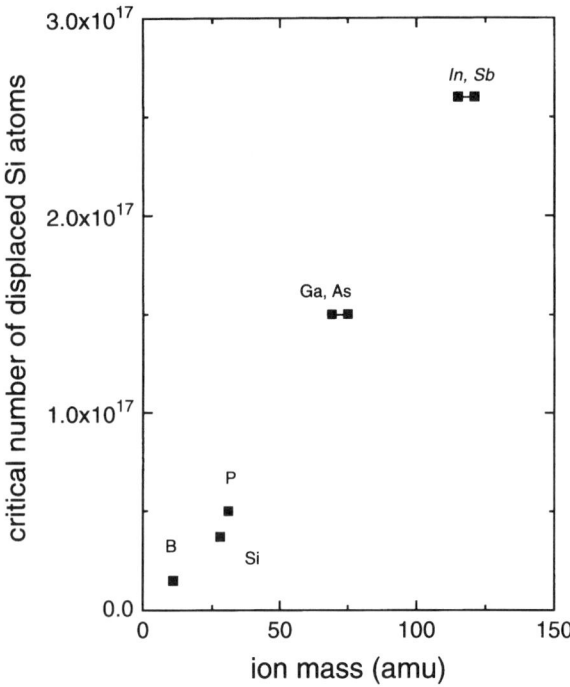

FIG. 7. Critical number of displaced silicon (Si) atoms needed for the formation of extended defects during annealing treatments. In, indium; Ga, gallium; As, arsenic; P, phosphorus; B, boron. (Adapted from Table 3 of Schreutelkamp, Custer, Liefting, Lu, and Saris (1991). "Preamorphization Damage in Ion-Implanted Silicon," *Material Science Reports* **6**, 275–366.)

profile, shown as a heavy line, is completely free from defects after annealing. Since the level of the low-energy (heavy line) profile overcomes that due to the high-energy (light line) implantation process, it is demonstrated that the integral of the primary damage, more than the local defect concentration, is responsible for the nucleation of the secondary damage and that the Si atoms participating in the formation of the extended defects must be mobile in all the implanted volume. An increase in the ion mass produces an increase in the critical number of displaced Si atoms needed for secondary defect formation, as measured by RBS (Fig. 7). This is explained by the fact that only the mobile part (Si interstitials) of the Si disorder participates in extended defect formation. An increase in the ion mass implies that an increased number of displaced Si atoms are trapped in highly disordered or amorphous zones, and consequently the number of mobile Si interstitials decreases. For heavier ions, a continuous amorphous layer can be formed

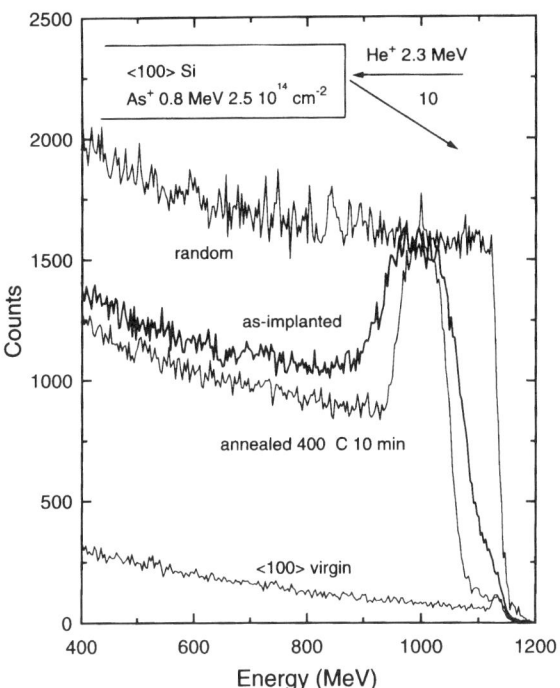

FIG. 8. ⟨100⟩-axial Rutherford backscattering spectrometry–channeling spectra of a ⟨100⟩ silicon (Si) sample implanted with As$^+$ (arsenic ions) at 0.8 MeV energy, $2.5 \cdot 10^{14}$ As$^+$/cm^{-2} fluence, and furnace-annealed at 400°C for 10 min in N$_2$ ambient. The random and aligned spectra of a virgin sample are shown for comparison. C, carbon; H, helium.

before the critical number of mobile Si interstitials for the formation of Category I damage is reached. Depending on the ion species and energy, the doping level corresponding to the critical damage may be too low for device applications; however, Schreutelkamp *et al.* (1991) has demonstrated that an implantation process producing damage below the critical value followed by annealing can be iterated up, yielding the desired doping level with a negligible formation of secondary defects. These results are of both practical and fundamental importance because they show how the secondary defects nucleate, even if the formation of a specific type of defect remains unexplained.

Figure 8 compares the RBS-channeling spectra of a buried amorphous layer before and after a low-temperature annealing treatment, 400°C for 10 min in N$_2$ ambient. The amorphous layer was obtained by implanting As$^+$ at 0.8 MeV and $2.5 \cdot 10^{14}$ cm^{-2} fluence. The random and aligned spectra of

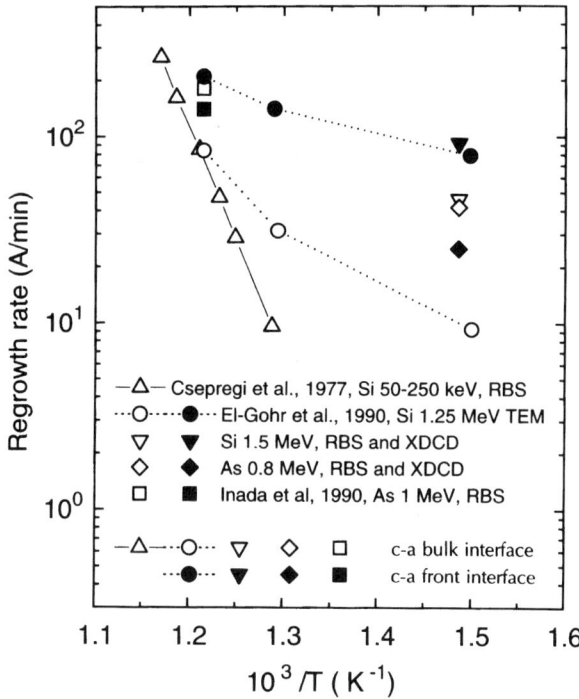

FIG. 9. Arrhenius plot of the solid-phase epitaxial regrowth rate at the crystalline-amorphous (c-a) interface for buried and near-surface amorphous layers obtained by ion implantation. Data from different authors and different diagnostic techniques are compared. Original data are also included. RBS, Rutherford backscattering spectrometry; As, arsenic; Si, silicon; XDCD, X-ray double-crystal diffractometry.

a virgin sample are also shown for comparison. After annealing, the Si signal from the buried amorphous layer decreases at both c-a interfaces, while keeping the random yield. This indicates that the crystal layer at the sample surface, as well as the bulk crystal, act as seeds for a solid-phase epitaxial process, whose magnitude is unusually high if compared with the solid-phase epitaxial regrowth of a surface amorphous layer produced by keV implantation. Moreover, the front and bulk c-a interfaces seem to have different displacement rates. Similar results were reported in the literature for MeV Si and arsenic (As) implants and are summarized in the Arrhenius plot of Figure 9, together with the historical data of Csepregi et al. (1977) for keV self-implanted Si. Data obtained with different diagnostic techniques — RBS (Inada et al., 1990), TEM (El-Gohr et al., 1990), and X-ray double-crystal diffractometry (XDCD) — are compared. Furthermore, some original RBS and XDCD data have been added. At MeV implantation energy the c-a interfaces are broad, with crystal inclusions in the amorphous phase and amorphous inclusions in the crystal. Observations by TEM reveal that the regrowth cannot be ascribed to low-temperature rearrangement of these rough c-a interfaces. Despite the differences in the absolute values of the measured regrowth rates, all the data seem to confirm that the solid-phase regrowth at the interfaces of a buried amorphous layer is a different phenomenon from that originally observed by Csepregi (1977) and later by Olson and Roth (1988). In addition, front and bulk interfaces react with different rates to the same thermal treatment. The crystallization at the front interface is faster than that at the bulk interface for Si implantation, the opposite being true for As implantation. This difference can be explained by the hypothesis that As enhances the solid-phase regrowth process (Csepregi et al., 1977; Lietoila et al., 1982; Olson and Roth, 1988). In fact, the near-bulk regrowth interface crosses depths at higher As concentration values than those of the front interface. Which parameters are characteristic of this enhanced epitaxial regrowth and why it is observed for damage profiles of large straggling, that is, for the MeV but not for the KeV implantation, is not yet clear. Holland et al. (1990) studied the phenomenon in MeV self-implanted Si at liquid nitrogen temperature. In these samples, small amorphous regions are the dominant defects over the damage distribution. These regions exhibited distinct annealing stages at 300 and 400°C. Holland et al. (1990) proposed a transient mechanism at the lowest temperature and a regrowth enhancement by the stress at the c-a interface at the highest temperature. This is a quite new argument that requires further investigation.

2. SILICON CARBIDE

Among the modern materials, SiC is probably the most interesting because it is used in a wide range of technical applications due to its extreme hardness, stability at high temperatures, high thermal conductivity, chemical inertness, and high-temperature semiconducting properties. So far, the ion implantation processing of SiC substrates has been investigated with two aims: (1) to modify the surface hardness of the material and eventually facilitate the sintering of SiC powder (Föhl, Emrick, and Carstanjen, 1992) and (2) to introduce electrical dopants into the SiC crystal to exploit its semiconducting properties (Hirano and Inada, 1995). The requirements for the thermal stability of the ion-induced damage are opposite in the two cases. The SiC powder sintering is a high-temperature ($\approx 2000°C$) process needing additives that can be substituted by the ion damage. At high

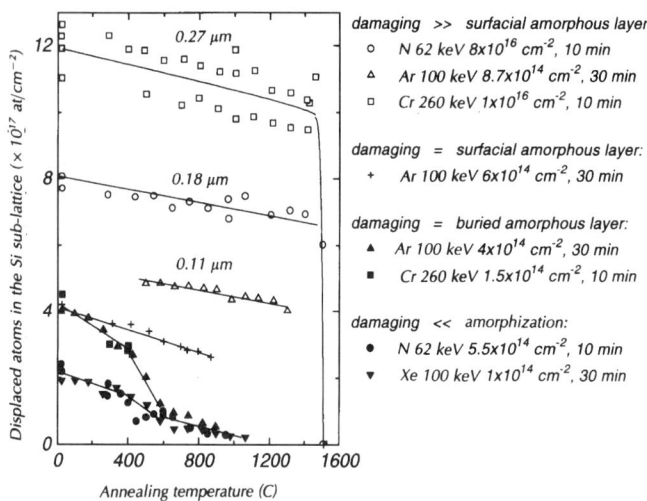

FIG. 10. Damage recovery versus the annealing temperature in 6H-SiH crystals as measured by Rutherford backscattering spectrometry–channeling. Different levels of the as-implantation damage are taken into account. Open symbols refer to samples implanted at fluences much higher than the threshold values to obtain a continuous amorphous layer from the sample surface up to a given depth that is a function of the ion mass and energy. Cross symbols refer to a sample implanted at the fluence threshold just described. Closed symbols refer to samples with amorphous—and not buried—damaged layers. N, nitrogen; Ar, argon; Cr, chromium; Xe, xenon; Si, silicon; SiC, silicon carbide.

temperatures the large mobility of the point defects produced by ion implantation can favor the sintering process by enhancing the volume and the grain boundary diffusion. This effect persists as long as the ion damage is not annealed out. The damage stability requirements for microelectronics applications are the opposite. In fact, the crystal recovery is necessary to electrically activate the implanted dopant. Moreover, the annealing temperature must be as low as possible to avoid broadening of the doping profile. In the following, a review is given of the most recent RBS studies on the implanted and annealed 6H-SiC polytype. Among the more studied SiC polytypes (4H, 6H and 3C), we chose 6H because it seems the most promising material for applications to microelectronics.

Figure 10 reviews the RBS results on the damage recovery in a large number of 6H-SiC samples. The data are by Campbell et al. (1974), Bohn et al. (1987). Föhl, Emrick, and Carstanjen (1992), and Rao et al. (1995). The annealing behaviors of samples implanted at different fluences are compared. Fluences were used to produce damage below the amorphization threshold, buried amorphous layers, and amorphous layers appearing at the surface. The annealing treatments were isochronal and the annealing times were equal to 10 or 30 min. The comparison between annealing treatments of different durations of time is often used in the literature for SiC. Some authors justify such comparisons, demonstrating that temperature rather than time seems to be the important parameter. The data in Figure 10 are the integral of the residual lattice disorder as measured by the Si signal (Section II) of RBS-channeling spectra versus the annealing temperature. The open symbols refer to the samples implanted at fluences well above the formation of an amorphous surface layer. These samples show just a small displacement of the c-a interface toward the sample surface for temperatures below 1400°C, whereas after 1500°C a sudden alignment with the bulk crystal is measured for the signal from the damaged region. Such kinetics are typical of any randomized SiC film, independent of its thickness, between 0.11 and 0.27 μm. The explosive crystallization at a threshold temperature might suggest that at lower temperatures a spontaneous nucleation of small crystal grains in the damaged volume is favored. Observations by TEM did not detect polycrystalline structures at temperatures near to <900°C (Bohn et al., 1987); however, optical images of a sample treated at 1040°C for 24 h show brownish amorphous islands alternating with a region of high transparency (Addamiano et al., 1972). The kinetics of the recrystallization of the SiC amorphous phase are not yet understood and for the moment the slope of the best fit of the experimental data below 1400°C is assumed as a measure of the solid-phase epitaxial regrowth of the 6H-SiC phase. This value ranges between 15 and 30 nm/°C, depending on the implanted species. The regrowth rate of the surface amorphous layer just described (cross symbols in Fig. 10) seems slightly faster than, although not very different from, these values. Completely different is the behavior of buried damaged layers — amorphous or not — the data of which are shown as closed symbols in Figure 10. At lower temperatures, it seems that the two types of implantation damage recover with different rates, converging to a single trend above 600°C. In both cases, the ion damage is considered completely annealed out above 1000°C, because the channeling spectra reproduced that of a virgin crystal. This does not exclude the presence of secondary damage that must be checked by other structural techniques (e.g., TEM). The Arrhenius plots (not shown) of the recovery rates of these data show the presence of different activation energies below and above approximately 600°C. However, variations in the data are such that a more accurate investigation is necessary before any detailed hypothesis about medium-to-low damage recovery in SiC is formulated. Even if not reported in Figure 10 (because of the lack of a correct scale factor), the data by Spitznagel et al. (1989, and related previous work) agree with those presented in Figure 10. This is because the completely amorphous surface layer, the buried amorphous layer, and the slightly damaged layer have different kinetic rates, depending to first-order on the concentration of the primary damage, and to second-order on the type of implanted species. Taking the results of Figure 10 into account, the semiconductor properties of SiC polytypes doped by ion implantation can be exploited as long as the damage produced by the implantation process is kept below or—at maximum—equal to the formation of a buried amorphous layer.

REFERENCES

Addamiano, A., Anderson, G. W., Comas, J., Hughens, H. L., and Lucke, W. (1972). "Ion Implantation Effects of Nitrogen, Boron, and Aluminum in Hexagonal Silicon Carbide," *J. Electrochem. Soc.* **119**, 135–1362.

Albertazzi, E., Bianconi, M., Lulli, G., Nipoti, R., and Cantiano, M. (1996). "Different Methods for the Determination of Damage Profiles in Si from RBS-Channeling Spectra: A Comparison," *Nucl. Instrum. Meth.* **B118**, 128–132.

Bentini, G. G., Bianconi, M., and Servidori, M. (1987). "Comparative Analysis of Extended Defect Depth Profiles in Silicon by Rutherford Backscattering, X-Ray and Electron Microscopy," *Nucl. Instrum. Meth.* **B18**, 145–152.

Bianconi, M., Nipoti, R., Cantiano, M., Gasparotto, A., and Sambo, A. (1994). "RBS-Channeling Spectra: Simulation of As-Implanted Si Samples Through an Empirical Formula for $\langle 100 \rangle$ Axial Dechanneling of He in Silicon," *Nucl. Instrum. Meth.* **B84** 507–511.

Bohn, H. G., Williams, J. M., McHargue, C. J., and Begun, G. M. (1987). "Recrystallization of Ion-Implanted α-SiC," *J. Material Res.* **2**, 107–116.

Breese, M. B. H., King, P. J. C., Smulders, P. J. M., and Grime, G. W. (1995). "Dechanneling of MeV Protons by 60° Dislocation," *Phys. Rev. B* **51**, 2742–2750.

Campbell, A. B., Shewchun, J., Thompson, D. A., Davies, J. A., and Michel, J. B. (1974). "Nitrogen Implantation in SiC: Lattice Disorder and Foreign-Atom Location." In: *Ion Implantation in Semiconductors* (Susumu Namba, ed.), Plenum Press, New York, pp. 291–298.

Chu, W. K., Mayer, J. W., and Nicolet, M. A. (1978). *Back-scattering Spectrometry*, Academic Press, London.

Csepregi, L., Kennedy, E. F., Gallagher, T. J., Mayer, J. W., and Sigmon, T. W. (1977). "Reordering of Amorphous Layers of Si Implanted with ^{31}P, ^{75}As, and ^{11}B Ions," *J. Appl. Phys.* **48**, 4234–4240.

El-Gohr, M. K., Holland, O. W., White, C. W., and Pennycook, S. J., (1990). "Structural Characterization of Damage in Si (100) Produced by MeV Si$^+$ Ion Implantation and Annealing," *J. Material Res.* **5**, 352–359.

Feldman, C., Mayer, J. W., and Picraux, S. T. (1982). *Materials Analysis by Ion Channeling*, Academic Press, London.

Föhl, A., Emrick, R. M., and Carstanjen, H. D. (1992). "A Rutherford Back-Scattering Study of Ar- and Xe-Implanted Silicon Carbide," *Nucl. Instrum. Meth.* **B65**, 335–340.

Gärtner, K., Hehl, K., and Schlotzhauer, G. (1983). "Axial Dechanneling: I. Perfect Crystal," *Nucl. Instrum. Meth.* **216**, 275–286.

Gärtner, K., Hehl, K., and Schlotzhauer, G. (1984a). "Axial Dechanneling: II. Point Defects," *Nucl. Instrum. Meth.* **4**, 55–62.

Gärtner, K., Hehl, K., and Schlotzhauer, G. (1984b). "Axial Dechanneling: III, Dislocations," *Nucl. Instrum. Meth.* **4**, 63–71.

Goldberg, R. D., Simpson, T. W., Mitchel, I. V., and Shultz, P. J. (1994). "Studies of Dislocation Formation in Annealed Self-Ion Irradiated Silicon," *Material Research Society Symposium Proceedings* **316**, 39–44.

Heera, V., Kögler, R., Skorupa, W., and Stomenos, J. (1994). "Amorphization and Recrystallization of 6H-SiC by Ion Beam Irradiation," *Material Research Society Symposium Proceedings* **339**, 197–202.

Hirano, Y., and Inada, T. (1995). "Nitrogen Implantation in (100)-β-SiC Layers Grown on Si Substrates," *J. Appl. Phys.* **77**, 1020–1028.

Holland, O. W., and White, C. W. (1991). "Ion-Induced Damage Amorphization in Si," *Nucl. Instrum. Meth.* **B59/60**, 353–362.

Holland, O. W., White, C. W., El-Gohr, M. K., and Budai, J. D. (1990). "MeV, Self-Ion Implantation in Si at Liquid Nitrogen Temperature; A Study of Damage Morphology and Its Anomalous Annealing Behavior," *J. Appl. Phys.* **68**, 2081–2086.

Inada, T., Nishida, A., Kanazawa, M., and Hasebe, H. (1990). "Annealing Characteristics and Electrical Properties of 1-MeV Arsenic-Ion-Implanted Layers in Silicon," *J. Appl. Phys.* **68**, 5555–5563.

Jones, K. S., Prussin, S., and Weber, E. R. (1988). "A Systematic Analysis of Defects in Ion-Implanted Silicon," *Appl. Phys.* **A45**, 1–34.

King, P. J. C., Breese, M. B. H., Smulders, P. J. M., Wilshaw, P. R., and Grime, G. W. (1995a). "Observation of a Blocking to Channeling Transition for MeV Protons at Stacking Faults in Silicon," *Phys. Rev. Lett.* **74**, 411–414.

King, P. J. C., Breese, M. B. H., P. J. M., Wilshaw, P. R., and Grime, G. W. (1995b). "Stacking-Fault Imaging Using Transmission Ion Channeling," *Phys. Rev. B* **51**, 2732–2741.

Lietoila, A., Wakita, A., Sigmon, T. W., and Gibbons, J. F. (1982). "Epitaxial Regrowth of Intrinsic, 31P-Doped and Compensated (^{31}P + ^{11}B-Doped) Amorphous Si," *J. Appl. Phys.* **53**, 4399–4405.

Mazzer, M., Drigo, A. V., and Romanato, F. (1992). "Dechanneling by Misfit Dislocation in III-V Semiconductor Heterostructures," *Nucl. Instrum. Meth.* **B64**, 103–107.

Olson, G. L., and Roth, J. A. (1988). "Kinetics of Solid Phase Crystallization in Amorphous Silicon," *Material Science Reports* **3**, 1–77.

Picraux, S. T., Follstaedt, D. M., Baeri, P., Campisano, S. U., Foti, G., and Rimini, E. (1980). "Depth Profile Studies of Extended Defects Induced by Ion Implantation in Si and Al," *Radiation Effects* **49**, 75–80.

Rao, M. V., Griffiths, P., Holland, O. W., Kelner, G., Freitas, J. A., Simons, D. S., Chi, P. H., and Ghezzo, M. (1995). "Al and B Ion-Implantation in 6H- and 3C-SiC," *J. Appl. Phys.* **77**, 2479–2485.

Santos, J. H. R., Grande, P. L., Boudinov, H., and Behar, M. (1995). "Stopping Power and Charge Equilibration Process for Channeled He Ions Along $\langle 100 \rangle$ and $\langle 110 \rangle$ Directions of Si Crystal," *Proceedings of the Ion Implantation Technology-94 Conference*, North Holland, Amsterdam, pp. 711–715.

Santos, J. H. R., Grande, P. L., Boudinov, H., Behar, M., Stoll, R. Klatt, C., and Kalbitzer, S. (1995). "Electronic Stopping Power of $\langle 100 \rangle$ Axial Channeled He-Ions in Si Crystal." *Nucl. Instrum. Meth.* **B106**, 51–54.

Schreutelkamp, R. J., Custer, J. S., Liefting, J. R., Lu, W. X., and Saris, F. W. (1991). "Preamorphization Damage in Ion-Implanted Silicon," *Material Science Reports* **6**, 275–366.

Sigmund, P., Winterbon, K. B. (1974). "Small-Angle Multiple Scattering of Ions in the Screened Coulomb Region," *Nucl. Instrum. Meth.* **119**, 541–557.

Spitznagel, J. A., Wood, S., Choyke, W. J., Devaty, R. P., and Ruan, J. (1989). "Amorphization and Annealing of 6H SiC Implanted with n-Type, p-Type or Isovalent Dopants," *Material Research Society Symposium Proceedings* **147**, 113–118.

Zaumseil, P., Winter, U., Cembali, F., Servidori, M., and Sourek, Z. (1987). "Determinations of Dislocation Loop Size and Density in Ion Implanted and Annealed Silicon by Simulation of Triple Crystal X-ray Rocking Curves," *Phys. Stat. Sol. (a)* **100**, 95–104.

Ziegler, J. F. (1971). "Determination of Lattice Disorder Profiles in Crystals by Nuclear Backscattering," *J. Appl. Phys.* **43**, 2973–2981.

Ziegler, J. F. (1977). *The Stopping and Ranges of Ions in Matter*, Pergamon Press, New York.

CHAPTER 9

X-Ray Diffraction Techniques

P. Zaumseil

INSTITUTE FOR SEMICONDUCTOR PHYSICS
FRANKFURT(ODER), GERMANY

I. INTRODUCTION 261
II. BASIC ASPECTS OF CRYSTAL LATTICE MODIFICATION DUE TO ION IMPLANTATION . 262
   1. *Implantation-Induced Variation of the Lattice Parameters* 263
   2. *Dopant Incorporation* 265
   3. *Defect Generation During Annealing* 266
III. X-RAY METHODS TO ANALYZE IMPLANTED SAMPLES 268
   1. *Simulation of the Reflection Curve of a Disturbed Crystal* 268
   2. *Experimental Techniques* 271
   3. *Limitations of X-ray Diffraction Techniques* 273
IV. EXAMPLES OF SPECIAL APPLICATIONS 274
   1. *Determination of Dislocation Loop Size and Density in Implanted and Annealed Silicon* 274
   2. *Arsenic Implantation and Annealing* 276
   3. *Characterization of Boron-Implanted and Annealed Silicon* 278
V. SUMMARY 280
   References 281

I. Introduction

X-ray diffraction techniques are widely used, in general, in the characterization of semiconductor substrate materials and of the features of layer structures, and they are successfully used, in particular, for the investigation of ion-implanted layers. Starting in the 1960s these techniques received an enormous impetus due to the increasing quality of semiconductor materials, and especially the requirements of semiconductor technology stimulated the development of new and more precise characterization methods. Based on the dynamical theory of X-ray diffraction (von Laue, 1960; Pinsker, 1978), the fundamental measuring and analyzing methods (Burgeat and Taupin, 1968) were developed until the beginning of the 1970s, during which time the investigation of diffusion layers was dominant. During the 1980s, the development was characterized by further refinement of existing methods due to increasing demands regarding sensitivity and accuracy, especially for the investigation of ion-implanted layers.

The starting point of all the methods discussed is the diffraction of X-rays at the regularly arranged lattice of a crystal. In the least, a nearly perfect single crystal as substrate is the necessary condition to be able to measure how a layer with modified structure near the surface generated by ion implantation changes the diffraction features of this substrate. It is a fundamental aspect of X-ray techniques, which is of importance for the appropriate analyzing algorithms, that a separation between "perfect" substrate and implanted (disturbed) layer is done, even if there generally exists a continuous transition between both. This makes it clear that it is impossible to investigate the implantation into polycrystalline or amorphous material with the discussed methods. Even an amorphous surface layer generated by implantation is invisible to X-ray diffraction techniques and therefore cannot be investigated by them.

Part II will show how implantation-induced modifications of the crystal structure act on the diffraction behavior of X-rays. In Part III the commonly used measuring and analyzing methods will be discussed in detail, and in Part IV some examples will demonstrate the successful application of X-ray diffraction techniques for the study of implanted layers.

II. Basic Aspects of Crystal Lattice Modification due to Ion Implantation

Any analysis by X-ray diffraction of ion-implanted layers is based on the modifications of the lattice parameters of the former perfect single crystalline substrate due to the implantation process. The term *perfect* refers to the detection limits of lattice defects of the used method, and in this case it means that the measured reflection curve of such a substrate corresponds to that calculated by the dynamical theory of X-ray diffraction. The crystal lattice can vary in its netplane distances or more generally in its perfection.

The relative change of netplane distance $\Delta d/d$ results starting from the Bragg equation

$$2d \sin \Theta_B = n\lambda$$

that describes the diffraction process, where d is the distance of the diffracting netplanes, Θ_B is the Bragg angle, n is the order of diffraction ($n = 1,2,3...$), and λ is the wavelength of the used X radiation, in a variation of the diffraction angle by

$$\Delta\Theta_B = -\frac{\Delta d}{d}\tan\Theta_B$$

The perfection of a lattice can be described by the statistical random displacement of the atoms from their position (disorder). Every defect distorts the lattice over a certain finite region, and the accumulation of defects increases the mean square displacement of the atoms. Consequently, the amplitude of the diffracted waves, proportional to the mean structure factor F_g^o, is reduced according to the relation (Speriosu, 1981)

$$\langle F_g \rangle = F_g^o \exp(-W)$$

where W (static Debye-Waller factor) depends on the random displacement standard deviation U and can be expressed under the assumption of a statistical distribution of the displacement of any atom according to a spherically symmetrical Gaussian form by

$$W = (8\pi^2/\lambda^2)\sin^2\Theta_B \times U^2$$

The perfection P of a crystal lattice is described by $\exp(-W)$. It goes to zero in the case of amorphization.

1. Implantation-Induced Variation of the Lattice Parameters

Independently of the implantation species, the implantation process itself causes such a variation of the lattice structure that the mean displacement of the atomic planes from their positions in the perfect crystal (strain $\varepsilon = \Delta d/d$) is positive. Consequently, the lattice constant of the diffraction planes in the implanted layer is larger than that of the substrate, and an additional diffraction peak occurs on the low-angle side relative to the substrate peak. Typically, the maximum strain increases with increasing implantation dose. Simultaneously, the disorder of the lattice also increases, which leads to a decreasing diffraction power of the disturbed layer, until the amorphous state is reached and the layer peak disappears.

Figure 1 demonstrates this dose-dependence of diffraction features with the example of a silicon (Si) implantation, at 100 keV and at room temperature, into a (100) Si substrate. Relative to the double-crystal diffractometer (DCD) rocking curve (RC) of the perfect substrate, the reflec-

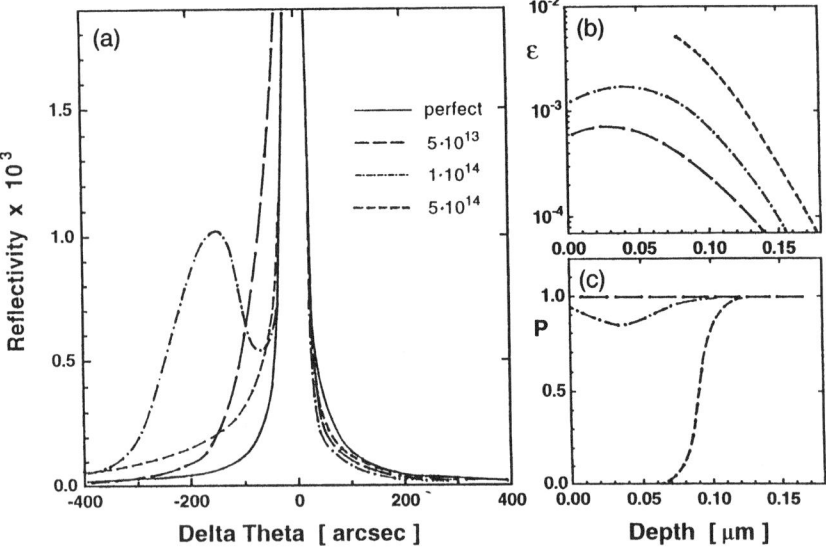

FIG. 1. (a) Double-crystal diffractometer rocking curves, silicon-(400)-reflection with CuK$_\alpha$-radiation, after silicon (Si) implantation with 100 keV of energy at various doses ($5 \cdot 10^{13}$, $1 \cdot 10^{14}$, and $5 \cdot 10^{14}$ Si$^+$/cm^{-2}) compared with the rocking curve of a perfect Si crystal. (b) Corresponding depth profile of lattice strain. (c) Profile of lattice perfection.

tivity on the low-angle side is clearly increased due to the increased lattice constant in the implanted layer (Fig. 1(a)). At the dose $D = 1 \cdot 10^{14}$ cm^{-2}, a separated layer peak is found at $\Delta\Theta = -150$ arcsec, which disappears again at $D = 5 \cdot 10^{14}$ cm^{-2} because of amorphization. However, even for doses above the amorphization limit, the RC remains unsymmetrical (increased reflectivity on the low-angle side), because there still exists a thin layer with increased lattice constant at the transition from the amorphous to the crystalline state. Figures 1(b) and 1(c) show the simulated depth profiles of strain and perfection (based on a simplified Gaussian shape) that were obtained by fitting the calculations to the experimental curves. The thickness of the amorphous layer at $D = 5 \cdot 10^{14}$ cm^{-2} of about 0.07 µm cannot be measured directly. It is the result of a rough extrapolation of perfection profiles at lower doses, and its error might be on the order of ±0.02 µm.

This example demonstrates that X-ray methods can be used to investigate the implantation process itself regarding the implantation dose and energy, substrate material and its orientation, implantation temperature, and so on. Speriosu et al. (1982) investigated, for example, the dose-dependence of Si implantation into Si, germanium (Ge), and gallium arsenide (GaAs). Qualitative and quantitative differences were found between the damage in GaAs, on the one hand, and in Si and Ge on the other, due to the creation of extended defects in GaAs. The nonlinear production of strain in (100) GaAs by room temperature implantation of neon (Ne), Si, and tellurium (Te) ions has been studied in more detail by Paine and Speriosu (1987) using X-ray diffraction in comparison with Rutherford backscattering spectrometry (RBS) measurements. A comparison of the maximum strain ε^{max} and of the integrated strain in self-implanted Si with the fractional displaced atoms and the concentration of displaced atoms, respectively, calculated by Monte Carlo methods, was shown by Servidori (1987) in its dependence on the implantation dose.

2. DOPANT INCORPORATION

If implanted ions are incorporated into lattice sites by any kind of annealing, as is usual for ion doping, an additional process varies the parameters of the crystal lattice. Since the tetrahedral radius of doping atoms r_d differs, in general, from that of the substrate material r_s, the lattice constant a_o is changed according to the Vegard law (Servidori, Zani, and Garulli, 1982)

$$\frac{\Delta a}{a_o} = \left(1 - \frac{r_d}{r_s}\right) K C_d N^{-1} = \beta_d K C_d$$

where N is the number of substrate atoms per unit volume ($N = 5.0 \cdot 10^{22}$ cm^{-3} for Si substrate) and C_d is the concentration of the substitutionally incorporated dopants. The factor K takes into account the fact that only the lattice strain normal to the wafer surface is different from zero ($\varepsilon_\perp \neq 0$), whereas the in-plane strain is zero ($\varepsilon_\parallel = 0$). This is realized as long as the doping level is lower than the critical level for the generation of misfit dislocations or for the occurrence of sample curvature. For typical substrate materials with cubic crystal lattice, K can be expressed by the elastic stiffness coefficients for the two important substrate orientations (Fukahara and Takano, 1977) as

$$K = 1 + \frac{2C_{12}}{C_{11}} \quad \text{for (100)}$$

$$K = 3\left(1 + \frac{2C_{12}}{C_{11}}\right) \bigg/ \left(1 + \frac{2C_{12}}{C_{11}} + \frac{4C_{44}}{C_{11}}\right) \quad \text{for (111)}$$

The fact that $\varepsilon_\parallel = 0$ for layers with sufficiently low dopant concentration to avoid the generation of misfit dislocations was experimentally confirmed by Golovin, Imamov, and Kondrashkina (1985) using an X-ray diffraction method under total external reflection that directly detects the in-plane strain.

As demonstrated, X-ray diffraction techniques are not able to measure the dopant concentration directly as can secondary ion mass spectroscopy (SIMS), for example. Therefore, the general problem is to know the relation between the measured strain $\varepsilon_\perp$ and C_d. As long as all dopant atoms are substitutionally incorporated into the substrate lattice, this is well known for the most-used species. The situation becomes more complicated in the cases of only partly substitutional incorporation. Then, a combination of X-ray and other methods can be useful to distinguish between the different states. Another problem arises when two different mechanisms cause a lattice strain—for example, in an incompletely annealed lattice expansion due to implantation process overlapping with the strain profile of the incorporated dopant atoms, or the profiles of two different kinds of implanted ions overlapping each other—then only the resulting strain effect can be measured, and it is often difficult or even impossible to distinguish between the different origins.

3. Defect Generation During Annealing

Different from the situation described above, that more or less pointlike defects (implantation-induced defects, dopant atoms) are generating a strained but nevertheless macroscopically perfect lattice, it is possible that two- or three-dimensional defects are generated especially after annealing of ion-implanted layers. These defects can be, for example, dislocations, stacking faults, or precipitates, and they can influence the X-ray diffraction behavior of the layer. The local displacement field in the area surrounding the defects produces diffuse scattering of X-rays that can be analyzed using triple-crystal diffractometer arrangements (Part III, Section 1), or the defects change the averaged lattice parameters and the RC is modified by analogy with the situations described in Sections 1 and 2 of Part II. Generally, a detection of defects by the measurement of diffuse scattering is relatively easy; however, a more detailed characterization should be the task for other techniques, for example, transmission electron microscopy (TEM).

A typical example is the generation of a layer of perfect interstitial dislocation loops during the annealing of an amorphous layer at the original crystalline-amorphous (a-c) interface. These loops, with concentration C_L and radius R, according to the theory of diffuse scattering of X-rays (Dederichs, 1973), cause an averaged lattice strain $\varepsilon_\perp$ and a Debye-Waller factor W (Servidori, 1987):

$$\varepsilon_\perp = \frac{K}{3} \frac{b\pi R^2 C_L}{V_C}$$

$$W = \frac{1}{2} (hb)^{3/2} \frac{R^3 C_L}{V_C}$$

where b is the modulus of the loop Burgers vector, C_L is the loop concentration, V_C is the volume per atom, and h is the modulus of the reciprocal lattice vector. Here a determination of the strain and perfection profile allows the determination of the radius and the concentration of the dislocation loops (Zaumseil et al., 1987).

Figure 2 demonstrates an example from bipolar complementary metal-oxide semiconductor (BICMOS) technology in which several implantations through a 30-nm-thick oxide were used to prepare shallow junctions. A (100)-oriented Si wafer was implanted with 70 keV, $5 \cdot 10^{15}$ As$^+$/cm^{-2}; 100 keV, $5 \cdot 10^{13}$ P$^+$/cm^{-2}; and 30 keV, $1 \cdot 10^{13}$ B$^+$/cm^{-2} and then furnace-annealed at 900°C for 75 min in N_2. The reflection curve measured by a triple-crystal diffractometer technique (Part III, Section 2b) is compared

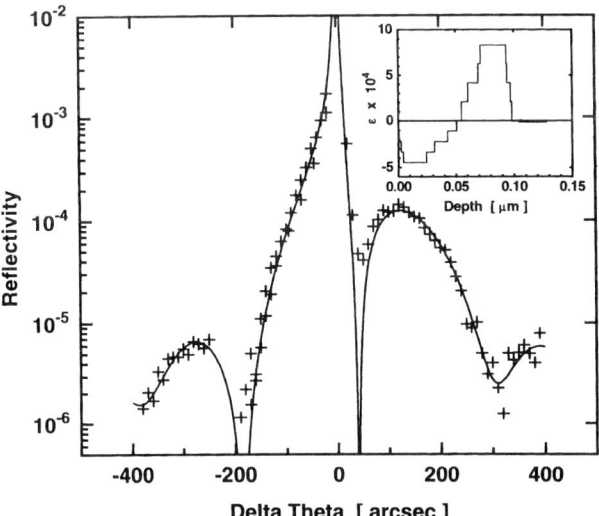

FIG. 2. Triple-crystal diffractometer rocking curve (+) of a silicon crystal after multiple ion implantation (see the text for details) and annealing at 900°C for 75 min, and best-fitted calculated curve (line). The inset shows the corresponding depth profile of lattice strain.

with the curve calculated by the depth profile of strain shown in the inset of Fig. 2 (the details of the calculations are discussed in Part III). The approximately 30-nm-thick layer with positive strain and a perfection $P \simeq 0.70$ results from small dislocation loops at the former c-a interface. The negatively strained layers can be explained by the phosphorus (P) and boron (B) incorporation according to the Vegard law (r_P and r_B are smaller than is r_{Si}). The substitutionally incorporated As does not significantly influence the strain profile, since the tetrahedral radius of As is nearly equal to that of Si. We must emphasize that here positive and negative strain components overlap, and only the resulting strain profile can be obtained.

Other examples of structural defects in implanted and annealed layers that have been investigated by X-ray techniques are misfit dislocations in highly P-doped or B-doped (Zaumseil and Winter, 1990) samples or the perfection of As-implanted Si after pulsed laser irradiation (Zaumseil, Winter, and Galler, 1984).

III. X-Ray Methods to Analyze Implanted Samples

It was shown, in principle, that the implantation-induced profile of strain and imperfection and structural defects are able to influence the X-ray reflection curve of a sample. The first problem is how this modification can be described theoretically, and the second is how the reflection curve can be measured. The reflection curve is a theoretical description of the diffraction of a plane, monochromatic X-ray wave, depending on its angle of incidence relative to the diffracting netplane. In reality, however, an ideal plane, monochromatic wave does not exist, and following, it is impossible to measure the reflection curve of a sample without any influence on the experimental arrangement. Consequently, the influence of the different measuring techniques and the possible loss of information also must be taken into account.

1. SIMULATION OF THE REFLECTION CURVE OF A DISTURBED CRYSTAL

a. Dynamical Theory

The general theory of X-ray diffraction, which properly accounts for normal absorption and extinction (coherent scattering or diffraction) of wave fields in a crystal medium, is called the dynamical theory of X-ray diffraction. When the homogeneity of the crystal specimen varies in only one direction—that is, in depth—with no lateral variation, the change in the X-ray complex amplitudes in depth is well described by the Takagi-Taupin equation, which was originally developed by Takagi (1962) and independently by Taupin (1964). Following the notation of Wie, Tombrello, and Vreeland (1986) the Takagi-Taupin equation is

$$i\frac{\lambda}{\pi}\beta_o \cdot \nabla D_o(r) = \psi_o D_o(r) + \psi_h D_h(r)$$

$$i\frac{\lambda}{\pi}\beta_h \cdot \nabla D_o(r) = \psi_o D_h(r) + \psi_h D_o(r) - \alpha_h D_h(r)$$

where $D_o(r)$ and $D_h(r)$ are the complex amplitudes of the incident and diffracted waves in the crystal, β_o and β_h are the wave vectors of the incident and diffracted waves in the crystal, $\psi_o = -(e^2/mc^2)(\lambda^2/\pi)(F_o/V)$, $\psi_h = -(e^2/mc^2)(\lambda^2/\pi)(F_h/V)$, F_o and F_h are the structure factors for the incident and diffracted waves, V is the cell volume, and $\alpha_h \approx - 2(\Theta - \Theta_B)\sin 2\Theta_B$. More useful for th/e study of implanted layers is the equation in the following form:

$$i\frac{dX}{dA} = (1 + ik)X^2 - 2(y + ig)X + (1 + ik)$$

where $X = D_h(r)\sqrt{b}\,D_o(r)$ is the scattering amplitude, $b = |\gamma_o/\gamma_h|$, and $\gamma_{o,h}$ are the direction cosines of the incident and diffracted waves with respect to the inward surface normal. A is a normalized depth coordinate, $g = (1 + b)\psi_o''/2|\psi_h'|\sqrt{b}$, $k = \psi_h''/\psi_h'$, and the depth distribution of lattice strain is included in the parameter $y = [(1 + b)\psi_o' - b\alpha_h]/2|\psi_h'|\sqrt{b}$ with $\alpha_h = -2(\Theta - \Theta_B)\sin 2\Theta_B - (c_1\epsilon_1 + c_2\epsilon_2)$, where $c_{1,2}$ are geometrical factors and ϵ_1 and ϵ_2 are the strain components perpendicular and parallel to the surface, respectively (for a detailed explanation of the parameters, see Wie, Tombrello, and Vreeland, 1986). Calculation of the reflection curve of the disturbed crystal requires integration of this differential equation starting at a certain depth, at which the crystal can be supposed to be perfect, up to the surface.

b. Semikinematic Approximation

A semikinematic approximation given by the first iteration of the Takagi-Taupin equation was presented by Kyutt, Petrashen, and Sorokin (1980). For the calculation the damaged layer is divided into n lamellae of thickness

Δ_k normalized by the extinction distance of the used reflection and strain ε_k ($k = 1, 2, \ldots, n$). Then the integration yields

$$R(\eta) = R_o(\eta) \left| 1 - 2i\eta \sum_{k=1}^{n} \exp(i\varphi_k) \frac{\sin(\eta - f_k)\Delta_k}{\eta - f_k} \right|^2$$

where

$$\varphi_k = -2 \sum_{k'=k+1}^{n} \Delta_{k'}(\eta - f_{k'}) - \Delta_k(\eta - f_k)$$

$$f_k = \frac{\lambda\sqrt{\gamma_o/|\gamma_o|}}{2\pi C |\psi_h|} \cdot \frac{2\sin\Theta_B|\gamma_h|}{\lambda} \cdot \varepsilon_{\perp k}$$

Here η is proportional to the angular distance between the diffraction angle and the Bragg angle, $R_o(\eta)$ is the reflection curve of the perfect crystal calculated by the dynamical theory, and C is the polarization factor. For simplicity, we suppose that the diffracting netplanes are parallel to the sample surface. Physically, this relation shows the interference of the amplitudes of reflections from the substrate and from the thin layers, with the phases determined by the geometrical position of the layer and by the value of the shift of this layer from its position in an ideal crystal.

Calculations of the reflection curve of a disturbed crystal based on this algorithm are much faster than are dynamical calculations, but are restricted to thin layers (thin compared with the extinction distance). This is fulfilled for most of the implanted layers as long as energies below 1 MeV are used.

c. Other Calculation Techniques

Additional approaches to simulate the reflection curves of implanted layers were based on a pure kinematic algorithm (see, e.g., Speriosu, Glass, and Kobayashi, 1979). Another method was proposed by Kohn et al. (1981) to obtain integral characteristics of an implanted layer such as the effective layer thickness and the averaged lattice strain directly from the measured RC. This information is often enough for many purposes of the comparison of different implantation parameters.

Meanwhile, not only was the one-dimensional (depth depending) problem treated, but first attempts were made to investigate two-dimensional deformation fields with a periodical implantation structure in the surface (see, e.g., Goureev et al., 1992).

2. Experimental Techniques

a. Double-Crystal Diffractometry

Double-crystal diffractometry (DCD) is the most-used X-ray technique to measure the diffraction behavior of a crystal with an ion-implanted layer. Figure 3 shows, in principle, the so-called (n,-n) arrangement. A first, perfect crystal is collimating, and in combination with slit systems monochromatizing the radiation, which is falling on the sample. Depending on the angular position of the sample Θ, typically expressed relative to the Bragg angle $\Delta\Theta = \Theta - \Theta_B$, the so-called rocking curve $R_c(\Delta\Theta)$ is measured, which is obtained for nonpolarized radiation as the convolution of the reflection curves of collimator crystal C_1 and sample C_2 (Pinsker, 1978)

$$R_c(\Delta\Theta) \sim \int_{-\infty}^{\infty} C_1^\sigma(x)C_2^\sigma(x - \Delta\Theta) + C_1^\pi(x)C_2^\pi(x - \Delta\Theta)\,dx$$

where σ and π describe the two polarization states of radiation perpendicular and parallel to the diffraction plane, respectively. In the tails of the RC ($|\Delta\Theta|$ greater than about five times the full width at half maximum RC) the more simple expression

$$R_c(\Delta\Theta) = \frac{C_2^\sigma(\Delta\Theta)I_1^\sigma + C_2^\pi(\Delta\Theta)I_1^\pi}{I_1^\sigma + I_1^\pi}$$

can be used with sufficient accuracy (Servidori and Cembali, 1988). Here C_2^σ, C_2^π and I_1^σ, I_1^π are the reflection curves of the sample and the integrated reflecting powers of the collimator, respectively, for the two components of polarization.

An important problem of the DCD technique is that due to the use of an open window scintillation counter to measure the intensity diffracted by the sample, diffuse scattering is also collected. This results in an angle depending

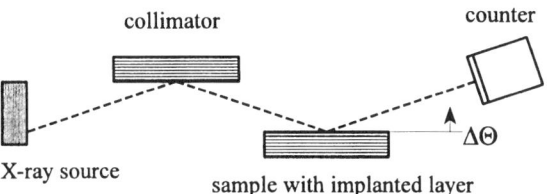

FIG. 3. Double-crystal diffractometer arrangement in parallel (n,-n) setting $\Delta\Theta$, rotation angle relative to the Bragg position.

intensity background that is commonly not included in the calculation procedures, which can lead to wrong depth profiles of deformation. Cembali et al. (1985) tried to overcome this problem by a semi-empirical correction of the RC, which works well as long as only Compton and thermal diffuse scattering act, but fails completely when structural defects in the layer generate an extra-diffuse scattering.

b. *Triple-Crystal Diffractometry*

To overcome this problem and to increase the sensitivity of the measurement, in general, triple-crystal diffractometry (TCD) in parallel (n,-n,n) setting (Iida and Kohra, 1979), which can be realized in different arrangements (Zaumseil and Winter, 1982), was successfully used (see, e.g., Cembali et al., 1986; Zaumseil et al., 1987; Boussey-Said et al., 1992). Figure 4 shows the principle sketch of a TCD. It looks like a DCD arrangement with a third, perfect crystal added behind the sample that acts as an analyzer crystal. Now, the two angular positions of sample α and analyzer $\Delta\Theta$ are free parameters to measure the intensity given here for the simplest case of symmetrical Bragg case reflections at all crystals, as shown in Figure 4 (both polarization states must be summed up as in the similar expression for the DCD arrangement):

$$I(\alpha, \Delta\Theta) \sim \int_{-\infty}^{\infty} C_1^{\sigma,\pi}(x - 2\alpha) C_2^{\sigma,\pi}(x - \alpha) C_3^{\sigma,\pi}(x - \Delta\Theta) \, dx$$

This allows determination of the two-dimensional intensity distribution in the surroundings of the reciprocal lattice point in the diffraction plane

$$(q_x, q_y) = [\Delta\Theta \cos \Theta_B, (2\alpha - \Delta\Theta)\sin \Theta_B]/\lambda$$

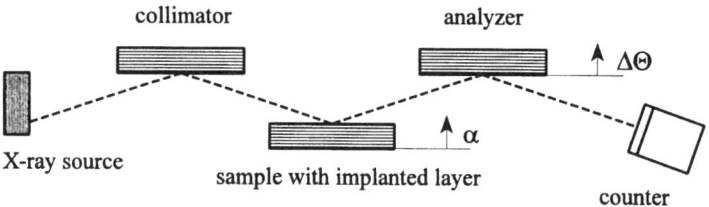

FIG. 4. Triple-crystal diffractometer arrangement in parallel (n,-n,n) setting.

where **q** is the scattering vector (see, e.g., Zaumseil et al., 1987). Consequently, TCD allows a differential measurement contrary to the integral measurement of the DCD.

The main advantage of TCD is the angular separation of dynamical and kinematic (diffuse) intensities (Iida and Kohra, 1979). To measure the dynamical reflection curve of the sample with high accuracy, the so-called main-peak analysis method was proposed (Zaumseil, 1985). It requires the step-by-step measurement of the main-peak intensity at sample position α and analyzer position $\Delta\Theta_{MP} = 2\alpha$. Similar to the DCD technique, the relation between the reflection curve in both polarizations and the measured main-peak reflectivity can be expressed in first approximation by

$$R(\alpha, \Delta\Theta_{MP}) = \frac{C_2^\sigma(\alpha)I_1^{\sigma^2} + C_2^\pi(\alpha)I_1^{\pi^2}}{I_1^\sigma + I_1^\pi}$$

Rocking curves are provided with much more details with TCD than with DCD, which leads to better results in the deformation profile analysis Furthermore, in some cases, TCD is the only method to obtain an RC that can be used for profile simulations (Zaumseil et al., 1987).

c. Other X-Ray Techniques

In addition to the DCD and TCD techniques that are mainly used, there are other X-ray diffraction methods for special purposes. X-ray topography is able to show the inhomogeneous distribution of strain fields in or around an implanted area (Wieteska, 1981) or the perfection of recrystallized implanted layers (Zaumseil, Winter, and Galler, 1984). X-ray diffraction under total external reflection conditions (Golovin, Imamov, and Kondrashkina, 1985) provides information about the strain parallel to the sample surface, and even amorphous layers can be characterized according to their densities and thicknesses. Time-resolved X-ray diffraction measurements of the lattice strain during pulsed laser annealing of B-implanted Si have been made with nanosecond resolution by using synchrotron radiation (Larson et al., 1983).

3. LIMITATIONS OF X-RAY DIFFRACTION TECHNIQUES

Some of the main advantages of X-ray techniques are that they are nondestructive, they require no special sample preparation, and the technique is relatively simple. However, their use is limited in four different ways:

1. The lateral resolution is relatively low. Usually, the X-ray beam size is about $1 \times 1\,\mathrm{mm}^2$ in order to have enough intensity as long as a conventional X-ray source is used. This requires appropriately large areas with homogeneous implantation. By using synchrotron radiation and focusing X-ray optics, first attempts have been made to increase the resolution to some micrometers (Iberl et al., 1995).
2. The minimum detectable lattice strain—and if this is caused by a dopant, the minimum dopant concentration—depends on the X-ray diffraction conditions and the thickness of the layer. Under the assumption of a Gaussian-shaped deformation profile $\varepsilon_\perp(z) = \varepsilon_o \exp\{-(z/2L)^2\}$, Zaumseil and Winter (1983) have calculated the relation between ε_o and L for six different symmetrical and asymmetrical Bragg case reflections at Si, which leads to a relative change of 5% at any point of the reflection curve. For $L = 100\,\mathrm{nm}$, a strain of $\varepsilon_o = 1\ldots 4 \cdot 10^{-4}$ is necessary to fulfill this criterion. Supposing a sufficiently thick implanted layer, the minimum dopant concentration can be estimated by a comparison of the dopant-induced change of the diffraction angle $\Delta\Theta_B = -\varepsilon_\perp \tan\Theta_B = -\beta_d K C_d \tan\Theta_B$ and the half-width of the dynamical reflection curve $\Delta\Theta_{1/2} = 2|\psi_h|/\sin 2\Theta_B$. For the Si-(400)-reflection with CuK_α-radiation, the minimum detectable concentrations of P and B are about $1.5 \cdot 10^{19}\,\mathrm{P/cm}^{-3}$ and $2 \cdot 10^{18}\,\mathrm{B/cm}^{-3}$, respectively.
3. The substitutional incorporation of As in Si does not significantly change the lattice parameters. Consequently, it is impossible to investigate the doping profile of As in Si by X-ray diffraction techniques.
4. The combination of deformation profile analysis and characterization of structural defects requires the application of TCD. A general statement about the detection limits of defects is impossible because they depend on too many parameters (e.g., defect type and size, layer thickness, and deformation profile in the surroundings). The final criterion is that the extra diffuse scattering of the defects must be stronger than is the thermal diffuse scattering of the sample.

IV. Examples of Special Applications

1. DETERMINATION OF DISLOCATION LOOP SIZE AND DENSITY IN IMPLANTED AND ANNEALED SILICON

The presence of perfect interstitial dislocation loops just below the original amorphous-crystalline interface is well known as the main structural feature determined by high-temperature treatments of implanted (100)

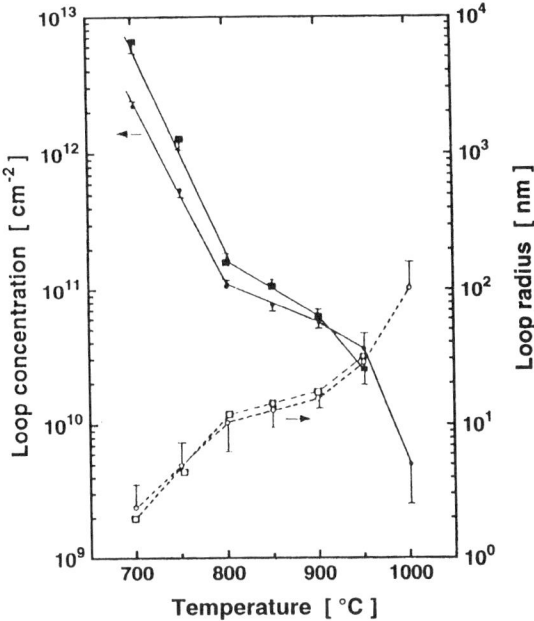

FIG. 5. Temperature evolution of loop mean radius R and concentration C obtained by triple-crystal diffractometer rocking curve simulation (□, ■) and transmission electron microscopy (○, ●). (Reprinted from Zaumseil, Winter, Cembali, Servidori, and Šourek (1987). "Determination of Dislocation Loop Size and Density in Ion Implanted and Annealed Silicon of Triple Crystal X-Ray Rocking Curves," *Phys. Stat. Sol.* (a) **100**, 95, with permission of the author and Akademie Verlag, Berlin.)

silicon wafers. These defects are commonly observed by TEM both in plane-view and cross-section view, so that their size, density, and depth position can be determined. Zaumseil et al. (1987) proposed an alternative method based on computer simulation of X-ray RCs taken in triple-crystal geometry. The TCD technique was preferred to the simpler DCD technique because it offers the possibility to discriminate the influence of diffuse scattering that increases with the defect size and density.

Samples of (100) silicon wafers were implanted with $10^{16}\,\mathrm{Si^+/cm^{-2}}$, at 100 keV ions, with a tilt angle of 8 degrees from the ⟨100⟩ direction. After implantation, each sample was furnace-annealed at a single temperature for 30 min. The temperatures of the different annealings ranged from 700 to 1000°C in increments of 50°C. Using the symmetrical (400)-reflection with CuK_α-radiation, RCs were obtained. The depth profiles of strain and disorder were calculated by means of the semikinematic approximation of the dynamical theory.

The profiles show a sharp peak of positive strain and decreased perfection at a depth of about 250 nm that is related to the buried layer of dislocation loops. Using the relation given in part II, section 3, the loop radius R and the density C_L were determined from the measured strain and disorder. These values and those obtained by TEM plane-views are reported in Figure 5. These data agree surprisingly well, indicating the validity of the model used. The temperature evolution of C_L and R follows the trend of decreasing density and increasing size, where three regions can be distinguished: (1) $T < 800°C$; (2) $800°C < T < 900°C$; and (3) $T > 900°C$.

2. ARSENIC IMPLANTATION AND ANNEALING

During the last few years, many experimental studies have been carried out in an attempt to understand the mechanisms of disorder production and crystalline-to-amorphous transition in ion-implanted Si. Similarly, many

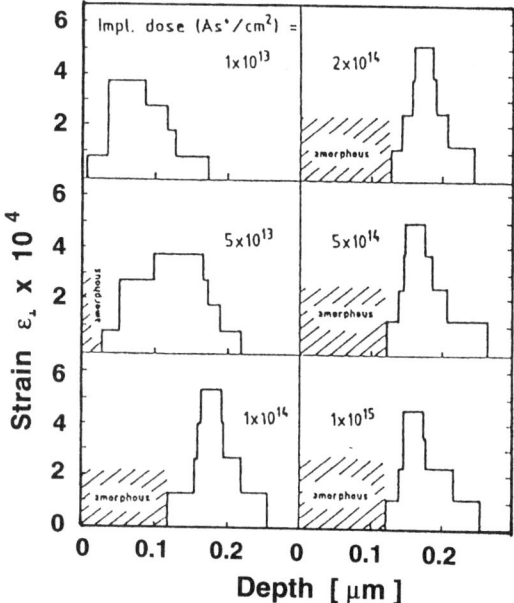

FIG. 6. Depth profiles of lattice strain obtained by triple-crystal diffractometer measurements from samples implanted (Impl.) with As^+ at an energy of 200 keV at various doses ($10^{13}...10^{15}$ As^+/cm^{-2}) and annealed for 1 h at low temperature (450°C). As, arsenic. (Reprinted from Boussey-Said, Ghibaudo, Stoemenos, and Zaumseil (1992). "Electrical and Structural Properties of Silicon Layers Heavily Damaged by Ion Implantation," *J. Appl. Phys.* **72**, 61, with permission from the author.)

works deal with the optimization of the postimplantation thermal annealing conditions for a better doping activation, while minimizing the junction depth. A proper combination of electrical and physical characterizations was used by Boussey-Said et al. (1992) to analyze in detail the effect of thermal annealing on the electrical and structural properties of high-dose ion-implanted Si. The TCD results obtained for As implantation, in particular, are shown here.

Samples of (100) silicon were implanted with As^+ at a tilt angle of 7 degrees relative to the $\langle 100 \rangle$ direction, at room temperature an energy of 200 keV, with doses ranging from 10^{13} to 10^{15} As^+/cm^{-2}. Isochronal annealings (1 h) were made in a thermal furnace under flowing nitrogen at temperatures ranging from 200 to 1100°C. The RCs were measured by TCD and the simulations are based on the semikinematic model.

Figure 6 shows depth profiles of the lattice strain $\varepsilon_\perp$ obtained at various doses after annealing at low temperature (450°C) for 1 h. A partial surface amorphization is detected after a $5 \cdot 10^{13}$ As^+/cm^{-2} implantation. For higher doses, amorphization occurs at the surface, and the deeper strained region is not significantly changed by further increasing of the implantation dose. The depth position of the positive strained layer was extrapolated from

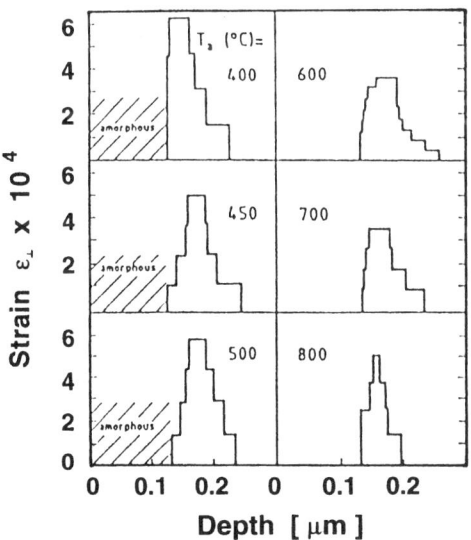

FIG. 7. Depth profiles of lattice strain, $D = 2 \cdot 10^{14}$ As^+/cm^{-2}, isochronally annealed (1 h) at different temperatures (400...800°C). As, arsenic; T_a, annealing temperature. (Reprinted from Boussey-Said, Ghibaudo, Stoemenos, and Zaumseil (1992). "Electrical and Structural Properties of Silicon Layers Heavily Damaged by Ion Implantation," J. Appl. Phys. 72, 61, with permission from the author.)

the case of high-temperature annealing for which recrystallization already has occurred. Nevertheless, there remains an uncertainty in the thickness of this amorphous layer of about $\pm 0.02\,\mu$m.

In order to study the evolution of structures with implantation doses near the amorphization threshold, cross-section TEM (XTEM) and TCD analyses have been performed on annealed samples under various conditions. Figure 7 shows the strain profiles obtained from the As-implanted sample ($2\cdot 10^{14}\,\text{As}^+/\text{cm}^{-2}$) and isochronally annealed (1 h) at 400, 450, 500, 600, 700, and 800°C. The significant modification of the near-surface layer, related to the amorphous-crystalline transition, occurs between 500 and 600°C. The remaining layer with positive strain is caused by dislocation loops that require very high temperature annealing (up to 1000°C) to be removed.

The comparison of these structural investigations with electrical measurements showed that a low temperature (< 450°C) electrical activation process is taking place in the surface layer, which is still amorphous (see Chapter 7 of this book.

3. Characterization of Boron-Implanted and Annealed Silicon

Boron implantation is a typical example in which, at high doses, the critical concentration for the generation of misfit dislocations can be exceeded. Depending on implantation dose, energy, and annealing conditions, the defect spectrum varies between individual dislocation loops and a narrow dislocation network. The possibility of using X-ray triple-crystal diffractometry (TCD) to analyze the depth profile of lattice distortion under such conditions was demonstrated by Zaumseil and Winter (1990).

One set of (100) Si samples was implanted with B^+ at 85 keV and doses of $8\cdot 10^{14}$, $2\cdot 10^{15}$, $1\cdot 10^{16}$, and $2\cdot 10^{16}\,B^+/\text{cm}^{-2}$ and then furnace-annealed at 1100°C for 30 min in dry nitrogen. The equal-intensity curves shown in Figure 8 were constructed from sets of TCD curves measured at different analyzer positions $\Delta\Theta$ and rotation of the sample. The contours of a perfect crystal (Zaumseil and Winter, 1982) in Fig. 8(a) are often called the resolution function of the given TCD arrangement. The sample with $8\cdot 10^{14}\,B^+/\text{cm}^{-2}$ shows nearly the same contours as does the perfect crystal, and therefore is not presented separately here. Contrary to this, the samples with implantation doses of $2\cdot 10^{15}$ up to $2\cdot 10^{16}\,B^+/\text{cm}^{-2}$ exhibit an increasingly diffuse scattering, which is shifted in its center in the direction of the reciprocal lattice vector. This unusual behavior of diffuse scattering is

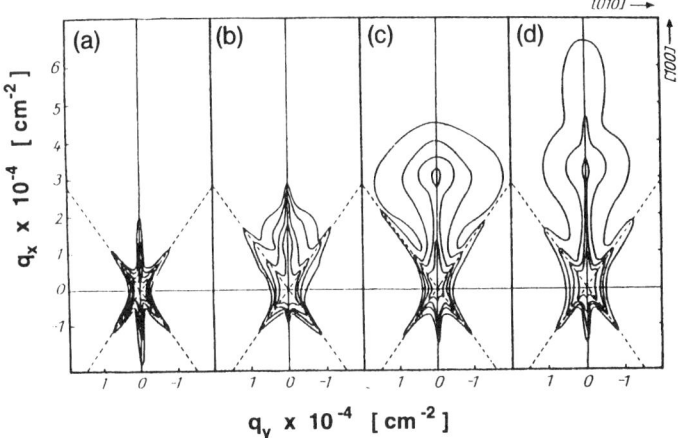

FIG. 8. Equal-intensity contours near the (400) reciprocal lattice point: (a) undisturbed crystal and $D = 8 \cdot 10^{14}$, (b) $2 \cdot 10^{15}$, (c) $1 \cdot 10^{16}$, and (d) $2 \cdot 10^{16}$ B$^+$/cm^{-2}. (Reprinted from Zaumseil and Winter (1990). "Characterization of Boron Implanted Silicon by X-Ray Triple-Crystal Diffractometry," *Phys. Stat. Sol. (a)* **120**, 67, with permission from the author and Akademie Verlag, Berlin.)

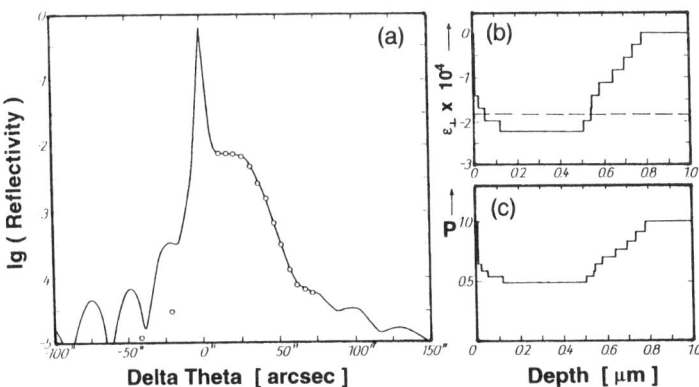

FIG. 9. (a) Comparison of corrected experimental triple-crystal diffractometer rocking curve (O) and best-fitted calculated curve (line), $D = 2 \cdot 10^{15}$ B$^+$/cm^{-2}. (b) Corresponding depth profile of lattice strain. (c) Profile of lattice perfection, P. (Reprinted from Zaumseil and Winter (1990). "Characterization of Boron Implanted Silicon by X-Ray Triple-Crystal Diffractometry," *Phys. Stat. Sol. (a)* **120**, 67, with permission from the author and Akademie Verlag, Berlin.)

related to the fact that the scattering defects are located in a matrix with a strained crystal lattice due to the incorporation of B atoms. Therefore, the center of diffuse scattering is shifted in the reciprocal space by $\Delta q_x = -\varepsilon_\perp \cdot h$, where h is the modulus of the reciprocal lattice vector.

The determination of the depth profile of strain and disorder from a RC requires TCD measurements, and even here the main-peak values must be corrected by the surrounding very strong diffuse scattering. Simulations of DCD RCs would fail completely. Figure 9 shows the comparison between experimental points obtained by corrected TCD measurements and the best-fitted simulated RC, and the corresponding profiles of strain and perfection for the sample with $2 \cdot 10^{15}$ B^+/cm^{-2}. Even if the experimental points are relatively few and the fit is good only for $\Delta\Theta > 0$, this is the only method for obtaining reliable profiles of strain and disorder in this case. The dashed line in Figure 9(b) indicates the $\varepsilon_\perp$ value corresponding to the center of diffuse scattering in Figure 8(b). The depth position of the structural defects can be estimated from these data to be approximately 0.4 to 0.6 μm, which was confirmed by TEM investigations. A determination of the B profile itself would lead to an underestimation in this example because the structural defects are generated simply to reduce the vertical strain.

V. Summary

X-ray diffraction is a powerful tool for investigating ion-implanted layers in single crystalline substrates. The primary information that can be obtained is the structural perfection of the layer expressed by modification of either the spacing of the diffracting netplanes or the statistical random displacement of the atoms (disorder), or both. Different measuring techniques are available, and TCD provides higher sensitivity and more information than does the simpler DCD, however, at the price of longer measuring times. The theoretical models used to simulate reflection curves of disturbed crystals are well established. Typical examples for the application of X-ray techniques are the investigation of the implantation process itself, characterization of dopant profiles after annealing, and detection and study of extended structural defects. Meanwhile, so many papers are published in this field, that only a tiny fraction could be mentioned here, especially related to implantation into Si. The different X-ray techniques have their limitations as does any diagnostic technique. In most cases, the combination with other methods is the best way to achieve optimal results.

REFERENCES

Boussey-Said, J., Ghibaudo, G., Stoemenos, I., and Zaumseil, P. (1992). "Electrical and Structural Properties of Silicon Layers Heavily Damaged by Ion Implantation," *J. Appl. Phys.* **72**, 61–68.

Burgeat, J., and Taupin, D. (1968). "Application de la théorie dynamique de la diffraction X a l'étude de la diffusion du bore et du phosphore dans les cristaux de silicium," *Acta Cryst.* **A24**, 99–103.

Cembali, F., Servidori, M., Gabilli, E., and Lotti, R. (1985). "Effects of Diffuse Scattering in the Strain Profile Determination by Double Crystal X-Ray Diffraction," *Phys. Stat. Sol. (a)* **87**, 225–233.

Cembali, F., Servidori, M., Solmi, S., Šourek, Z., Winter, U., and Zaumseil, P. (1986). "Structural and Electrical Characterization of Boron Implanted in Preamorphized Silicon Layers," *Phys. Stat. Sol. (a)* **98**, 511–516.

Dederichs, P. H. (1973). "The Theory of Diffuse X-Ray Scattering and Its Application to the Study of Point Defects and their Clusters," *J. Phys.* **F3**, 471–493.

Fukahara, A., and Takano, Y. (1977). "Determination of Strain Distributions from X-Ray Bragg Reflection by Silicon Single Crystals," *Acta Cryst.* **A33**, 137–142.

Golovin, A. L., Imamov, R. M., and Kondrashkina, E. A. (1985). "Potentialities of New X-Ray Diffraction Methods in Structural Studies of Ion-Implanted Silicon Layers," *Phys. Stat. Sol. (a)* **88**, 505–514.

Goureev, T. E., Nikulin, A. Yu., and Petrashen, P. V. (1992). "X-Ray Diagnostics of 2D Deformation Profiles in Crystals with Periodically Distorted Near-Surface Region," *Phys. Stat. Sol. (a)* **130**, 263–271.

Iberl, A., Schuster, M., Göbel, H., Meyer, A., Baur, B., Matz, R., Snigirev, A., Snigireva, I., Freund, A., Lengeler, B., and Heinecke, H. (1995). "Micro-Focus Double-Crystal X-Ray Diffractometry on III–V Heterostructures Grown by Selective-Area Epitaxy," *J. Phys.* **D28**, A200–A205.

Iida, A., and Kohra, K. (1979). "Separate Measurement of Dynamical and Kinematical X-Ray Diffractions from Silicon Crystals with a Triple Crystal Diffractometer," *Phys. Stat. Sol. (a)* **51**, 533–542.

Kohn, V. G., Kovalchuk, M. V., Imamov, R. M., and Lobanovich, E. F. (1981). "The Method of Integral Characteristics in X-Ray Diffraction Studies of the Structure of the Surface Layers of Single Crystals," *Phys. Stat. Sol. (a)* **64**, 435–442.

Kyutt, R. N., Petrashen, P. V., and Sorokin, L. M. (1980). "Strain Profiles in Ion-Doped Silicon Obtained from X-Ray Rocking Curves," *Phys. Stat. Sol. (a)* **60**, 381–389.

Larson, B. C., White, C. W., Noggle, T. S., Barhorst, J. F., and Mills, D. M. (1983). "Time-Resolved x-Ray Diffraction Measurement of the Temperature and Temperature Gradients in Silicon During Pulsed Laser Annealing," *Appl. Phys. Lett.*, **42**, 282–284.

Paine, B. M., and Speriosu, V. S. (1987). "Nonlinear Effects in Ion-Implanted GaAs," *J. Appl. Phys.* **62**(5), 1704–1709.

Pinsker, Z. G. (1978). *Dynamical Scattering of X-Rays in Crystals*, Springer-Verlag, Berlin, Heidelberg, New York.

Servidori, M. (1987). "Characterization of Lattice Damage in Ion Implanted Silicon by Multiple Crystal X-Ray Diffraction," *Nucl. Instru. Meth. Phys. Res.* **B19/20**, 443–449.

Servidori, M., and Cembali, F. (1988). "Accuracy in X-Ray Rocking Curve Analysis as a Necessary Requirement for Revealing Vacancies and Interstitials in Regrown Silicon Layers Amorphized by Ion Implantation," *J. Appl. Cryst.* **21**, 176–181.

Servidori, M., Zani, A., and Garulli, G. (1982). "Residual Lattice Disorder in Self-Implanted Silicon after Pulsed Laser Irradiation," *Phys. Stat. Sol. (a)* **70**, 691–701.

Speriosu, V. S. (1981). "Kinematical X-Ray Diffraction in Nonuniform Crystalline Films: Strain and Damage Distributions in Ion-Implanted Garnets," *J. Appl. Phys.* **52**, 6094–6103.

Speriosu, V. S., Glass, H. L., and Kobayashi, T. (1979). "X-ray Determination of Strain and Damage Distributions in Ion-Implanted Layers," *Appl. Phys. Lett.*, **34**(9), 539–542.

Speriosu, V. S., Paine, B. M., Nicolet, M.-A., and Glass, H. L. (1982). "X-Ray Rocking Curve Study of Si-Implanted GaAs, Si, and Ge," *Appl. Phys. Lett.*, **40**(7), 604–606.

Takagi, S. (1962). "Dynamical Theory of Diffraction Applicable to Crystals with Any Kind of Small Distortions," *Acta Cryst.* **15**, 1311–1312.

Taupin, D. (1964). "Théorie dynamique de la diffraction des rayons x par les cristaux déformés," *Bull. Soc. Franc. Minéral. Crist.* **87**, 469–511.

von Laue, M. (1960). *Röntgenstrahlinterferenzen*, Akademische Verlagsgesellschaft, Frankfurt.

Wie, C. R., Tombrello, T. A., and Vreeland, Jr., T. (1986). "Dynamical x-Ray Diffraction from Nonuniform Crystalline Films: Application to X-Ray Rocking Curve Analysis," *J. Appl. Phys.* **59**, 3743–3746.

Wieteska, K. (1981). "X-Ray Diffraction Investigations of High Energy α-Particle Damage in Silicon," *Phys. Stat. Sol.* (a) **68**, 179–185.

Zaumseil, P. (1985). "On the Increased Sensitivity of X-ray Rocking Curve Measurements by Triple-Crystal Diffractometry," *Phys. Stat. Sol.* (a) **91**, K31–K33.

Zaumseil, P., and Winter, U. (1982). "Triple Crystal Diffractometer Investigations of Silicon Crystals with Different Collimator-Analyzer Arrangements," *Phys. Stat. Sol.* (a) **70**, 497–505.

Zaumseil, P., and Winter, U. (1983). "Investigation of Shallow Strain Distributions in Silicon Single Crystals with X-Ray Bragg Reflection," *Crystal Res. Technol.* **18**, 219–228.

Zaumseil, P., and Winter, U. (1990). "Characterization of Boron Implanted Silicon by X-Ray Triple-Crystal Diffractometry," *Phys. Stat. Sol.* (a) **120**, 67–75.

Zaumseil, P., Winter, U., Cembali, F., Servidori, M., and Šourek, Z. (1987). "Determination of Dislocation Loop Size and Density in Ion Implanted and Annealed Silicon by Simulation of Triple Crystal X-Ray Rocking Curves," *Phys. Stat. Sol.* (a) **100**, 95–104.

Zaumseil, P., Winter, U., and Galler, R. (1984). "X-Ray Triple-Crystal Diffractometer Investigation of Arsenic Implanted Silicon after Laser Irradiation," *Crystal Res. Technol.* **19**, 633–641.

Index

A

Active dopant concentration, 189. *See also* dopant incorporation
Amorphization threshold, 242
Amorphization, 87, 204, 240, 256, 276
Amorphous semiconductors, 85–120
Amorphous Silicon, 85–120
  ion beam production of, 87
  plasma deposited amorphous silicon, 106
Amorphous-Crystal interface, 206
Annealing kinetics, 115, 143
  activation energy, 151
  relaxation time, 147
Annealing of defects. *See* Annihilation of implantation induced defects. *See also* Annealing
Annealing, 86, 115, 144, 181, 248, 276
  laser annealing, 86,
  of implantation damage, 115, 210
  thermal annealing, 86, 143, 175
  rapid thermal annealing, 181
Annihilation of implantation induced defects, 115, 144, 181, 206, 248, 276
Anodic stripping, 166
Arsenic emitter, 23
Arsenic implanted silicon, 143, 276
Atomic displacement, 240, 276
Axial channeling, 69

B

Bipolar devices, 22
  current gain, 25
Bipolar transistors. *See* Bipolar devices
Boron base, 24
Boron fluorine-implanted silicon, 152, 175

Boron implanted silicon, 152, 278
Brandt-Kitagawa model, 42
  projectile, 42
  stopping number, 43

C

Carrier density profiling, 172
Carrier mobility profiling, 172
Channeling effect, 69, 240
Chemical composition, 199
Chemical compound stopping power, 69
Collinear four-probe method, 131
Configurational changes, 121. *See also* structural changes
Coulomb explosion, 47
Crystalline-amorphous interface, 144, 204, 240
Crystallization, 99, 205. *See also* Solid phase epitaxial regrowth
Crystallographic defects, 201. *See also* defects
  dislocations, 217
  EOR defects, 204
  extended defects, 202
  interstitials, 203
  rodlike defects (RLD), 203
  stacking faults, 218
  vacancies, 203
Crystallographic structure, 199

D

Defects,
  carbon precipitates, 213
  cluster, 201

283

Defects (*continued*)
  dislocation, 217
  dislocation loop, 249
  dissolution, 223
  extended defect distribution, 249
  generation of, 115, 266
  interstitials, 201, 249
  precipitates, 225
  microtwin, 249
  stacking faults, 219
  vacancies, 201
Dielectric function, 34
  polarizability, 34
Diffraction dynamical theory, 268
Dislocation loop, 274. *See also* defects
Disorder, 86
  crystalline silicon target, 87
  depth profile of, 240
  impact of disorder on doping, 87
  impact of disorder on optical properties, 88
  impact of disorder on electronic transport, 91
Displaced atom number, 252–256, 276–280
Dopant incorporation, 189, 265
Doping mechanism, 110. *See also* dopant incorporation
Double crystal diffractometry, 263, 271
Drain engineering, 20
  LATID, 14, 20
  LDD, 20
  silicide, 21

E

Effective stopping charge, 44
Electrical characterization, 87, 129, 165, 208
Electrical doping activation, 99, 143, 189. *See also* dopant incorporation
Electron gas model, 33
Electronic stopping power, 55–80
  Bragg rule, 71
  combination rule, 60, 71
Electronic transport, 87, 129, 165
End-of-range (EOR) defects, 204
  carbon implantation, 212
  Electrical properties of, 208
  kinetics of EOR, 208
  implantation condition inhibiting EOR

  formation, 212
Energetic ions, 31-50

F

Fluorination, 105
Furnace annealing, 115, 143, 175, 248, 274

G

Gas phase doping, 106

H

Hall effect, 166
  basic theory of, 166
Hall scattering factor, 172
Heavy ions, 42
Hydrogenated material, 110, 122
Hydrogenation, 105

I

Impact parameter, 64
Implantation damage, 115, 204, 240, 276
  residual damage, 246
Implantation, 5,
  acceleration, 8
  ion beam, 2
  ion irradiation, 2
  high temperature implantation, 213
  mass separation, 8
Implanter trends, 14
Implanter, see Ion implanter
Ion beam doping, 106
Ion beam induced epitaxial crystallization (IBIEC), 230
Ion beam synthesis, 216
Ion beam, 85–120
Ion bombardment, 86
Ion implantation 1–280. *See also* implantation
Ion implantation doping, 106
Ion implanter, 5
  accelerators, 6
  beam scanning, 11
  contamination, 13
  development of, 5

INDEX

dosimetry, 11–12
linear accelerator (LINAC), 8
 preacceleration, 9
 principle of, 8–11
 postacceleration, 9
SIMOX, 13,
Ion range, 32

L

Lattice parameter modification, 262
 implantation induced lattice parameter variation, 263
Lattice strain profile, 276
Local density approximation (LDA), 35, 58
 LDA for binary collisions, 58
LSS theory, 2, 56

M

Metal-Oxide-Semiconductor devices, 15
 MOS transistor, 15
 CMOS transistor, 15
Microtwin, 249
Mobility measurement, 170
Molecular ions, 47, 210

N

Nuclear stopping power, 59
 numerical calculations of, 36

O

Oen-Robinson model, 64
Optical properties, 87
Ordering phenomena, 95, 99

P

Physical characterization, 201, 239, 261
Polycrystalline Silicon, 189
Precipitates, 249

R

Radiation damage, 32, 201, 248, 263, 276
Random phase approximation (RPA), 32

Range profile, 72-75
 computer simulations, 72
 Monte carlo simulations, 76
Rapid thermal annealing (RTA), 181
RBS channeling, 240
RBS-channeling characterization, 240
RBS studies, 248
Recrystallization velocity, 206. *See also* Regrowth velocity
Regrowth velocity, 144
Resistivity measurement, 138, 168
Resistivity profiles, 138–155
Rocking curve, 262–280
Rod-like defect, 249. *See also* defects
Rutherford Backscattering Spectrometry (RBS), 239–258

S

Self-aligned gate process, 15
Semiconductors, 1–280
Sheet resistance, 129–140
 sheet resistance measurement, 130
Silicon carbide, 227, 248, 255
 RBS studies, 248
 TEM studies, 227
Silicon Separation by Implanted Oxygen (SIMOX), 216
Silicon, 1–280
 RBS studies, 248
 TEM studies, 195
 Electrical studies, 85, 129, 165
 X-ray studies, 261
Solid effect on electronic stopping power, 68, 76
Solid phase crystallization (SPC), 206
Solid phase epitaxial regrowth, 144, 175, 206, 254
Solid phase solubility, 186,
Solid state target atoms, 58
Spreading resistance profiling, 135
 principle of, 136
Spreading resistance, 129–150
Stopping power, 31–50
 electronic stopping power, 33
 electronic stopping power for heavy ions, 42
 electronic stopping power for molecular ions, 47

Stopping power (*continued*)
    electronic stopping power for protons, 36
    fitted formula, 38
    low and high velocity approximation, 37
    numerical calculations of, 36
Stripping Hall effect, 165–191
    errors and limitations, 174
Structural changes, 121, 196, 248, 274

T

TEM characterization, 201–230
Thermal crystallization, 99, 206, 248
Threshold adjust, 16
Threshold voltage, 16
Transmission electron microscopy (TEM), 195–234
    principle of, 196
    imaging ray path, 196
    phase contrast, 197
    diffraction contrast, 198
    specimen preparation for, 199
Triple crystal diffractometry, 267, 272

V

Vacancy, 249. *See also* defects
Vicinage effect, 49

W

Well doping, 18

X

X-ray diffraction, 261–280
    principle of, 262
    dynamical theory, 268
    Semikinematic approximation, 269
    limitations of 273

Contents of Volumes in This Series

Volume 1 Physics of III–V Compounds

C. Hilsum, Some Key Features of III–V Compounds
Franco Bassani, Methods of Band Calculations Applicable to III–V Compounds
E. O. Kane, The k-p Method
V. L. Bonch-Bruevich, Effect of Heavy Doping on the Semiconductor Band Structure
Donald Long, Energy Band Structures of Mixed Crystals of III–V Compounds
Laura M. Roth and Petros N. Argyres, Magnetic Quantum Effects
S. M. Puri and T. H. Geballe, Thermomagnetic Effects in the Quantum Region
W. M. Becker, Band Characteristics near Principal Minima from Magnetoresistance
E. H. Putley, Freeze-Out Effects, Hot Electron Effects, and Submillimeter Photoconductivity in InSb
H. Weiss, Magnetoresistance
Betsy Ancker-Johnson, Plasma in Semiconductors and Semimetals

Volume 2 Physics of III–V Compounds

M. G. Holland, Thermal Conductivity
S. I. Novkova, Thermal Expansion
U. Piesbergen, Heat Capacity and Debye Temperatures
G. Giesecke, Lattice Constants
J. R. Drabble, Elastic Properties
A. U. Mac Rae and G. W. Gobeli, Low Energy Electron Diffraction Studies
Robert Lee Mieher, Nuclear Magnetic Resonance
Bernard Goldstein, Electron Paramagnetic Resonance
T. S. Moss, Photoconduction in III–V Compounds
E. Antoncik ad J. Tauc, Quantum Efficiency of the Internal Photoelectric Effect in InSb
G. W. Gobeli and I. G. Allen, Photoelectric Threshold and Work Function
P. S. Pershan, Nonlinear Optics in III–V Compounds
M. Gershenzon, Radiative Recombination in the III–V Compounds
Frank Stern, Stimulated Emission in Semiconductors

Volume 3 Optical of Properties III–V Compounds

Marvin Hass, Lattice Reflection
William G. Spitzer, Multiphonon Lattice Absorption
D. L. Stierwalt and R. F. Potter, Emittance Studies
H. R. Philipp and H. Ehrenveich, Ultraviolet Optical Properties
Manuel Cardona, Optical Absorption above the Fundamental Edge
Earnest J. Johnson, Absorption near the Fundamental Edge
John O. Dimmock, Introduction to the Theory of Exciton States in Semiconductors
B. Lax and J. G. Mavroides, Interband Magnetooptical Effects
H. Y. Fan, Effects of Free Carries on Optical Properties
Edward D. Palik and George B. Wright, Free-Carrier Magnetooptical Effects
Richard H. Bube, Photoelectronic Analysis
B. O. Seraphin and H. E. Bennett, Optical Constants

Volume 4 Physics of III–V Compounds

N. A. Goryunova, A. S. Borschevskii, and D. N. Tretiakov, Hardness
N. N. Sirota, Heats of Formation and Temperatures and Heats of Fusion of Compounds $A^{III}B^{V}$
Don L. Kendall, Diffusion
A. G. Chynoweth, Charge Multiplication Phenomena
Robert W. Keyes, The Effects of Hydrostatic Pressure on the Properties of III–V Semiconductors
L. W. Aukerman, Radiation Effects
N. A. Goryunova, F. P. Kesamanly, and D. N. Nasledov, Phenomena in Solid Solutions
R. T. Bate, Electrical Properties of Nonuniform Crystals

Volume 5 Infrared Detectors

Henry Levinstein, Characterization of Infrared Detectors
Paul W. Kruse, Indium Antimonide Photoconductive and Photoelectromagnetic Detectors
M. B. Prince, Narrowband Self-Filtering Detectors
Ivars Melngalis and T. C. Harman, Single-Crystal Lead-Tin Chalcogenides
Donald Long and Joseph L. Schmidt, Mercury-Cadmium Telluride and Closely Related Alloys
E. H. Putley, The Pyroelectric Detector
Norman B. Stevens, Radiation Thermopiles
R. J. Keyes and T. M. Quist, Low Level Coherent and Incoherent Detection in the Infrared
M. C. Teich, Coherent Detection in the Infrared
F. R. Arams, E. W. Sard, B. J. Peyton, and F. P. Pace, Infrared Heterodyne Detection with Gigahertz IF Response
H. S. Sommers, Jr., Macrowave-Based Photoconductive Detector
Robert Sehr and Rainer Zuleeg, Imaging and Display

Volume 6 Injection Phenomena

Murray A. Lampert and Ronald B. Schilling, Current Injection in Solids: The Regional Approximation Method
Richard Williams, Injection by Internal Photoemission
Allen M. Barnett, Current Filament Formation

R. Baron and J. W. Mayer, Double Injection in Semiconductors
W. Ruppel, The Photoconductor-Metal Contact

Volume 7 Application and Devices
Part A

John A. Copeland and Stephen Knight, Applications Utilizing Bulk Negative Resistance
F. A. Padovani, The Voltage-Current Characteristics of Metal-Semiconductor Contacts
P. L. Hower, W. W. Hooper, B. R. Cairns, R. D. Fairman, and D. A. Tremere, The GaAs Field-Effect Transistor
Marvin H. White, MOS Transistors
G. R. Antell, Gallium Arsenide Transistors
T. L. Tansley, Heterojunction Properties

Part B

T. Misawa, IMPATT Diodes
H. C. Okean, Tunnel Diodes
Robert B. Campbell and Hung-Chi Chang, Silicon Junction Carbide Devices
R. E. Enstrom, H. Kressel, and L. Krassner, High-Temperature Power Rectifiers of $GaAs_{1-x}P_x$

Volume 8 Transport and Optical Phenomena

Richard J. Stirn, Band Structure and Galvanomagnetic Effects in III–V Compounds with Indirect Band Gaps
Roland W. Ure, Jr., Thermoelectric Effects in III–V Compounds
Herbert Piller, Faraday Rotation
H. Barry Bebb and E. W. Williams, Photoluminescence I: Theory
E. W. Williams and H. Barry Bebb, Photoluminescence II: Gallium Arsenide

Volume 9 Modulation Techniques

B. O. Seraphin, Electroreflectance
R. L. Aggarwal, Modulated Interband Magnetooptics
Daniel F. Blossey and Paul Handler, Electroabsorption
Bruno Batz, Thermal and Wavelength Modulation Spectroscopy
Ivar Balslev, Piezopptical Effects
D. E. Aspnes and N. Bottka, Electric-Field Effects on the Dielectric Function of Semiconductors and Insulators

Volume 10 Transport Phenomena

R. L. Rhode, Low-Field Electron Transport
J. D. Wiley, Mobility of Holes in III–V Compounds
C. M. Wolfe and G. E. Stillman, Apparent Mobility Enhancement in Inhomogeneous Crystals
Robert L. Petersen, The Magnetophonon Effect

Volume 11 Solar Cells

Harold J. Hovel, Introduction; Carrier Collection, Spectral Response, and Photocurrent; Solar Cell Electrical Characteristics; Efficiency; Thickness; Other Solar Cell Devices; Radiation Effects; Temperature and Intensity; Solar Cell Technology

Volume 12 Infrared Detectors (II)

W. L. Eiseman, J. D. Merriam, and R. F. Potter, Operational Characteristics of Infrared Photodetectors
Peter R. Bratt, Impurity Germanium and Silicon Infrared Detectors
E. H. Putley, InSb Submillimeter Photoconductive Detectors
G. E. Stillman, C. M. Wolfe, and J. O. Dimmock, Far-Infrared Photoconductivity in High Purity GaAs
G. E. Stillman and C. M. Wolfe, Avalanche Photodiodes
P. L. Richards, The Josephson Junction as a Detector of Microwave and Far-Infrared Radiation
E. H. Putley, The Pyroelectric Detector—An Update

Volume 13 Cadmium Telluride

Kenneth Zanio, Materials Preparations; Physics; Defects; Applications

Volume 14 Lasers, Junctions, Transport

N. Holonyak, Jr. and M. H. Lee, Photopumped III–V Semiconductor Lasers
Henry Kressel and Jerome K. Butler, Heterojunction Laser Diodes
A Van der Ziel, Space-Charge-Limited Solid-State Diodes
Peter J. Price, Monte Carlo Calculation of Electron Transport in Solids

Volume 15 Contacts, Junctions, Emitters

B. L. Sharma, Ohmic Contacts to III–V Compounds Semiconductors
Allen Nussbaum, The Theory of Semiconducting Junctions
John S. Escher, NEA Semiconductor Photoemitters

Volume 16 Defects, (HgCd)Se, (HgCd)Te

Henry Kressel, The Effect of Crystal Defects on Optoelectronic Devices
C. R. Whitsett, J. G. Broerman, and C. J. Summers, Crystal Growth and Properties of $Hg_{1-x}Cd_xSe$ alloys
M. H. Weiler, Magnetooptical Properties of $Hg_{1-x}Cd_xTe$ Alloys
Paul W. Kruse and John G. Ready, Nonlinear Optical Effects in $Hg_{1-x}Cd_xTe$

Volume 17 CW Processing of Silicon and Other Semiconductors

James F. Gibbons, Beam Processing of Silicon
Arto Lietoila, Richard B. Gold, James F. Gibbons, and Lee A. Christel, Temperature Distribu-

tions and Solid Phase Reaction Rates Produced by Scanning CW Beams
Arto Leitoila and James F. Gibbons, Applications of CW Beam Processing to Ion Implanted Crystalline Silicon
N. M. Johnson, Electronic Defects in CW Transient Thermal Processed Silicon
K. F. Lee, T. J. Stultz, and James F. Gibbons, Beam Recrystallized Polycrystalline Silicon: Properties, Applications, and Techniques
T. Shibata, A. Wakita, T. W. Sigmon, and James F. Gibbons, Metal-Silicon Reactions and Silicide
Yves I. Nissim and James F. Gibbons, CW Beam Processing of Gallium Arsenide

Volume 18 Mercury Cadmium Telluride

Paul W. Kruse, The Emergence of $(Hg_{1-x}Cd_x)Te$ as a Modern Infrared Sensitive Material
H. E. Hirsch, S. C. Liang, and A. G. White, Preparation of High-Purity Cadmium, Mercury, and Tellurium
W. F. H. Micklethwaite, The Crystal Growth of Cadmium Mercury Telluride
Paul E. Petersen, Auger Recombination in Mercury Cadmium Telluride
R. M. Broudy and V. J. Mazurczyck, (HgCd)Te Photoconductive Detectors
M. B. Reine, A. K. Soad, and T. J. Tredwell, Photovoltaic Infrared Detectors
M. A. Kinch, Metal-Insulator-Semiconductor Infrared Detectors

Volume 19 Deep Levels, GaAs, Alloys, Photochemistry

G. F. Neumark and K. Kosai, Deep Levels in Wide Band-Gap III–V Semiconductors
David C. Look, The Electrical and Photoelectronic Properties of Semi-Insulating GaAs
R. F. Brebrick, Ching-Hua Su, and Pok-Kai Liao, Associated Solution Model for Ga-In-Sb and Hg-Cd-Te
Yu. Ya. Gurevich and Yu. V. Pleskon, Photoelectrochemistry of Semiconductors

Volume 20 Semi-Insulating GaAs

R. N. Thomas, H. M. Hobgood, G. W. Eldridge, D. L. Barrett, T. T. Braggins, L. B. Ta, and S. K. Wang, High-Purity LEC Growth and Direct Implantation of GaAs for Monolithic Microwave Circuits
C. A. Stolte, Ion Implantation and Materials for GaAs Integrated Circuits
C. G. Kirkpatrick, R. T. Chen, D. E. Holmes, P. M. Asbeck, K. R. Elliott, R. D. Fairman, and J. R. Oliver, LEC GaAs for Integrated Circuit Applications
J. S. Blakemore and S. Rahimi, Models for Mid-Gap Centers in Gallium Arsenide

Volume 21 Hydrogenated Amorphous Silicon Part A

Jacques I. Pankove, Introduction
Masataka Hirose, Glow Discharge; Chemical Vapor Deposition
Yoshiyuki Uchida, di Glow Discharge
T. D. Moustakas, Sputtering
Isao Yamada, Ionized-Cluster Beam Deposition
Bruce A. Scott, Homogeneous Chemical Vapor Deposition

Frank J. Kampas, Chemical Reactions in Plasma Deposition
Paul A. Longeway, Plasma Kinetics
Herbert A. Weakliem, Diagnostics of Silane Glow Discharges Using Probes and Mass Spectroscopy
Lester Gluttman, Relation between the Atomic and the Electronic Structures
A. Chenevas-Paule, Experiment Determination of Structure
S. Minomura, Pressure Effects on the Local Atomic Structure
David Adler, Defects and Density of Localized States

Part B

Jacques I. Pankove, Introduction
G. D. Cody, The Optical Absorption Edge of a-Si:H
Nabil M. Amer and Warren B. Jackson, Optical Properties of Defect States in a-Si:H
P. J. Zanzucchi, The Vibrational Spectra of a-Si:H
Yoshihiro Hamakawa, Electroreflectance and Electroabsorption
Jeffrey S. Lannin, Raman Scattering of Amorphous Si, Ge, and Their Alloys
R. A. Street, Luminescence in a-Si:H
Richard S. Crandall, Photoconductivity
J. Tauc, Time-Resolved Spectroscopy of Electronic Relaxation Processes
P. E. Vanier, IR-Induced Quenching and Enhancement of Photoconductivity and Photoluminescence
H. Schade, Irradiation-Induced Metastable Effects
L. Ley, Photoelectron Emission Studies

Part C

Jacques I. Pankove, Introduction
J. David Cohen, Density of States from Junction Measurements in Hydrogenated Amorphous Silicon
P. C. Taylor, Magnetic Resonance Measurements in a-Si:H
K. Morigaki, Optically Detected Magnetic Resonance
J. Dresner, Carrier Mobility in a-Si:H
T. Tiedje, Information about band-Tail States from Time-of-Flight Experiments
Arnold R. Moore, Diffusion Length in Undoped a-Si:H
W. Beyer and J. Overhof, Doping Effects in a-Si:H
H. Fritzche, Electronic Properties of Surfaces in a-Si:H
C. R. Wronski, The Staebler-Wronski Effect
R. J. Nemanich, Schottky Barriers on a-Si:H
B. Abeles and T. Tiedje, Amorphous Semiconductor Superlattices

Part D

Jacques I. Pankove, Introduction
D. E. Carlson, Solar Cells
G. A. Swartz, Closed-Form Solution of I–V Characteristic for a a-Si:H Solar Cells
Isamu Shimizu, Electrophotography
Sachio Ishioka, Image Pickup Tubes

P. G. LeComber and W. E. Spear, The Development of the a-Si:H Field-Effect Transistor and Its Possible Applications
D. G. Ast, a-Si:H FET-Addressed LCD Panel
S. Kaneko, Solid-State Image Sensor
Masakiyo Matsumura, Charge-Coupled Devices
M. A. Bosch, Optical Recording
A. D'Amico and G. Fortunato, Ambient Sensors
Hiroshi Kukimoto, Amorphous Light-Emitting Devices
Robert J. Phelan, Jr., Fast Detectors and Modulators
Jacques I. Pankove, Hybrid Structures
P. G. LeComber, A. E. Owen, W. E. Spear, J. Hajto, and W. K. Choi, Electronic Switching in Amorphous Silicon Junction Devices

Volume 22 Lightwave Communications Technology
Part A

Kazuo Nakajima, The Liquid-Phase Epitaxial Growth of InGaAsP
W. T. Tsang, Molecular Beam Epitaxy for III–V Compound Semiconductors
G. B. Stringfellow, Organometallic Vapor-Phase Epitaxial Growth of III–V Semiconductors
G. Beuchet, Halide and Chloride Transport Vapor-Phase Deposition of InGaAsP and GaAs
Manijeh Razeghi, Low-Pressure Metallo-Organic Chemical Vapor Deposition of $Ga_xIn_{1-x}As P_{1-y}$ Alloys
P. M. Petroff, Defects in III–V Compound Semiconductors

Part B

J. P. van der Ziel, Mode Locking of Semiconductor Lasers
Kam Y. Lau and Ammon Yariv, High-Frequency Current Modulation of Semiconductor Injection Lasers
Charles H. Henry, Special Properties of Semiconductor Lasers
Yasuharu Suematsu, Katsumi Kishino, Shigehisa Arai, and Fumio Koyama, Dynamic Single-Mode Semiconductor Lasers with a Distributed Reflector
W. T. Tsang, The Cleaved-Coupled-Cavity (C^3) Laser

Part C

R. J. Nelson and N. K. Dutta, Review of InGaAsP InP Laser Structures and Comparison of Their Performance
N. Chinone and M. Nakamura, Mode-Stabilized Semiconductor Lasers for 0.7–0.8- and 1.1–1.6-μm Regions
Yoshiji Horikoshi, Semiconductor Lasers with Wavelengths Exceeding 2 μm
B. A. Dean and M. Dixon, The Functional Reliability of Semiconductor Lasers as Optical Transmitters
R. H. Saul, T. P. Lee, and C. A. Burus, Light-Emitting Device Design
C. L. Zipfel, Light-Emitting Diode-Reliability
Tien Pei Lee and Tingye Li, LED-Based Multimode Lightwave Systems
Kinichiro Ogawa, Semiconductor Noise-Mode Partition Noise

Part D

Federico Capasso, The Physics of Avalanche Photodiodes
T. P. Pearsall and M. A. Pollack, Compound Semiconductor Photodiodes
Takao Kaneda, Silicon and Germanium Avalanche Photodiodes
S. R. Forrest, Sensitivity of Avalanche Photodetector Receivers for High-Bit-Rate Long-Wavelength Optical Communication Systems
J. C. Campbell, Phototransistors for Lightwave Communications

Part E

Shyh Wang, Principles and Characteristics of Integrable Active and Passive Optical Devices
Shlomo Margalit and Amnon Yariv, Integrated Electronic and Photonic Devices
Takaoki Mukai, Yoshihisa Yamamoto, and Tatsuya Kimura, Optical Amplification by Semiconductor Lasers

Volume 23   Pulsed Laser Processing of Semiconductors

R. F. Wood, C. W. White, and R. T. Young, Laser Processing of Semiconductors: An Overview
C. W. White, Segregation, Solute Trapping, and Supersaturated Alloys
G. E. Jellison, Jr., Optical and Electrical Properties of Pulsed Laser-Annealed Silicon
R. F. Wood and G. E. Jellison, Jr., Melting Model of Pulsed Laser Processing
R. F. Wood and F. W. Young, Jr., Nonequilibrium Solidification Following Pulsed Laser Melting
D. H. Lowndes and G. E. Jellison, Jr., Time-Resolved Measurement During Pulsed Laser Irradiation of Silicon
D. M. Zebner, Surface Studies of Pulsed Laser Irradiated Semiconductors
D. H. Lowndes, Pulsed Beam Processing of Gallium Arsenide
R. B. James, Pulsed CO_2 Laser Annealing of Semiconductors
R. T. Young and R. F. Wood, Applications of Pulsed Laser Processing

Volume 24   Applications of Multiquantum Wells, Selective Doping, and Superlattices

C. Weisbuch, Fundamental Properties of III–V Semiconductor Two-Dimensional Quantized Structures: The Basis for Optical and Electronic Device Applications
H. Morkoc and H. Unlu, Factors Affecting the Performance of (Al,Ga)As/GaAs and (Al,Ga)As/InGaAs Modulation-Doped Field-Effect Transistors: Microwave and Digital Applications
N. T. Linh, Two-Dimensional Electron Gas FETs: Microwave Applications
M. Abe et al., Ultra-High-Speed HEMT Integrated Circuits
D. S. Chemla, D. A. B. Miller, and P. W. Smith, Nonlinear Optical Properties of Multiple Quantum Well Structures for Optical Signal Processing
F. Capasso, Graded-Gap and Superlattice Devices by Band-Gap Engineering
W. T. Tsang, Quantum Confinement Heterostructure Semiconductor Lasers
G. C. Osbourn et al., Principles and Applications of Semiconductor Strained-Layer Superlattices

Volume 25 Diluted Magnetic Semiconductors

W. Giriat and J. K. Furdyna, Crystal Structure, Composition, and Materials Preparation of Diluted Magnetic Semiconductors
W. M. Becker, Band Structure and Optical Properties of Wide-Gap $A^{II}_{1-x}Mn_xB^{IV}$ Alloys at Zero Magnetic Field
Saul Oseroff and Pieter H. Keesom, Magnetic Properties: Macroscopic Studies
Giebultowicz and T. M. Holden, Neutron Scattering Studies of the Magnetic Structure and Dynamics of Diluted Magnetic Semiconductors
J. Kossut, Band Structure and Quantum Transport Phenomena in Narrow-Gap Diluted Magnetic Semiconductors
C. Riquaux, Magnetooptical Properties of Large-Gap Diluted Magnetic Semiconductors
J. A. Gaj, Magnetooptical Properties of Large-Gap Diluted Magnetic Semiconductors
J. Mycielski, Shallow Acceptors in Diluted Magnetic Semiconductors: Splitting, Boil-off, Giant Negative Magnetoresistance
A. K. Ramadas and R. Rodriquez, Raman Scattering in Diluted Magnetic Semiconductors
P. A. Wolff, Theory of Bound Magnetic Polarons in Semimagnetic Semiconductors

Volume 26 III–V Compound Semiconductors and Semiconductor Properties of Superionic Materials

Zou Yuanxi, III–V Compounds
H. V. Winston, A. T. Hunter, H. Kimura, and R. E. Lee, InAs-Alloyed GaAs Substrates for Direct Implantation
P. K. Bhattacharya and S. Dhar, Deep Levels in III–V Compound Semiconductors Grown by MBE
Yu. Yu. Gurevich and A. K. Ivanov-Shits, Semiconductor Properties of Supersonic Materials

Volume 27 High Conducting Quasi-One-Dimensional Organic Crystals

E. M. Conwell, Introduction to Highly Conducting Quasi-One-Dimensional Organic Crystals
I. A. Howard, A Reference Guide to the Conducting Quasi-One-Dimensional Organic Molecular Crystals
J. P. Pouquet, Structural Instabilities
E. M. Conwell, Transport Properties
C. S. Jacobsen, Optical Properties
J. C. Scott, Magnetic Properties
L. Zuppiroli, Irradiation Effects: Perfect Crystals and Real Crystals

Volume 28 Measurement of High-Speed Signals in Solid State Devices

J. Frey and D. Ioannou, Materials and Devices for High-Speed and Optoelectronic Applications
H. Schumacher and E. Strid, Electronic Wafer Probing Techniques
D. H. Auston, Picosecond Photoconductivity: High-Speed Measurements of Devices and Materials
J. A. Valdmanis, Electro-Optic Measurement Techniques for Picosecond Materials, Devices, and Integrated Circuits.
J. M. Wiesenfeld and R. K. Jain, Direct Optical Probing of Integrated Circuits and High-Speed Devices
G. Plows, Electron-Beam Probing
A. M. Weiner and R. B. Marcus, Photoemissive Probing

Volume 29 Very High Speed Integrated Circuits: Gallium Arsenide LSI

M. Kuzuhara and T. Nazaki, Active Layer Formation by Ion Implantation
H. Hasimoto, Focused Ion Beam Implantation Technology
T. Nozaki and A. Higashisaka, Device Fabrication Process Technology
M. Ino and T. Takada, GaAs LSI Circuit Design
M. Hirayama, M. Ohmori, and K. Yamasaki, GaAs LSI Fabrication and Performance

Volume 30 Very High Speed Integrated Circuits: Heterostructure

H. Watanabe, T. Mizutani, and A. Usui, Fundamentals of Epitaxial Growth and Atomic Layer Epitaxy
S. Hiyamizu, Characteristics of Two-Dimensional Electron Gas in III–V Compound Heterostructures Grown by MBE
T. Nakanisi, Metalorganic Vapor Phase Epitaxy for High-Quality Active Layers
T. Nimura, High Electron Mobility Transistor and LSI Applications
T. Sugeta and T. Ishibashi, Hetero-Bipolar Transistor and LSI Application
H. Matsueda, T. Tanaka, and M. Nakamura, Optoelectronic Integrated Circuits

Volume 31 Indium Phosphide: Crystal Growth and Characterization

J. P. Farges, Growth of Discoloration-free InP
M. J. McCollum and G. E. Stillman, High Purity InP Grown by Hydride Vapor Phase Epitaxy
T. Inada and T. Fukuda, Direct Synthesis and Growth of Indium Phosphide by the Liquid Phosphorous Encapsulated Czochralski Method
O. Oda, K. Katagiri, K. Shinohara, S. Katsura, Y. Takahashi, K. Kainosho, K. Kohiro, and R. Hirano, InP Crystal Growth, Substrate Preparation and Evaluation
K. Tada, M. Tatsumi, M. Morioka, T. Araki, and T. Kawase, InP Substrates: Production and Quality Control
M. Razeghi, LP-MOCVD Growth, Characterization, and Application of InP Material
T. A. Kennedy and P. J. Lin-Chung, Stoichiometric Defects in InP

Volme 32 Strained-Layer Superlattices: Physics

T. P. Pearsall, Strained-Layer Superlattices
Fred H. Pollack, Effects of Homogeneous Strain on the Electronic and Vibrational Levels in Semiconductors
J. Y. Marzin, J. M. Gerárd, P. Voisin, and J. A. Brum, Optical Studies of Strained III–V Heterolayers
R. People and S. A. Jackson, Structurally Induced States from Strain and Confinement
M. Jaros, Microscopic Phenomena in Ordered Suprlattices

Volume 33 Strained-Layer Superlattices: Materials Science and Technology

R. Hull and J. C. Bean, Principles and Concepts of Strained-Layer Epitaxy
William J. Schaff, Paul J. Tasker, Marc C. Foisy, and Lester F. Eastman, Device Applications of Strained-Layer Epitaxy

S. T. Picraux, B. L. Doyle, and J. Y. Tsao, Structure and Characterization of Strained-Layer Superlattices
E. Kasper and F. Schaffer, Group IV Compounds
Dale L. Martin, Molecular Beam Epitaxy of IV–VI Compounds Heterojunction
Robert L. Gunshor, Leslie A. Kolodziejski, Arto V. Nurmikko, and Nobuo Otsuka, Molecular Beam Epitaxy of II–VI Semiconductor Microstructures

Volume 34 Hydrogen in Semiconductors

J. I. Pankove and N. M. Johnson, Introduction to Hydrogen in Semiconductors
C. H. Seager, Hydrogenation Methods
J. I. Pankove, Hydrogenation of Defects in Crystalline Silicon
J. W. Corbett, P. Deák, U. V. Desnica, and S. J. Pearton, Hydrogen Passivation of Damage Centers in Semiconductors
S. J. Pearton, Neutralization of Deep Levels in Silicon
J. I. Pankove, Neutralization of Shallow Acceptors in Silicon
N. M. Johnson, Neutralization of Donor Dopants and Formation of Hydrogen-Induced Defects in n-Type Silicon
M. Stavola and S. J. Pearton, Vibrational Spectroscopy of Hydrogen-Related Defects in Silicon
A. D. Marwick, Hydrogen in Semiconductors: Ion Beam Techniques
C. Herring and N. M. Johnson, Hydrogen Migration and Solubility in Silicon
E. E. Haller, Hydrogen-Related Phenomena in Crystalline Germanium
J. Kakalios, Hydrogen Diffusion in Amorphous Silicon
J. Chevalier, B. Clerjaud, and B. Pajot, Neutralization of Defects and Dopants in III–V Semiconductors
G. G. DeLeo and W. B. Fowler, Computational Studies of Hydrogen-Containing Complexes in Semiconductors
R. F. Kiefl and T. L. Estle, Muonium in Semiconductors
C. G. Van de Walle, Theory of Isolated Interstitial Hydrogen and Muonium in Crystalline Semiconductors

Volume 35 Nanostructured Systems

Mark Reed, Introduction
H. van Houten, C. W. J. Beenakker, and B. J. van Wees, Quantum Point Contacts
G. Timp, When Does a Wire Become an Electron Waveguide?
M. Büttiker, The Quantum Hall Effects in Open Conductors
W. Hansen, J. P. Kotthaus, and U. Merkt, Electrons in Laterally Periodic Nanostructures

Volume 36 The Spectroscopy of Semiconductors

D. Heiman, Spectroscopy of Semiconductors at Low Temperatures and High Magnetic Fields
Arto V. Nurmikko, Transient Spectroscopy by Ultrashort Laser Pulse Techniques
A. K. Ramdas and S. Rodriguez, Piezospectroscopy of Semiconductors
Orest J. Glembocki and Benjamin V. Shanabrook, Photoreflectance Spectroscopy of Microstructures
David G. Seiler, Christopher L. Littler, and Margaret H. Wiler, One- and Two-Photon Magneto Optical Spectroscopy of InSb and $Hg_{1-x}Cd_xTe$

Volume 37 The Mechanical Properties of Semiconductors

A.-B. Chen, Arden Sher and W. T. Yost, Elastic Constants and Related Properties of Semiconductor Compounds and Their Alloys
David R. Clarke, Fracture of Silicon and Other Semiconductors
Hans Siethoff, The Plasticity of Elemental and Compound Semiconductors
Sivaraman Guruswamy, Katherine T. Faber and John P. Hirth, Mechanical Behavior of Compound Semiconductors
Subhanh Mahajan, Deformation Behavior of Compound Semiconductors
John P. Hirth, Injection of Dislocations into Strained Multilayer Structures
Don Kendall, Charles B. Fleddermann, and Kevin J. Malloy, Critical Technologies for the Micromachining of Silicon
Ikuo Matsuba and Kinji Mokuya, Processing and Semiconductor Thermoelastic Behavior

Volume 38 Imperfections in III/V Materials

Udo Scherz and Matthias Scheffler, Density-Functional Theory of sp-Bonded Defects in III/V Semiconductors
Maria Kaminska and Eicke R. Weber, El2 Defect in GaAs
David C. Look, Defects Relevant for Compensation in Semi-Insulating GaAs
R. C. Newman, Local Vibrational Mode Spectroscopy of Defects in III/V Compounds
Andrzej M. Hennel, Transition Metals in III/V Compounds
Kevin J. Malloy and Ken Khachaturyan, DX and Related Defects in Semiconductors
V. Swaminathan and Andrew S. Jordan, Dislocations in III/V Compounds
Krzysztof W. Nauka, Deep Level Defects in the Epitaxial III/V Materials

Volume 39 Minority Carriers in III–V Semiconductors: Physics and Applications

Niloy K. Dutta, Radiative Transitions in GaAs and Other III–V Compounds
Richard K. Ahrenkiel, Minority-Carrier Lifetime in III–V Semiconductors
Tomofumi Furuta, High Field Minority Electron Transport in p-GaAs
Mark S. Lundstrom, Minority-Carrier Transport in III–V Semiconductors
Richard A. Abram, Effects of Heavy Doping and High Excitation on the Band Structure of GaAs
David Yevick and Witold Bardyszewski, An Introduction to Non-Equilibrium Many-Body Analyses of Optical Processes in III–V Semiconductors

Volume 40 Epitaxial Microstructures

E. F. Schubert, Delta-Doping of Semiconductors: Electronic, Optical, and Structural Properties of Materials and Devices
A. Gossard, M. Sundaram, and P. Hopkins, Wide Graded Potential Wells
P. Petroff, Direct Growth of Nanometer-Size Quantum Wire Superlattices
E. Kapon, Lateral Patterning of Quantum Well Heterostructures by Growth of Nonplanar Substrates
H. Temkin, D. Gershoni, and M. Panish, Optical Properties of $Ga_{1-x}In_xAs/InP$ Quantum Wells

Volume 41 High Speed Heterostructure Devices

F. Capasso, F. Beltram, S. Sen, A. Pahlevi, and A. Y. Cho, Quantum Electron Devices: Physics and Applications
P. Solomon, D. J. Frank, S. L. Wright, and F. Canora, GaAs-Gate Semiconductor–Insulator-Semiconductor FET
M. H. Hashemi and U. K. Mishra, Unipolar InP-Based Transistors
R. Kiehl, Complementary Heterostructure FET Integrated Circuits
T. Ishibashi, GaAs-Based and InP-Based Heterostructure Bipolar Transistors
H. C. Liu and T. C. L. G. Sollner, High-Frequency-Tunneling Devices
H. Ohnishi, T. More, M. Takatsu, K. Imamura, and N. Yokoyama, Resonant-Tunneling Hot-Electron Transistors and Circuits

Volume 42 Oxygen in Silicon

F. Shimura, Introduction to Oxygen in Silicon
W. Lin, The Incorporation of Oxygen into Silicon Crystals
T. J. Schaffner and D. K. Schroder, Characterization Techniques for Oxygen in Silicon
W. M. Bullis, Oxygen Concentration Measurement
S. M. Hu, Intrinsic Point Defects in Silicon
B. Pajot, Some Atomic Configurations of Oxygen
J. Michel and L. C. Kimerling, Electical Properties of Oxygen in Silicon
R. C. Newman and R. Jones, Diffusion of Oxygen in Silicon
T. Y. Tan and W. J. Taylor, Mechanisms of Oxygen Precipitation: Some Quantitative Aspects
M. Schrems, Simulation of Oxygen Precipitation
K. Simino and I. Yonenaga, Oxygen Effect on Mechanical Properties
W. Bergholz, Grown-in and Process-Induced Effects
F. Shimura, Intrinsic/Internal Gettering
H. Tsuya, Oxygen Effect on Electronic Device Performance

Volume 43 Semiconductors for Room Temperature Nuclear Detector Applications

R. B. James and T. E. Schlesinger, Introduction and Overview
L. S. Darken and C. E. Cox, High-Purity Germanium Detectors
A. Burger, D. Nason, L. Van den Berg, and M. Schieber, Growth of Mercuric Iodide
X. J. Bao, T. E. Schlesinger, and R. B. James, Electrical Properties of Mercuric Iodide
X. J. Bao, R. B. James, and T. E. Schlesinger, Optical Properties of Red Mercuric Iodide
M. Hage-Ali and P. Siffert, Growth Methods of CdTe Nuclear Detector Materials
M. Hage-Ali and P Siffert, Characterization of CdTe Nuclear Detector Materials
M. Hage-Ali and P. Siffert, CdTe Nuclear Detectors and Applications
R. B. James, T. E. Schlesinger, J. Lund, and M. Schieber, $Cd_{1-x}Zn_xTe$ Spectrometers for Gamma and X-Ray Applications
D. S. McGregor, J. E. Kammeraad, Gallium Arsenide Radiation Detectors and Spectrometers
J. C. Lund, F. Olschner, and A. Burger, Lead Iodide
M. R. Squillante, and K. S. Shah, Other Materials: Status and Prospects
V. M. Gerrish, Characterization and Quantification of Detector Performance
J. S. Iwanczyk and B. E. Patt, Electronics for X-ray and Gamma Ray Spectrometers
M. Schieber, R. B. James, and T. E. Schlesinger, Summary and Remaining Issues for Room Temperature Radiation Spectrometers

Volume 44 II–IV Blue/Green Light Emitters: Device Physics and Epitaxial Growth

J. Han and R. L. Gunshor, MBE Growth and Electrical Properties of Wide Bandgap ZnSe-based II–VI Semiconductors
Shizuo Fujita and Shigeo Fujita, Growth and Characterization of ZnSe-based II–VI Semiconductors by MOVPE
Easen Ho and Leslie A. Kolodziejski, Gaseous Source UHV Epitaxy Technologies for Wide Bandgap II–VI Semiconductors
Chris G. Van de Walle, Doping of Wide-Band-Gap II–VI Compounds — Theory
Roberto Cingolani, Optical Properties of Excitons in ZnSe-Based Quantum Well Heterostructures
A. Ishibashi and A. V. Nurmikko, II–VI Diode Lasers: A Current View of Device Performance and Issues
Supratik Guha and John Petruzello, Defects and Degradation in Wide-Gap II–VI-based Structures and Light Emitting Devices

Volume 45 Effect of Disorder and Defects in Ion-Implanted Semiconductors: Electrical and Physiochemical Characterization

Heiner Ryssel, Ion Implantation into Semiconductors: Historical Perspectives
You-Nian Wang and Teng-Cai Ma, Electronic Stopping Power for Energetic Ions in Solids
Sachiko T. Nakagawa, Solid Effect on the Electronic Stopping of Crystalline Target and Application to Range Estimation
G. Müller, S. Kalbitzer and G. N. Greaves, Ion Beams in Amorphous Semiconductor Research
Jumana Boussey-Said, Sheet and Spreading Resistance Analysis of Ion Implanted and Annealed Semiconductors
M. L. Polignano and G. Queirolo, Studies of the Stripping Hall Effect in Ion-Implanted Silicon
J. Stoemenos, Transmission Electron Microscopy Analyses
Roberta Nipoti and Marco Servidori, Rutherford Backscattering Studies of Ion Implanted Semiconductors
P. Zaumseil, X-ray Diffraction Techniques